Economic Analysis and Policies in the Energy Sector

Economic Analysis and Policies in the Energy Sector

Guest Editor

George Halkos

MDPI

Basel • Beijing • Wuhan • Barcelona • Belgrade • Novi Sad • Cluj • Manchester

Guest Editor
George Halkos
Department of Economics
University of Thessaly
Volos
Greece

Editorial Office
MDPI AG
Grosspeteranlage 5
4052 Basel, Switzerland

This is a reprint of the Special Issue, published open access by the journal *Energies* (ISSN 1996-1073), freely accessible at: https://www.mdpi.com/journal/energies/special_issues/L3314488L8.

For citation purposes, cite each article independently as indicated on the article page online and as indicated below:

Lastname, A.A.; Lastname, B.B. Article Title. *Journal Name* **Year**, *Volume Number*, Page Range.

ISBN 978-3-7258-5011-2 (Hbk)
ISBN 978-3-7258-5012-9 (PDF)
https://doi.org/10.3390/books978-3-7258-5012-9

Contents

Contents

About the Editor

George Halkos

George Halkos is a Professor in Economics of Natural Resources at the University of Thessaly. He is an economist who graduated from the Department of Economics at the University of Piraeus (Greece). He holds an M.Sc in Project Analysis, Finance and Investment and a D.Phil in Cost–Benefit Analysis with the use of Mathematical Models and Game Theory and with application to Environmental Economics from the Economics Department of the University of York (England). Prof. George Halkos has taught various postgraduate courses across international institutions as a visiting professor. He has worked as a team leader and research fellow in various research and academic institutions for research projects financed by the European Union (PHARE programs), the Hellenic Republic Ministry of Development, the National Statistical Service of Greece, EUROSTAT, the Ministry of Employment, etc. His research interests are in the fields of Applied Statistics and Econometrics, Simulations of Economic Modeling, Natural Resource and Environmental Economics, Applied Micro-economics with emphasis in Welfare Economics, Air Pollution, Game Theory, and Mathematical Models (Non-Linear Programming). In recognition of his scholarly impact, Professor Halkos has been included in Stanford University's list of the world's top 2% of highly cited scientists for five consecutive years, up to his most recent inclusion in 2024.

Preface

This Reprint addresses the growing research interest in tackling environmental and energy challenges through advanced empirical and theoretical economic analysis, with a particular focus on *energy economics*, *renewable energy*, and the *energy transition*. Furthermore, it seeks to feature innovative approaches for assessing and enhancing sustainable energy policies, integrating both theoretical insights and empirical applications.

A key focus is on methodological advancements that capture spatio-temporal patterns of energy markets, resource and energy use, clean energy adoption, and associated environmental impacts, highlighting implications for *decarbonization* and the *just energy transition*. Essentially, the contributions in this Reprint have explored frameworks that support green technologies and sustainable energy policy development, proposed initiatives generating socio-economic benefits, identified obstacles to sustainability, and illuminated the complex interactions between economic and environmental systems.

Ultimately, this Reprint provides actionable insights into the formulation of effective energy policy by setting clear targets, presenting models for sustainable growth, and analyzing policy synergies and trade-offs. Overall, the policy implications in this Reprint highlight the core research advances in energy economics, renewable energy, and green technologies.

George Halkos
Guest Editor

energies

MDPI

Editorial

Economic Analysis and Policies in the Energy Sector

George Halkos

Laboratory of Operations Research, Department of Economics, University of Thessaly, 28hs Octovriou 78, 38333 Volos, Greece; halkos@uth.gr

Abstract

The aim of this Special Issue is to consider economic analysis in terms of the most up-to-date and advanced empirical and theoretical methods applied to energy problems. The main purpose of this Special Issue is to feature the theoretical and empirical practice of sustainable policy performance measurement. The progress of the green economy includes methodological issues in order to indicate and present spatio-temporal patterns of resource and energy use and associated pollution. Results will be discussed in support of sustainable energy policies. This Special Issue seeks the methodological framework to contribute to sustainable energy policy development, provide energy policy initiatives targeted to socio-economic goods/benefits, to capture sustainability obstacles and negative environmental impacts, and highlight links and interactions between economic and environmental systems. The expected outcome is to set targets, propose models for sustainable growth and energy policies, and analyze policy interactions.

Keywords: energy economics; renewable energy; energy transition; energy markets; clean energy; just energy transition; green technologies; sustainable energy; decarbonization; energy policy

1. Introduction

The energy sector is currently facing, worldwide, a variety of concerns that are impeding progress towards a future that is less polluting. It is the traditional dependence on fossil fuels that remains a challenge, as it generates more greenhouse gas emissions, causing further climate change and environmental deterioration. Additionally, issues such as energy security, price volatility, ageing infrastructure, and uneven accessibility to commodities and services can all hinder the sector's capacity to satisfy rising demand. As a consequence, energy systems under enormous strain continue to develop, but the route to transition is plagued with technological, monetary, and political hurdles. The above challenges not only impede the mainstreaming of greener innovations but also reduce the decarbonization of key industries for economic development.

The current energy transformation, which aims to shift towards renewables and enhance energy efficiency, is inextricably linked to meeting the United Nations Sustainable Development Goals (SDGs), notably SDG7 (Affordable and Clean Energy) and SDG13 (Climate Action). However, difficulties in the energy industry threaten to halt or undermine growth. Furthermore, the widespread adoption of renewable energy solutions is hindered by the significant upfront investment requirements, ambiguous regulatory frameworks, and the demand for extensive grid upgrades. Additionally, in the absence of a just transition framework that guarantees equitable opportunities and safeguards for workers and marginalized groups, the transition may deepen existing social disparities. Overall, addressing these obstacles is essential to ensure that the energy transition supports sustainable development by harmonizing economic progress, environmental responsibility, and social equity.

Therefore, bearing in mind the above issues in energy sectors, it is important to amass potential methodological advancements in energy-related sectors that might lead the path for sustainable growth. The aim of this Special Issue is to review the updated and robust empirical and theoretical economic methods that have been utilized in diverse energy-related challenges; hence, the main purpose of this Special Issue is to feature practical solutions to actualize sustainability-related performance in policymaking.

2. State-of-the-Art Energy Policy Practices and Instruments

Sustainability is inextricably linked to the progressive disengagement of modern societies from conventional fossil fuel energy production [1]. Halkos and Tsirivis [2] revealed that the distinctive attributes of individual electricity markets raise the complexity of strategic planning and implementation of multinational pro-environmental initiatives, making horizontal energy and environmental policymaking practically inefficient. The study initially affirmed the pivotal influence of electricity prices and energy taxation on renewable deployment in EU member states, yet the direction and strength of this impact significantly varied even between neighboring countries, thereby favoring the implementation of national energy strategies in an effort to fulfill SDGs.

Ji et al. [3] revealed how energy storage can provide power supply reliability and facilitate the sustainable energy transition. This study monitored an electricity supply chain and compared the equilibrium results under four different scenarios influenced by the Stackelberg game theory. The empirical results showed that two factors are essential: the discharge subsidy and the investment subsidy, as they can ameliorate energy storage technology quality. Moreover, especially during the early stages, the investment subsidy can influence positively both technology levels and electricity demand. Nevertheless, during the later stages, the discharge subsidies are rendered more advantageous for technological advancements and electricity demand. Essentially, the paper is suitable for finding profitable solutions by enhancing technological standards and satisfying electricity demand.

Considering the nature and special features of the Mexican energy market, Flores et al. [4] determined that combining the withdrawal of all state subsidies for energy generation with a new carbon tax would yield substantial environmental and welfare gains for Mexican society. This study revealed necessary policy-related aspects such as the influence of taxes on welfare-oriented fiscal policies as it utilizes the Household Income and Expenditure Survey (ENIGH) to estimate the demand for fuels, specifically electricity, liquefied petroleum gas, and gasoline, showing that welfare losses would be regressive regarding the incremental rises in electricity price, whereas the opposite stands for the changes in gasoline prices. Essentially, the redistribution of the tax revenues that have been gathered through the removal of energy subsidies and the setting of a carbon tax would significantly benefit Mexican households, resulting even in welfare gains of up to 350% for the most vulnerable households.

Similarly, Basu et al. [5] verified a decoupling process of associated GHG emissions and capital formation in the Indian power sector, strongly arguing against the subsidization of both fossil fuel imports and RES investments. This study provides insights on thenet-zero economy by monitoring the economy-electricity-emissions (3E) relations, showing that in the long term, the presence of electricity-driven emissions decouples with capital formation, whereas inflation and economic development raise CO_2 emissions. The primary finding demonstrates that GDP and emissions are not directly related by highlighting the negative feedback loop of inflation to trade and of trade to emissions. Decoupling should be prevented by this long-term macroeconomic dynamic death spiral, imports of fossil fuels should not be subsidized for economic shock recovery, and risk hedging of energy transition investments should take place in the post-COVID-19 future.

Rather than relying on subsidy policies, Wan et al. [6] advocate the promotion of environmental, social, and governance (ESG) evaluation and rating of distinct power companies as an incentive to enhance their environmental performance profile and green transition progress. This study observes the companies of the Shanghai Stock Exchange in China through machine learning algorithms in order to reveal potential causal relationships between ESG ratings. Contributing to the existing literature, as ESG parameters are linked to companies in the energy sector, especially those with companies with high capital market performance.

Smart metropolitan agglomerations are expected, according to Hu and Han [7] to utterly change future urban planning; such megacities can boost energy efficiency of urban living while at the same time facilitating the modification of the industrial model so that energy consumption from urban economic activity is minimized. The empirical results show that the simulations demonstrate an "S"-shaped development curve for the "smart" and "green" elements, providing an important indication of a robust development model. Overall, this paper paves the way for long-term planning with periodic goals and stages, taking into account the unique characteristics of megacities and addressing outside energy-driven urban-focused pressures.

Nonetheless, the required time horizon for the complete decarbonization of modern societies raises the significance of prominent transitional fuels, such as natural gas. von Döllen and Schlüter [8] support that the large storage capacity of existing natural gas facilities confers a competitive advantage in the usage of the particular fossil fuel, since it offers energy planners a critical solution to the insurmountable challenge of energy storage. The authors assert that a wise power generation strategy must necessarily exploit all available technologies, including renewables, transitional fuels, and conventional fossil fuel power plants.

In harmony with the previous analysis, Navarro et al. [9] postulate that diminishing natural gas supply for power generation to below 50% of current pipeline full capacity would trigger a substantial domino effect on production expenses and supply security. This paper simulates the primary natural gas pipeline's capacity congestion, showing interesting policy-related interrelations between natural gas management and the electricity sector through the interlinkages of production costs and load flows. The empirical results can underpin policy-specific coordinated management based on these interlinked systems with economic optimization and holistic grids.

Considering the necessity for an uninterrupted power supply for both the industrial sector and households, Bambirra et al. [10] allege that, apart from the economic cost of energy facility appraisals, it is essential to integrate other critical aspects such as power system security, operational risks, and potential synergies among the various energy sources contributing to the generation fuel mix. This planning can be undertaken through meticulous asset management based on the NEWAVE model that is used to simulate potential scenarios of hydraulic production that can reorient forthcoming fiscal policies in revenue redistribution. In order, however, to deal with the shortcoming of the proposed methodology, the paper employs multicriteria decision-making tools to assess the performance of long-term energy assets. Henceforth, this empirical contribution can be used as a robust decision-making tool in energy sectors worldwide.

In this sense, the monitoring of how new energy policies can affect urban carbon equity, as for example in China, reveals that while policies promote energy transition, they also worsen carbon distribution inequity. For instance, Yang et al. [11] showed that cultural forces, particularly Confucianism, might help mitigate these inequities, especially in non-industrial and non-resource-dependent cities. The empirical findings highlight the need for culturally informed policy frameworks to achieve more equitable energy transitions.

Qudrat-Ullah [12] proposed a framework for nuclear peace and trust in order to strengthen nuclear energy governance by addressing trust deficits, access inequities, and regional instability. It emphasizes trust-building, equitable technology sharing, and inclusive governance, linking nuclear energy to sustainable development. Actionable steps include the need for oversight from the International Atomic Energy Agency, capacity-building, and training to support safe, cooperative nuclear advancement, especially in developing nations.

3. Conclusions

The present Special Issue shows that sustainability depends on reducing fossil fuel reliance, but variations in electricity markets complicate multinational energy policies, making national strategies more effective for meeting SDGs. Moreover, energy storage, supported by investment and discharge subsidies, plays a crucial role in improving technology and supporting renewable transitions, with investment subsidies being more impactful early on and discharge subsidies more beneficial later. In Mexico, removing energy subsidies and introducing a carbon tax can yield environmental gains, while in India, decoupling emissions from capital formation argues against subsidizing fossil fuels or renewables. Instead of subsidies, ESG evaluations are recommended to drive green transitions. Smart cities are expected to enhance energy efficiency and reshape industrial models. Despite decarbonization goals, transitional fuels, such as natural gas, remain vital due to storage advantages, necessitating a balanced use of renewables, transitional, and conventional energy sources. Finally, ensuring power supply security requires evaluating both economic and operational risks within the energy mix.

To recapitulate, this Special Issue focuses on using the latest empirical and theoretical methods to analyze energy-related economic challenges; furthermore, it aims to highlight best practices in measuring the performance of sustainable policies and address methodological issues related to tracking spatiotemporal patterns of resource use, energy consumption, and pollution. The goal is to support the development of sustainable energy policies that maximize socio-economic benefits while minimizing environmental impacts. By examining the connections between economic and environmental systems, the present Special Issue has proposed models for sustainable growth, set clear policy targets, and evaluated policy interactions.

Acknowledgments: The Editor would like to thank the support and help of Tiffany Lv and the authors that contributed to this Special Issue.

Conflicts of Interest: The author declares no conflicts of interest.

References

1. Halkos, G.E. Optimal Abatement of Sulphur Emissions in Europe. *Environ. Resour. Econ.* **1994**, *4*, 127–150. [CrossRef]
2. Halkos, G.E.; Tsirivis, A.S. Sustainable Development of the European Electricity Sector: Investigating the Impact of Electricity Price, Market Liberalization and Energy Taxation on RES Deployment. *Energies* **2023**, *16*, 5567. [CrossRef]
3. Ji, C.; Wang, X.; Zhao, W.; Wang, X.; Qian, W. The Impact of Environmental Policies on Renewable Energy Storage Decisions in the Power Supply Chain. *Energies* **2025**, *18*, 2152. [CrossRef]

4. Rosas Flores, J.A.; Morillón Gálvez, D.; Silva, R. Effects of Removing Energy Subsidies and Implementing Carbon Taxes on Urban, Rural and Gender Welfare: Evidence from Mexico. *Energies* **2024**, *17*, 2237. [CrossRef]
5. Basu, S.; Ishihara, K.; Ogawa, T.; Okumura, H. Structural Effects of Economic Shocks on the Macroeconomic Economy–Electricity–Emissions Nexus in India via Long-Term Cointegration Approach. *Energies* **2024**, *17*, 4354. [CrossRef]
6. Wan, J.; Wang, Y.; Wang, Y. Promoting or Hindering: The Impact of ESG Rating Differences on Energy Enterprises' Green Transformation—A Causal Test from Double Machine-Learning Algorithms. *Energies* **2025**, *18*, 464. [CrossRef]
7. Hu, R.; Han, X. Toward a "Smart-Green" Future in Cities: System Dynamics Study of Megacities in China. *Energies* **2023**, *16*, 6395. [CrossRef]
8. von Döllen, A.; Schlüter, S. Heat Pumps for Germany—Additional Pressure on the Supply–Demand Equilibrium and How to Cope with Hydrogen. *Energies* **2024**, *17*, 3053. [CrossRef]
9. Navarro, R.; Rojas, H.; Luyo, J.E.; Silva, J.L.; Molina, Y.P. Impacts of Natural Gas Pipeline Congestion on the Integrated Gas–Electricity Market in Peru. *Energies* **2024**, *17*, 4586. [CrossRef]
10. Bambirra, R.; Schiavo, L.; Lima, M.; Miranda, G.; Reis, I.; Cassemiro, M.; Andrade, A.; Laender, F.; Silva, R.; Vieira, D.; et al. Robust Multiobjective Decision Making in the Acquisition of Energy Assets. *Energies* **2023**, *16*, 6089. [CrossRef]
11. Yang, Z.; Yu, H.; Zhang, J. New Energy Policies and Informal Cultural Norms Promoting Carbon Equity in Chinese Cities: Synergistic Effects and Regional Heterogeneity. *Energies* **2025**, *18*, 2475. [CrossRef]
12. Qudrat-Ullah, H. The Q-NPT: Redefining Nuclear Energy Governance for Sustainability. *Energies* **2025**, *18*, 2784. [CrossRef]

energies

Article

New Energy Policies and Informal Cultural Norms Promoting Carbon Equity in Chinese Cities: Synergistic Effects and Regional Heterogeneity

Zixuan Yang [1], Huang Yu [1] and Jingqiu Zhang [2,*]

[1] School of International Trade and Economics, University of International Business and Economics, Beijing 100029, China; sarahyoungest@outlook.com (Z.Y.); jamespao520@outlook.com (H.Y.)
[2] College of Applied Arts and Science, Beijing Union University, Beijing 100191, China
* Correspondence: jqzhang4811@outlook.com

Abstract: In the era of energy transition, there is a lack of targeted research on the synergistic effects of new energy policies and informal institutions on carbon equity. This study examines how new energy policies influence urban carbon equity, with a focus on the mediating role of cultural forces. Utilizing panel data from 256 Chinese cities (2000–2021) and employing the New Energy Demonstration City (NEDC) policy as a quasi-natural experiment, this study adopts a staggered difference-in-differences (DID) approach to identify causal relationships. Key findings reveal: (1) China has been accompanied by a rise in carbon distribution inequity measured through the cumulative distribution patterns of carbon emissions and economic outcomes, highlighting the equity-efficiency trade-off. (2) The NEDC policy, while advancing energy transition, inadvertently exacerbates urban carbon inequity. The conclusion is robust to parallel trend tests, placebo analyses, and controls for concurrent policies. (3) Confucianism, as an informal institutional force, can effectively mitigate the urban policy-driven inequities. (4) Heterogeneity analysis finds that the synergistic effect of Confucianism and the policy is more significant in non-old industrial base cities and non-resource-dependent cities. Theoretically, this research bridges energy transition literature with institutional theory by revealing the compensatory role of cultural systems in formal policy frameworks. Practically, it advocates for culturally informed energy governance models, proposing Confucian principles of harmony and collective responsibility as design pillars for equitable sustainability transitions.

Keywords: carbon equity; energy transition; Confucianism; new energy demonstration city; differences-in-differences method

1. Introduction

The global imperative for climate mitigation demands a fundamental restructuring of energy systems [1], yet the equity implications of such transitions remain critically understudied. China's unique dual identity—as both the world's largest carbon emitter and a pioneering force in renewable deployment—provides a pivotal testing ground for reconciling decarbonization with distributive justice. Recent advances in hydrogen technologies—such as electrochemical production methods and advanced materials-based storage energy systems—offer scalable pathways to bridge energy transformation [2,3]. While fossil fuels still dominate the national energy matrix, strategic interventions like the New Energy Demonstration City (NEDC) policy have catalyzed localized energy transitions through multi-level governance innovations [4]. These measures have significantly elevated clean energy adoption rates, with pilot cities demonstrating accelerated renewable integration alongside sustained economic growth compared to non-participating counterparts [5]. However, the prevailing transition paradigm exhibits a critical tension: techno-economic efficiency metrics prioritize cost-optimal emission reductions [6,7], whereas equitable transitions require spatially and socially balanced allocation of decarbonization burdens and benefits. The equity-efficiency paradox is particularly pronounced in China's institutional context. Confucian governance traditions, which emphasize hierarchical responsibility allocation and culturally mediated social contracts [8,9], tend to underlie the political economy of carbon rights distribution. Consequently, developing transition frameworks that simultaneously optimize technical pathways and institutionalize carbon equity mechanisms emerges as both a scholarly imperative and a governance necessity.

Current research on carbon equity focuses on three primary domains: measurement and evaluation of carbon emission equity, analysis of factors affecting carbon emission equity, and design of carbon emission allocation and compensation mechanisms based on the principle of equity. First, in terms of measuring and evaluating the equity of carbon emissions, existing studies are mostly based on the core principles of economic capacity, responsibility allocation, interpersonal fairness and ecological balance [10], and use quantitative tools such as the Gini coefficient, the Theil index, and the Kakwani index to assess the equity of carbon emissions [11,12]. Second, in the research on the factors influencing carbon emission equity, most scholars regard the income level as the core driver of the phenomenon of carbon emission inequity or disparity [13]. However, comprehensive analyses incorporating multidimensional factors such as economic development, urbanization, energy systems, and industrial composition remain limited. While these studies have revealed the multi-faceted causes of carbon emission disparities [14], they often overlook the role of energy transition policies. Third, in the study of carbon emission allocation and compensation mechanisms grounded in equity principles, existing literature predominantly focuses on constructing equitable and efficient allocation schemes [15,16] but lacks a systematic quantitative analysis of the impacts of these schemes on the equity of carbon emission in practical application. This theory-practice disconnect is particularly pronounced in urban governance scenarios, when one-dimensional models encounter dynamic and complex urban systems, their efficacy tends to be drastically diminished [17]. Specifically, in-depth empirical analyses of how energy policy design and implementation impact carbon emission equity are still insufficient.

The complex economic and demographic dynamics of cities pose a great challenge to a one-size-fits-all approach. China's prefecture-level cities are characterized by differences in industrial structure and development endowments, which require governance frameworks to be embedded in local contexts. Notably, these contextual frameworks extend beyond mere socioeconomic and demographic parameters but are instead deeply rooted in enduring cultural norms [18]. Confucianism, as a deep cultural tradition, provides a unique perspective for understanding carbon equity issues [19]. Confucianism, with its ecological philosophy of "the unity of heaven and mankind" and ethical concepts centered on "benevolence", provides a deep cultural foundation for sustainable development [20]. This philosophical system closely integrates environmental issues with social, ethical, cultural, and economic dimensions, emphasizing the symbiosis and harmony between humans and nature [21]. In addition, "moderation", an important principle of Confucianism, advocates the pursuit of fairness and justice while tailoring governance strategies to specific contexts [22]. When applied to urban energy transition, this philosophical framework endows policy design with humanistic concerns and simultaneously emphasizes a strong sense of social responsibility [23]. This cultural logic promotes a delicate balance between the responsibility for carbon emission reduction and the right to development in practice, providing rationality for the realization of carbon equity. The neglect of traditional cultural regulatory mechanisms in existing research fundamentally undermines the cultural appropriateness of policy design, which constitutes the most notable research gap in the current body of theory.

Based on the above discussion, this study aims to address several key questions: first, can energy transition policies promote the realization of urban carbon equity? Second, do energy transition policies and Confucianism exhibit synergistic effects in promoting urban carbon equity? Finally, is there heterogeneity in the impact of the interaction of these two factors on urban carbon equity across cities with different industrial structures and resource endowments? To answer the above questions, this study empirically explores the impact of energy system transition on carbon equity based on unbalanced panel data from 256 Chinese cities from 2000 to 2021 using a progressive differential model, with a special focus on the unique role of Confucianism in this process, reinforcing the understanding of traditional cultural norms as key determinants of sustainable development paths.

Compared to previous studies, the main contributions of this study can be summarized in three aspects: first, by emphasizing the importance of carbon equity, this study proposes a perspective that complements and deepens the existing efficiency-centered emission reduction strategies, takes the right to development and the equity in the distribution of emission rights as the foundational principles, and quantifies urban carbon equity based on the cumulative distribution curves of carbon emissions and economic development at the district and county levels. Second, the energy transition is incorporated into the research framework of urban carbon equity, and the quasi-natural experiment is utilized to explore the theoretical and practical effects of the NEDC policy on urban carbon equity. Last, this study integrates economic and cultural perspectives, offering new insights into achieving carbon equity. This study advances an institutional synthesis framework that innovatively integrates formal regulatory mechanisms with Confucianism. The systematic analysis of the co-evolutionary dynamics demonstrates quantifiable scalability across both Confucian-heritage societies and institutional ecosystems.

2. Institutional Background and Theoretical Hypotheses

2.1. Institutional Background

Energy transition is an important policy tool for promoting sustainable development, in which the development and utilization of renewable energy is considered a key path to reducing environmental pollution and mitigating climate change. To address global climate change, the Kyoto Protocol proposes three flexible cooperation mechanisms to achieve greenhouse gas emission reduction targets through international cooperation, one of which is the Clean Development Mechanism (CDM) [24]. As one of the world's largest carbon emitters, China has prioritized clean energy development and positioned carbon reduction as a cornerstone of its national strategy. Intending to effectively address environmental and climate change issues and implement the goals of the 12th Five-Year Plan for Renewable Energy Development, the Chinese government introduced the concept of "New Energy Demonstration Cities" in 2012. This policy framework seeks to embed energy supply and consumption into the core of urban planning by deeply integrating new energy systems with urban construction. Its objectives are to foster the development of new energy industries and technological advancement, optimize the urban energy consumption structure, and gradually phase out the traditional "high-energy consumption, high-emission, high-pollution" economic model—all with the aim of enhancing cities' sustainable development capabilities [25]. It will gradually get rid of the traditional economic development model characterized by "high energy consumption, high emission, and high pollution" and enhance the sustainable development capability of the city. The declaration mechanism of new energy demonstration cities is characterized by both "top-down" and "bottom-up" approaches. Specifically, the declaration process is authorized by the central government, local governments apply independently according to their conditions, and finally, the central government reviews and approves the application. According to the "New Energy Demonstration City Evaluation Indicator System and Explanation (for Trial Implementation)", the declaring city needs to meet two basic conditions: firstly, the city's comprehensive capacity meets the standard, and secondly, the city's new energy utilization base meets the standard. These conditions are the prerequisites for the declaration work, ensuring that the construction of the demonstration city is scientific and feasible. The National Energy Administration (NEA) conducted a comprehensive assessment of the declared cities based on the basic conditions and evaluation index system, and in January 2014, it announced the list of the first batch of new energy demonstration cities and industrial parks [26]. The spatial distribution pattern of these demonstration cities is shown in Figure 1, covering representative cities in different regions of China, providing practical samples and policy references for the development of energy transition nationwide.

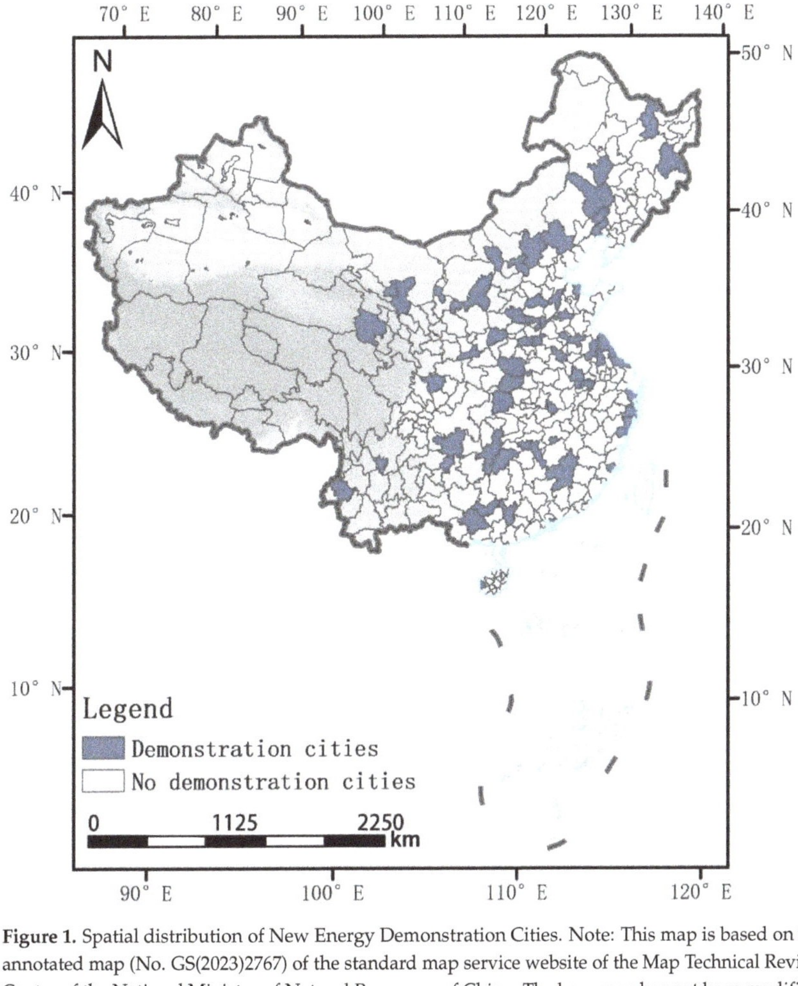

Figure 1. Spatial distribution of New Energy Demonstration Cities. Note: This map is based on the annotated map (No. GS(2023)2767) of the standard map service website of the Map Technical Review Center of the National Ministry of Natural Resources of China. The base map has not been modified.

2.2. Theoretical Hypotheses

In the process of promoting the transformation of urban energy systems, the new energy model city policy may not be conducive to the realization of carbon equity in cities. First, China's decentralized system is prone to the GDP catch-up and environmental "race to the bottom" effects that are common in different counties when weighing environmental and economic development [27]. The counties that rely heavily on traditional energy consumption will blindly accelerate the transformation of the secondary industry into a cleaner one without significantly reducing industrial energy dependence, which will easily lead to the energy consumption rebound effect, resulting in a large scale of energy consumption in the secondary industry in the early stage, increasing carbon emissions, deviating from the allocation of a reasonable amount of emissions, and exacerbating the inequity in carbon emissions in the city [28]. Secondly, from the perspective of constraints, the NEDC policy is a combination of environmental regulation relying on urban space, which explicitly puts forward energy consumption constraints and energy consumption intensity constraint indicators [29]. The "hard constraints" resulting from the command-type environmental regulation policy have a heterogeneous impact on counties with different energy structure endowments, and counties with a high degree of dependence on traditional fossil energy may be penalized for failing to complete the task in time, and thus be forced to take on more responsibility for carbon emissions [30]. Third, from the perspective of guidance, the NEDC policy aims to guide the promotion and utilization of clean technologies and explicitly requires local governments to set up special funds for support, such as cash incentives or tax breaks for new energy enterprises' technological innovation, transformation, and upgrading [31]. However, in the actual situation of decentralized financial governance in districts and counties, the allocation of resources is often affected by the economic strength of districts and counties and the bargaining power of local governments, which results in a regional lock-in effect of green technology innovation. This effect causes districts and counties with poorer resource endowments to lag behind in the application of technology, making it difficult to achieve rapid reductions in carbon emissions [32], thus exacerbating the imbalance in carbon emission levels between districts and counties. The social equity theory emphasizes the fair distribution of social resources and opportunities, and this principle is also applicable to the distribution of carbon emission rights. In the process of policy implementation, different regions may lead to uneven distribution of carbon emission responsibilities due to differences in resource endowment, economic structure, and development level. Therefore, the following hypotheses are proposed.

Hypothesis 1. *Urban energy transition inhibits the realization of urban carbon equity.*

Culture implicitly influences people's way of thinking, cognitive and practical activities [33]. In the context of the implementation of the NEDC policy, the theory of polycentric governance offers valuable insights into the complex interactions among various stakeholders involved, such as central and local governments, different departments, and social organizations. Polycentric governance theory underscores the significance of coordination, collaboration, and mutual checks and balances among these entities. However, in practice, local governments may face a dilemma between pursuing economic growth and achieving carbon emission reduction targets, which can result in deviations during policy implementation. It is here that the essence of Confucianism, with its social and ethical principles, becomes particularly relevant. As a cultural cornerstone deeply ingrained in Chinese society, Confucianism shapes behavioral norms and moral values. It provides a common ethical framework that can guide decision-making and resource allocation [34].

First, Confucianism emphasizes collectivism and believes that social resources should be allocated reasonably according to people's contributions and needs. Under the influence of the Confucian idea of "the way of neutrality", city managers will take into account the interests of the majority group when formulating income distribution policies [35]. In the energy system transition, this concept can guide policymakers to pay more attention to the fair distribution of carbon emission rights and reduce the uneven responsibility of carbon emissions caused by the difference in energy structure. Second, the absence of institutionalized trust mechanisms among stakeholders in carbon quota allocation triggers regulatory arbitrage and strategic gaming, not only impeding policy compliance but also exacerbating systemic resource misallocation through institutional exclusion and reverse selection effects The trust relationship is endogenous to institutions and cultures and is especially built on the foundations of responsibility and traditional culture [36], among others. Confucianism focuses on human relations and moral obligations, and such moral norms are the ideological basis of the credit system, which helps to enhance the trust between different interest groups in the process of carbon emission allocation [37]. In the energy system transition, this trust can promote cooperation and communication between districts and counties, share resources, technologies, and experiences more actively, reduce the waste of resources and conflicts caused by mistrust, and work together to meet the challenges encountered in the energy transition process, and ultimately promote the realization of carbon equity. Third, the unique taxation ideas in Confucianism provide a deep cultural soil and theoretical support for solving urban carbon equity issues. The Confucian principle of "giving to the rich, giving to the poor, and collecting from the poor" emphasizes the balance of income, expenditure, and welfare [38]. In the energy system transition, this principle elevates the societal responsibility of policymakers and implementers, preventing the neglect of carbon equity due to short-term economic gains. It simultaneously guides them to focus on the equitable distribution of financial resources, ensuring that districts and counties with fewer resources receive enhanced support. This dual focus fosters a more just allocation of carbon emissions. Therefore, the following hypotheses are proposed.

Hypothesis 2. *Confucianism can promote the realization of urban carbon equity in the process of energy transition.*

3. Materials and Methods

3.1. Model Setting

3.1.1. Baseline Model

In 2014, China launched the New Energy Demonstration Cities (NEDC) policy. Adhering to "clean, efficient, multi-energy complementarity", NEDC cities integrate urban construction and energy transformation. This paper uses NEDC to characterize energy system transformation [39], employing an empirical model:

$$Gini_{i,t} = \alpha + \beta_1 did_{i,t} + \sum \beta_k X_{i,t} + \sum YearFE + \sum CityFE + \sum (Province \times Year)FE + \varepsilon_{i,t} \quad (1)$$

$$did_{i,t} = Treat_i \times Post_t \quad (2)$$

$Treat_i$ = 1 for NEDC cities, 0 otherwise; $Post_t$ = 1 post-NEDC announcement, 0 otherwise. To mitigate potential confounding factors and ensure robustness, we included a set of control variables $X_{i,t}$. Additionally, city-fixed effects ∑CityFE, year-fixed effects ∑YearFE, and Province-Year fixed effects ∑(Province × Year)FE are included to address potential confounding factors.

3.1.2. Regulatory Effect Model

To verify the role of Confucianism on urban carbon equity in the process of energy system transformation, the following regulatory effect model is constructed:

$$Gini_{i,t} = \alpha + \gamma\, did_{i,t} \cdot \mathrm{Conf}_{i,t} + \beta_1 did_{i,t} + \beta_2 \mathrm{Conf}_{i,t} + \sum \beta_k X_{i,t} + \sum YearFE + \sum CityFE + \sum (Province \times Year)FE + \varepsilon_{i,t} \quad (3)$$

3.2. Variable Selection

3.2.1. Explained Variable

Carbon emission equity means that each unit in the city obtains equal development rights and carbon emission rights allocation space. County-level administrative units in China are grassroots units that combine economic development autonomy with environmental governance responsibilities (Counties refer to county-level administrative units in China, including municipal districts, counties, and county-level cities, which are the basic spatial units for analyzing intra-city developmental disparities). On the one hand, county-level governments have substantial decision-making power in economic activities such as investment attraction and industrial layout, directly affecting regional carbon emission intensity (such as the energy consumption structure of county-level industrial parks). On the other hand, national-level carbon emission policies (such as carbon peaking pilots) need to be implemented through county-level units. Therefore, taking county-level units as the analysis object can accurately capture the matching relationship between development rights and environmental burdens. However, due to significant differences in resource endowments and energy production and consumption structures across counties, coupled with high development fluidity between counties, a one-size-fits-all distribution mechanism is impractical. Additionally, dynamic monitoring through traditional development or emission processes is challenging [40]. Therefore, this paper uses the balance of cumulative distribution patterns of carbon emissions and economic development outcomes as a measure of carbon emission fairness. This approach reflects fairness in the distribution of results and addresses the challenges of dynamic monitoring in development and emission processes. The Carbon Lorentz curve is a manifestation of the development effect and environmental burden brought by the economy and carbon emissions. On the one hand, carbon emissions are essentially generated in the process of economic development of districts and counties; On the other hand, counties have the right to a fair share of carbon emissions commensurate with development. The carbon Lorentz curve at the urban scale quantifies the sorted cumulative distribution function of socioeconomic activities and carbon emission intensity across county-level administrative units. The Gini coefficient, calculated from the area between the Lorenz curve and the absolute fairness line, offers an objective measure of carbon emission distribution fairness, where values closer to 0 indicate greater equity. The formula is as follows [41]:

$$Gini_{i,t} = 1 - \sum_{n=1}^{j} \left(x_{j,t} - x_{j-1,t}\right)\left(y_{j,t} + y_{j-1,t}\right)(j = 1, 2, \ldots, m) \quad (4)$$

where j represents each district and county, $x_{j,t}$ is the reference factor the cumulative percentage of GDP, and $y_{j,t}$ is the cumulative percentage of carbon emissions.

3.2.2. Explanatory Variable

This study uses the New Energy Demonstration Cities policy to characterize energy transition and reveal the impact of energy transition on urban carbon equity. The promotion and deployment of this policy is the result of external forces at the national level. The policy has the characteristics of long-term stability, and the demonstration zone covers many regions in central and western China, providing a rich sample and comparison basis for identifying the effect of the policy. Specifically, the core explanatory variable *did* was set in the study, and the value was 1 after the policy took effect at the location of the policy, and 0 otherwise.

3.2.3. Moderating Variable

The measurement of Confucianism. The quantification of cultural ideologies, particularly Confucian value systems, with methodological approaches, remains a subject of debate. A central focus of this paper is the degree to which regions have been influenced by Confucianism [42], a construct typically quantified using historical data on Confucian temples, academies, scholarly centers, and the number of Confucian scholars.

In the macro-level study, Chen et al. (2021) measured the concentration of Confucianism in a region by the number of scholars in the Ming and Qing dynasties [43]. However, Chen et al. (2020) demonstrated that the number of scholars during the Ming and Qing periods significantly impacted contemporary education levels [44]. This suggests that the influence of scholar counts on other variables may primarily stem from regional education levels, rather than directly reflecting Confucianism. Regarding Wang and Chen (2025), this study employs the number of Confucian temples as a proxy for Confucianism [36], rather than relying on the number of Jinshi scholars. Confucian temples, as structures dedicated to commemorating Confucius, primarily serve sacrificial functions and thereby more purely reflect regional reverence for Confucian traditions. This measure captures the enduring influence of Confucianism, even after historical events such as the May Fourth Movement and the "Breaking the Four Olds" campaign. The current number of Confucian temples used in this analysis highlights the degree to which local societies value and preserve Confucian heritage, offering an indicator of contemporary cultural influence. To assess the influence, this study integrates it as a moderating factor in regression analyses. It is measured by the natural logarithm of the number of Confucian temples, incremented by one.

3.2.4. Control Variables

This paper accounts for a range of factors that could potentially influence both NEDC policy and urban carbon equity [44,45]. Control variables encompass several key economic indicators. Economic Advancement (Dev) is measured by the annual growth rate of gross regional product, indicating the rate of growth of the city's economic development. Income Level (Income), represented by the logarithmic transformation of per capita GDP. Fiscal Pressure (Fsc) is ascertained by the ratio of net financial revenue to fiscal revenue, which provides insight into the financial health and stability of a region. Industrial Structure (Str), quantified by the value added of the secondary industry relative to that of the tertiary industry, offers a measure of economic diversification and the balance between different sectors. Investment level (Invest) is the ratio of actual foreign investment utilization to GDP. The use of foreign capital (FDI), represented by the actual amount of foreign capital used in the current year, reflects technological spillover effects and their impact on urban carbon equity.

3.3. Data Specification

The study sample consists of 256 cities at the prefecture level and above in China, spanning the period from 2000 to 2021. Carbon emissions data were sourced from the latest version of the EDGAR v8.0 database (https://edgar.jrc.ec.europa.eu/report_2023, accessed on 8 September 2023), specifically utilizing its high-resolution fossil energy carbon emissions grid data at a spatial granularity of $0.1^{\circ} \times 0.1^{\circ}$. These grid-level data were aggregated to the city level based on the geographical coordinates of each region, ensuring precise alignment with urban boundaries. Urban characteristic data were obtained from multiple authoritative sources, including the China Urban Statistical Yearbook, relevant

emission, which reflects the ecologically fragile region's responsibility for the low-carbon development. This spatial differentiation highlights the urgency of conducting urban carbon equity research: on the one hand, the traditional efficiency-centered policy tools may lead to the "stronger the stronger, weaker the weaker" Matthew effect, exacerbating the imbalance of inter-regional emission responsibilities; on the other hand, as the core ethical dimension of sustainable development, the disconnection between carbon equity and efficiency may lead to social contradictions and resistance to governance in the low-carbon transition. As China is now in the overlapping period of energy system restructuring and accelerated urbanization, there is an urgent need to break the single evaluation paradigm of "efficiency first" and establish a multidimensional governance framework that takes into account both equity and efficiency.

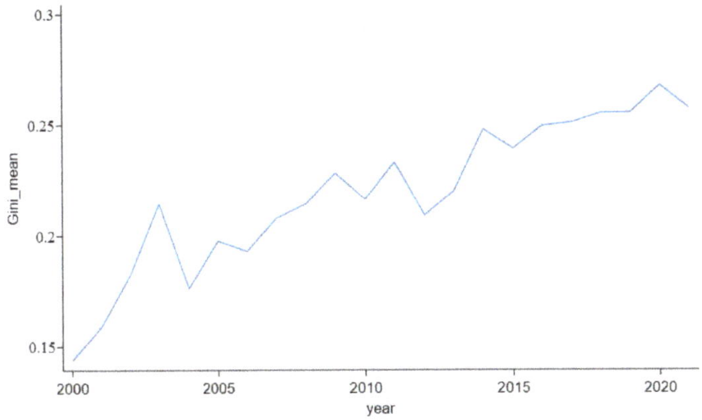

Figure 3. Annual Average Carbon Emission Gini Coefficients in China.

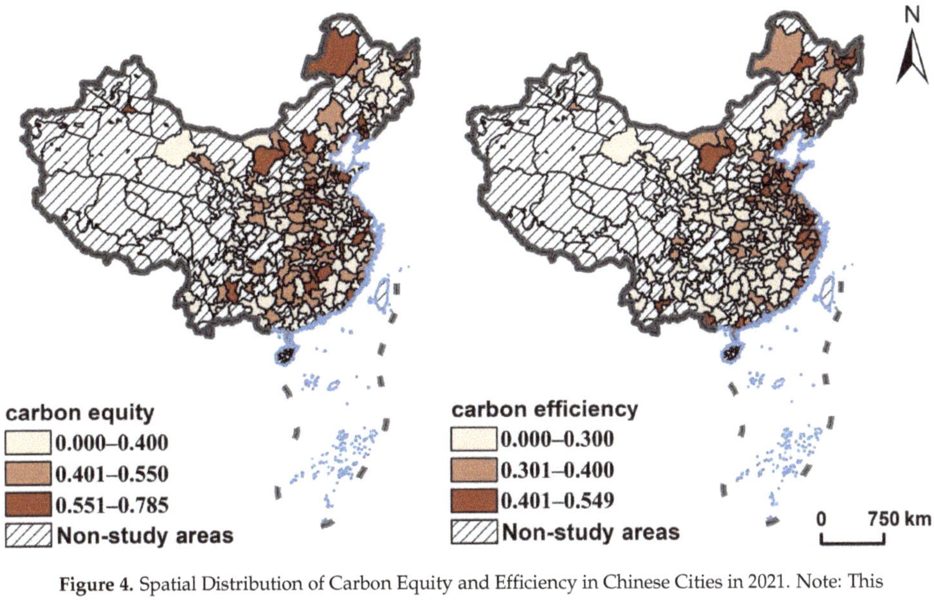

Figure 4. Spatial Distribution of Carbon Equity and Efficiency in Chinese Cities in 2021. Note: This map is based on the annotated map (No. GS(2023)2767) of the standard map service website of the Map Technical Review Center of the National Ministry of Natural Resources of China. The base map has not been modified.

4.2. Benchmark Regression

Columns (1) and (2) in Table 1 present the results of the data analysis based on the baseline regression model. In Column (1), the regression introduces only the policy variable did, while controlling for time and individual fixed effects. The estimated coefficient is 0.032, which is significant at the 5% significance level, suggesting that the transition of the energy system led to a worsening of carbon inequality. Comparing the statistics of the data within the sample, the mean value of the Gini coefficient within the sample is 0.344, which means that the implementation of the policy correspondingly brought about a rise in inequality within the region of about 10 percent. In Column (2), we further incorporate additional control variables and account for both time and individual fixed effects in the regression model. The coefficient on the core explanatory variable remains at 0.027, which is statistically significant at the 10% significance level. This result is highly consistent with the findings in Column (1), reinforcing the robustness of the relationship between the key variable and the outcome. These findings suggest that the transition may have overlooked the equity of carbon emission controls. Considering the common ecological impacts across regions, urban policies do not appear to be equitable in balancing economic development rights and environmental protection responsibilities within cities [46].

Table 1. Benchmark regression results.

	(1) Gini	(2) Gini
did	0.032 **	0.027 *
	(0.016)	(0.015)
Dev.		0.002 *
		(0.001)
Income		0.091 ***
		(0.013)
Str.		0.006
		(0.010)
Invest		−0.057 **
		(0.024)
FDI		−0.084 **
		(0.035)
Fsc.		0.358 ***
		(0.039)
Constant	0.341 ***	−0.465 ***
	(0.004)	(0.119)
Control variables	No	Yes
Year FE	Yes	Yes
City FE	Yes	Yes
Prov. × Year	Yes	Yes
Obs.	3678	3678
R2	0.071	0.165

Note: Standard errors in parentheses; *, **, *** denote $p < 0.10$ $p < 0.05$ $p < 0.01$, respectively.

4.3. Parallel Trend Hypothesis Testing

The validity of the DID approach relies on the assumption of parallel trends between the treatment and control groups. Therefore, this paper investigates the common trend between the experimental and control groups before and after the introduction of the policy. As shown in Figure 5, there is no significant difference in carbon equity between experimental and non-experimental areas before the policy was introduced, which is consistent with the parallel trend hypothesis.

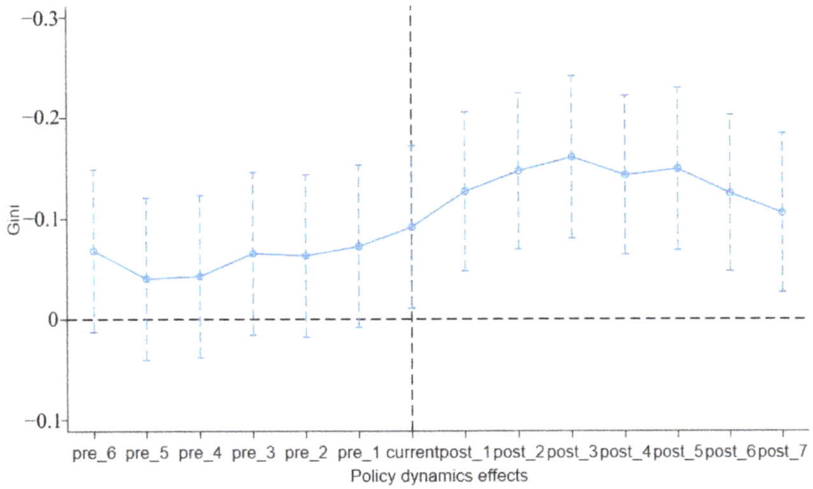

Figure 5. Parallel trend test results.

4.4. Robustness Test

4.4.1. Placebo Test

A placebo test was required to exclude the effects of other possible unobservable factors. A sample of 1000 was randomly selected from the samples, and each sample was randomly assigned to a demonstration city and a non-demonstration city. Figure 6a,b show the results for the treated group and $did_{i,t}$ is randomly selected 1000 times, and the kernel density of the regression coefficients is all around 0 and normally distributed [47].

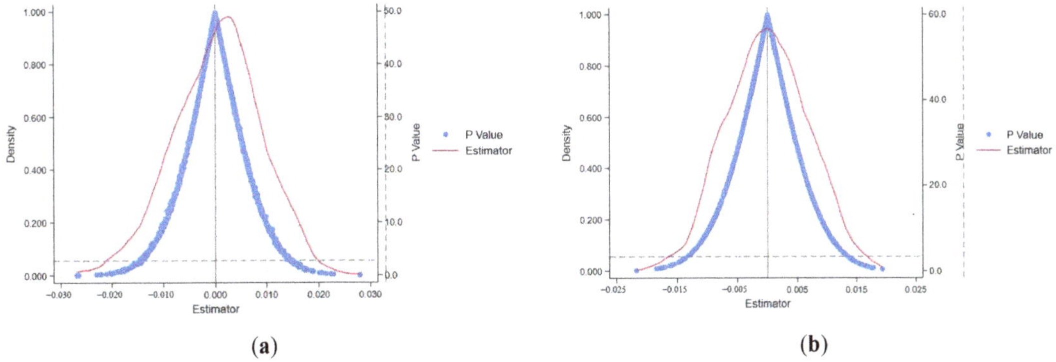

Figure 6. (**a**) Results of a random sampling of 1000 fitted distributions for the treated group cities. (**b**) Results of a random sampling of 1000 fitted distributions for the DID term.

4.4.2. Eliminate Concurrent Policy Interference

Considering that other policies during the sample period may interfere with the causal identification of this paper, this paper collects and organizes three policies that overlap with the study sample interval and may affect the equity of urban carbon emissions: Broadband China, Green Finance, and Energy Saving and Emission Reduction Fiscal Policies [48]. To ensure robust causal identification, we construct policy-specific dummy variables and incorporate them into the benchmark regression model. The results, as shown in Columns (1), (3), and (5) of Table 2, demonstrate that the regression findings remain consistent with the baseline conclusions even after excluding potential interferences from other policies.

Statistically, this indicates that the impact of this NEDC on urban carbon emissions is highly stable. The direction or degree of significance has not changed due to the control of other policies. Economically, this stabilizing negative coefficient implies that the studied policy still has a significant inhibitory effect on urban carbon equity.

Table 2. Excluding other policy effects.

	(1)	(2)	(3)	(4)	(5)	(6)
	Broadband		**Green Finance**		**Fiscity**	
did × Conf.		−0.030 **		−0.031 **		−0.030 **
		(0.013)		(0.013)		(0.013)
did	0.027 *	0.047 **	0.027 *	0.049 **	0.029 *	0.049 **
	(0.015)	(0.022)	(0.015)	(0.022)	(0.015)	(0.022)
Dev.	0.002 *	0.001	0.002 *	0.001	0.003 **	0.001
	(0.001)	(0.001)	(0.001)	(0.001)	(0.001)	(0.001)
Income	0.094 ***	0.095 ***	0.091 ***	0.099 ***	0.090 ***	0.100 ***
	(0.013)	(0.018)	(0.013)	(0.018)	(0.013)	(0.018)
Str.	0.005	−0.018 **	0.006	−0.018 **	0.008	−0.017 **
	(0.010)	(0.009)	(0.010)	(0.009)	(0.010)	(0.009)
Invest	−0.056 **	−0.070 ***	−0.057 **	−0.070 ***	−0.057 **	−0.070 ***
	(0.024)	(0.012)	(0.024)	(0.012)	(0.024)	(0.012)
FDI	−0.082 **	−0.004	−0.084 **	−0.002	−0.085 **	−0.002
	(0.035)	(0.025)	(0.035)	(0.025)	(0.035)	(0.025)
Fsc.	0.362 ***	−0.007	0.358 ***	−0.009	0.352 ***	−0.009
	(0.039)	(0.029)	(0.039)	(0.029)	(0.039)	(0.029)
did × Broadland	−0.030 *	−0.013				
	(0.016)	(0.009)				
did × GF			0.039	−0.002		
			(0.077)	(0.043)		
did × Fiscity					0.078 ***	0.042 ***
					(0.023)	(0.015)
Constant	−0.492 ***	−0.587 ***	−0.465 ***	−0.630 ***	−0.465 ***	−0.644 ***
	(0.120)	(0.179)	(0.119)	(0.176)	(0.119)	(0.176)
Control variables	Yes	Yes	Yes	Yes	Yes	Yes
Year FE	Yes	Yes	Yes	Yes	Yes	Yes
City FE	Yes	Yes	Yes	Yes	Yes	Yes
Prov. × Year	Yes	Yes	Yes	Yes	Yes	Yes
Obs.	3678	3678	3678	3678	3678	3678
R2	0.166	0.802	0.165	0.802	0.168	0.802

Note: Standard errors in parentheses; *, **, *** denote $p < 0.10$ $p < 0.05$ $p < 0.01$, respectively.

4.5. Further Analysis

4.5.1. The Moderating Effect of Confucianism

The organizational and cultural climates, as well as governance perspectives, underscore the interplay of cultural values and governance strategies in advancing sustainable energy transitions [49]. As shown in Columns (1) and (2) of Table 3, the coefficients of the interaction terms are consistently negative, indicating that Confucianism significantly promotes carbon equity. This moderating effect essentially reflects the synergistic governance efficacy of cultural institutions and formal policies—when policy shocks are embedded in a high-trust Confucian cultural environment, transaction costs in policy implementation can be reduced. This effect remains robust in Columns (3) and (4), where we include the interaction term between city classification variables and time. The moderating role of Confucianism in the policy effect of urban carbon emissions remains robust after controlling for other contemporaneous policy disturbances. Columns (2), (4), and (6) of Table 2 show that the coefficients of Confucianism and the policy interaction term did×Conf range from −0.030 to −0.031 and are all significant at the 5% level.

Table 3. Moderating effects of Confucianism.

	(1) Gini	(2) Gini	(3) Gini	(4) Gini
did × Conf.	−0.033 **	−0.031 **	−0.035 ***	−0.036 ***
	(0.013)	(0.013)	(0.013)	(0.013)
did	0.055 **	0.049 **	0.054 **	0.054 **
	(0.022)	(0.022)	(0.022)	(0.022)
Grade × year			0.006 ***	
			(0.001)	
Centre × year				0.005 ***
				(0.001)
Constant	0.344 ***	−0.630 ***	−1.678 ***	−1.690 ***
	(0.002)	(0.176)	(0.291)	(0.315)
Control variables	No	Yes	Yes	Yes
Year FE	Yes	Yes	Yes	Yes
City FE	Yes	Yes	Yes	Yes
Prov. × Year	Yes	Yes	Yes	Yes
Obs.	3678	3678	3653	3653
R2	0.798	0.802	0.803	0.802

Note: Standard errors in parentheses; **, *** denote $p < 0.05$ and $p < 0.01$, respectively.

Under its unique values and ethical framework, Confucianism offers a cultural foundation and innovative pathway for achieving urban carbon equity. This cultural drive promotes trust and cooperation among different stakeholders, concurrently enhancing policy implementation, optimizing resource allocation, and improving policy transparency and fairness [50]. These effects collectively facilitate a more equitable distribution of carbon emissions in the energy system transition. This corroborates theoretical hypothesis 2: informal norms can correct formal institutional implementation biases, especially in fiscal decentralization systems, and cultural capital can be transformed into a flexible regulatory tool for environmental governance.

In Chinese Confucian culture, "benevolence" and "righteousness" originate from family kinship (e.g., filial piety and fraternal duty) and gradually extend to social relationships [51]. Filial piety is a unique characteristic that emphasizes intergenerational emotional ties, distinguishing it from Western cultures. Based on data from the Chinese General Social Survey (CGSS) released in 2021, this paper measures regional Confucianism through the distribution of responses to the questionnaire "Should children bear the responsibility of supporting one's family?". The CGSS is a comprehensive, continuous, nationwide social survey organized by the Renmin University of China, using multistage stratified probability sampling, covering all provincial administrative units in China, with a sample size of about 12,000 per year [52]. The survey data span eight core observation years (2010–2013, 2015, 2017–2018, and 2021). For unsampled years within the 2010–2021 study period, missing values were imputed using the linear interpolation method, which calculates the arithmetic mean of adjacent observed years.

One question regarding aging: "Who do you think should be primarily responsible for the aging of elderly people with children?" Based on this question, we calculated the percentage of people in the region who believe that their children should be primarily responsible for their old age. According to Table 4, the moderating effect of Confucianism, which is measured by survey data, is consistent with that shown in Table 3. For regions with a dense Confucian culture, NEDC implementation has had a positive effect on urban carbon equity, as evidenced by a 0.125-unit decrease in the Gini coefficient (according to the coefficient in Column (2): 0.120−0.245 = −0.125). Confucianism has a significant positive effect on the realization of urban carbon equity.

Table 4. Robustness test of moderating effects.

	(1) Gini	(2) Gini	(3) Gini	(4) Gini
did × Conf. 1	−0.247 **	−0.245 **	−0.281 ***	−0.274 ***
	(0.097)	(0.097)	(0.097)	(0.097)
did	0.120 **	0.120 **	0.137 ***	0.133 ***
	(0.051)	(0.051)	(0.051)	(0.051)
Grade × year			0.010 ***	
			(0.003)	
Centre × year				0.008 ***
				(0.003)
Constant	0.361 ***	−0.667 ***	−2.316 ***	−2.399 ***
	(0.003)	(0.255)	(0.560)	(0.641)
Control variables	No	Yes	Yes	Yes
Year FE	Yes	Yes	Yes	Yes
City FE	Yes	Yes	Yes	Yes
Obs.	1846	1846	1846	1846
R2	0.797	0.800	0.801	0.801

Note: Standard errors in parentheses; **, *** denote $p < 0.05$ and $p < 0.01$, respectively.

4.5.2. Heterogeneity Analysis

The structural heterogeneity of cities engenders differential policy responsiveness, necessitating tailored governance frameworks for sustainable energy transitions. Cities' resource endowments and industrial configurations create distinct transition constraints. Accounting for the heterogeneity in cities' industrial structures and resource endowments enhances the understanding of Confucianism's role in urban energy transitions and how policy and cultural interventions contribute to achieving carbon equity. Old industrial bases and resource-dependent cities face significant transformation resistance: on the one hand, the capital mismatch and sunk cost effect of traditional industries prolong the lag period of the policy's effectiveness, leading to structural delay in carbon emission reduction; on the other hand, the "resource curse" and economic lock-in effect of resource-dependent cities may trigger the risk of trans-regional transfer of carbon emissions, and such cities tend to outsource high-carbon production activities under policy pressure, creating a "beggar-thy-neighbor" situation, which exacerbates the carbon burdens of neighboring cities and damage to regional ecosystem.

Based on the national urban resource types delineated in the National Sustainable Development Plan for Resource-dependent Cities (2013–2020), the sample cities are categorized into resource-dependent cities and non-resource-dependent cities [50]. As shown in columns (3) and (4) of Table 5, for non-resource-dependent cities, the synergy between Confucianism and NEDC policy significantly promotes urban carbon equity; for resource-dependent cities, the positive effect of Confucianism in promoting urban carbon equity in the energy transition process is not significant. The relatively diversified industrial structure of non-resource-dependent cities makes them more flexible and sustainable in their organizational scope and governance thinking, and better able to translate the equity concepts of Confucianism into practical policies. The long-term dependence of resource cities on specific resources and their mindset of "prioritizing growth over governance" limit the influence of Confucianism in energy transition [53].

Table 5. Heterogeneity analysis.

	(1) Non-Old Industrial Base	(2) Old Industrial Base	(3) Non-Resource- Dependent	(4) Resource- Dependent
did × Conf.	−0.097 ***	−0.030	−0.054 ***	−0.012
	(0.017)	(0.033)	(0.019)	(0.021)
did	0.149 ***	0.122 **	0.070 **	0.029
	(0.030)	(0.048)	(0.032)	(0.034)
Constant	0.365 *	−0.402	−0.237	−1.475 ***
	(0.214)	(0.273)	(0.232)	(0.334)
Control variables	Yes	Yes	Yes	Yes
Year FE	Yes	Yes	Yes	Yes
City FE	Yes	Yes	Yes	Yes
Prov. × Year	Yes	Yes	Yes	Yes
Obs.	2412	1209	2204	1414
R2	0.804	0.175	0.786	0.819

Note: Standard errors in parentheses; *, **, *** denote $p < 0.10$ $p < 0.05$ $p < 0.01$, respectively.

To examine whether the urban industrial system significantly influences the synergistic effect of energy transition policies and Confucianism, this paper focuses on the industrial system outlined in the "National Old Industrial Base Adjustment and Reform Plan (2013–2022)" [54], as shown in columns (1) and (2) of Table 5. In non-old industrial base cities, the synergy between Confucianism and the NEDC policy significantly enhances urban carbon equity. However, in old industrial-based cities, the positive role of Confucianism in promoting urban carbon equity during the energy transition process is less pronounced. Old industrial base cities have long been characterized by a single industrial structure and a high concentration of large state-owned enterprises. While the development model of "one industry alone" has brought huge economic benefits, its management mode and benefit distribution mechanism are often relatively rigid, and the endogenous motivation to seek paths for energy use transformation is relatively weak. In this highly administrative and planned economy context [55], the influence of Confucianism has been weakened, making it difficult to promote substantive energy transformation and carbon equity.

5. Discussion

5.1. Energy Transition and Carbon Equity

The energy transition is a key pathway to achieving the goal of carbon neutrality and involves a shift from traditional fossil energy sources to renewable energy sources, as well as improvements in energy efficiency. According to the theory of equity-efficiency trade-offs in economics, the energy transition aims to improve energy efficiency and reduce carbon emissions, but this process may exacerbate interregional inequities [56]. Carbon equity, as an organic unity of the three elements of environment, development, and equity, is a key issue affecting the low-carbon transition and social welfare. Scholars have extensively studied the conceptual connotations, measurement indicators, and evolutionary trends of carbon equity at the global, national, and regional scales [57], but there are relatively few empirical studies measuring carbon equity at the city level, especially lacking in the in-depth excavation of intra-city carbon inequality. Examined from an individual's behavioral perspective, the equity of urban carbon emissions constitutes a key factor influencing individual behavioral decisions. The equity of carbon emissions directly affects the behavioral dynamics of district and county governments. Any bias or unfairness in the distribution of carbon emission rights among different administrative units within the city may weaken the behavioral motivation of the district and county governments, which in turn constitutes an obstacle to the realization of low-carbon goals.

From a macro perspective, when promoting carbon emission reductions, district and county governments focus on local economic development and emission performance while concurrently monitoring the development outcomes and emission status of peer administrative regions. If districts and counties pursue carbon emission targets in isolation rather than synergistically, it often leads to an inefficient allocation of resources and fails to maximize the overall benefits of the city under the framework of welfare economics theory. This study quantifies urban carbon equity through the cumulative distribution curves of carbon emissions and economic development at the district and county levels and proposes an analytical framework based on the principles of the right to development and equity in the distribution of emission rights, which provides a complementary perspective to the existing efficiency-centered emission reduction strategies. The study finds that although the Gini coefficient of carbon emissions at the national level in China is within a relatively average range (Gini coefficient < 0.4), its fluctuating upward trend suggests that the inequity in the distribution of carbon emissions is increasing. In addition, China's new energy model city policy exacerbates the inequity of urban carbon emissions. This result is consistent with the theory of policy externalities in economics. Although the NEDC policy has achieved remarkable results in technology promotion and energy structure optimization, its implementation has exacerbated the unequal distribution of resources among regions. Early clarification of the impact of energy transition policy implementation on carbon equity in cities is essential for achieving sustainable development goals and promoting coordinated regional development.

5.2. New Energy Policy and Informal Cultural Norms

The realization of carbon equity, while relying on the upgrading of formal systems, technologies, and facilities at the "hardware" level, necessitates parallel advancement in informal systems, such as cultural and conceptual frameworks, at the "software" level [58]. From the perspective of behavioral economics, cultural informal institutions directly influence the implementation effectiveness of energy policies by shaping the decision-making preferences of individuals and organizations. In a cultural atmosphere that emphasizes collective responsibility and environmental friendliness, enterprises are more inclined to incorporate sustainable development into their strategic goals and proactively adjust their production technologies and energy structures to meet the requirements of energy policies. Such culture-driven corporate ethics can reduce the psychological costs for enterprises to implement environmental regulations and decrease resistance to policy implementation. Analyzing from the perspective of transaction cost theory, cultural informal institutions help build trust networks and reduce information asymmetry among enterprises, the government, and consumers. When trust relationships are established between enterprises and policymakers based on shared cultural values, the costs of policy communication and supervision significantly decrease. As a result, enterprises are more willing to cooperate in policy implementation, voluntarily disclose carbon emission data, and participate in the carbon trading market. At the government level, cultural informal institutions have a significant impact on the implementation of energy policies by influencing the administrative ethics and risk preferences of officials. Although innovations in formal institutions, including environmental regulations, are crucial for China's low-carbon economic transformation, under the unique Chinese government governance model, political promotion tournaments and fiscal decentralization lead to intense competition among local governments [59]. Local officials face information asymmetry and goal conflicts when implementing central environmental policies. In the absence of the reinforcement of public responsibility by cultural informal institutions, officials may, under the pressure of promotion tournaments, prioritize high-energy-consuming industrial projects that can quickly boost their political achievements, resulting in the weakening of environmental policy implementation.

Compared with existing studies, this paper particularly emphasizes the role of cultural informal institutions in urban carbon emissions and examines the synergistic effects and heterogeneities between energy transition policies and cultural informal institutions on carbon emission equity. The study finds that Confucianism, as a unique informal institution in China, interacts with formal energy policies and influences urban carbon equity. Its emphasis on "harmony without uniformity" and "benevolence towards others" helps create a cultural environment that emphasizes public interests and long-term development [60]. This changes the risk preferences of officials, making them more inclined to implement low-carbon development policies and achieve sustainable development goals by promoting industrial structure upgrading [61]. This culture-driven transformation of administrative ethics can effectively mitigate the opportunistic behavior of local governments in environmental policy implementation and enhance the effectiveness of energy policy implementation. In non-resource-based cities and non-old industrial bases, where the service industry accounts for a high proportion and the cost of industrial transformation is low, cultural informal institutions can more easily enhance public environmental awareness and the long-term governance willingness of the government. Therefore, Confucianism demonstrates significant synergistic effects with new energy demonstration city policies. However, in resource-dependent cities and old industrial bases, due to the significant lock-in effect of traditional industries, the promoting effects of Confucianism on enterprise transformation and the administrative thinking of the government are offset by asset specificity and sunk costs, resulting in insignificant policy synergistic effects [62]. Such heterogeneities call for the construction of differentiated policy synergy mechanisms. In non-resource-based cities, efforts should be made to strengthen the integration of cultural informal institutions and policy tools. Methods such as establishing a low-carbon cultural points system and incorporating environmental protection into enterprise credit evaluations can be adopted to form a positive cycle of "cultural guidance—policy incentives". In resource-dependent cities, it is necessary to improve the constraint mechanisms of formal institutions. Measures such as the construction of carbon emission trading markets and the normalization of environmental inspections should be taken. Only by compensating for regional differences in cultural factors through institutional design can the dual goals of energy transition and carbon emission equity be achieved.

5.3. Research Limitations and Future Prospects

While this study has yielded valuable insights into the impact of new energy model city policies on urban carbon equity and the moderating role of Confucianism, limitations persist. Quantitative methods, though precise, may be less effective in interpreting complex socio-cultural dynamics. Qualitative research methods, on the other hand, can dig deeper into the mechanisms and logic behind these phenomena, but the generalizability of their results may be limited. Future research could address these limitations by integrating quantitative and qualitative methodologies, thereby providing a more comprehensive perspective on energy transitions and carbon equity.

The spatial spillover effects of energy policies are another area of concern. Policies may have spillover effects on neighboring regions in the course of implementation, which are complex and heterogeneous. For example, as economic activities between regions become more frequent, demonstration cities may have demonstration and diffusion effects on the surrounding areas, which in turn affect the carbon emission behavior of neighboring cities. Future research can analyze the spatial spillover effects of energy policies and their impact on carbon equity by constructing spatial econometric models. Dynamic spatial panel models incorporating time-varying connectivity matrices could capture evolving inter-city interactions in economic, technological, and demographic dimensions. Researchers might employ spatial regime-switching models to identify critical thresholds in policy diffusion, complemented by difference-in-differences designs with spatial lags to disentangle demonstration effects from competitive behaviors.

Confucianism, as a cultural informal system, has a moderating role in the energy transition, but its specific transmission mechanism has not been quantitatively analyzed. The cultural informal system influences carbon emissions through channels such as social trust and governance thinking, but the specific paths of action and time lags of these channels need further study. This begins with tracing how Confucian cultural endowments influence mediating mechanisms like social trust networks and corporate governance practices, ultimately shaping emission-intensive production and consumption patterns. Crucially, future work should investigate institutional complementarity between Confucian norms and formal regulations through policy experiments in pilot zones, supported by longitudinal databases tracking cultural-institutional synergies across development stages.

Emerging frontiers call for reimagining carbon equity and energy justice through three lenses: (1) examining how smart city technologies mediate between cultural systems and carbon governance, particularly through AI-driven simulations incorporating cultural variables; (2) developing intergenerational equity models that embed Confucian filial ethics into long-term decarbonization commitments; and (3) constructing transnational frameworks analyzing how localized cultural systems interface with global climate regimes. Realizing these ambitions necessitates forming transdisciplinary consortia integrating energy engineering, cultural geography, and computational social science expertise—ultimately forging context-sensitive transition frameworks that harmonize technological pathways with cultural-institutional ecosystems.

6. Conclusions and Recommendations

6.1. Conclusions

Based on panel data of 256 cities across China from 2000 to 2021, this paper characterizes urban energy transition with NEDC policy, empirically examines the urban carbon equity effect of energy system transition using the DID method, and explores the moderating role of Confucianism in it. The findings of the study include the following:

(1) China's national carbon emission Gini coefficient, though currently below 0.4, reveals an upward trend from 2000 to 2021, indicating growing inequality in carbon distribution. (2) The NEDC policy appears to worsen urban carbon inequality, a finding that survives rigorous testing, including parallel trend and placebo tests, and remains robust after accounting for confounding policies and factors. (3) Confucianism positively moderates energy transitions, fostering urban carbon equity. (4) The impact of policy synergy between Confucianism and NEDC policy on urban carbon equity is characterized by typical heterogeneity. Specifically, the synergistic effect of Confucianism and NEDC policy is more significant in non-old industrial base cities and non-resource-dependent cities; the positive effect of Confucianism on promoting carbon equity is not obvious in resource-dependent cities and old industrial base cities.

6.2. Policy Recommendations

Optimizing NEDC policy according to local conditions. The policy framework should be adjusted in regions where carbon emissions are unevenly distributed or where the NEDC policy has proven ineffective, ensuring alignment with regional realities. For resource-dependent cities and old industrial base cities, where the positive impact of Confucianism on carbon equity is less pronounced, increased policy support and financial investment are critical. Targeted interventions such as subsidies for green technology adoption, incentives for economic diversification, and support for energy system transformation can help break path dependencies. Culturally resonant areas could amplify bottom-up community-driven transitions anchored in Confucian collectivism. This multi-layered approach—spatially differentiated regulation, cultural-policy synergy, redistributive market mechanisms, and adaptive governance—constitutes a coherent framework to reconcile decarbonization imperatives with interregional equity, aligning China's climate trajectory with its ecological civilization ethos.

Strengthen the organic integration of Confucianism promotion and energy transition policies. First, Confucianism's ecological and ethical principles should be embedded in policy communication to increase public awareness and acceptance of renewable energy, thereby strengthening the social foundation for policy implementation. Second, enterprises can be encouraged to adopt Confucian values of responsibility and integrity to enhance environmental stewardship. This can be achieved through incentives for green technological innovation and sustainable industrial upgrades. Finally, aligning Confucian ethical frameworks with modern governance practices can drive green innovation and support equitable energy transitions.

Promote cooperation and coordination of regional carbon emissions. To mitigate the persistent rise in national carbon inequality, a dual-track redistribution system should be established. First, a unified carbon trading platform must prioritize equitable initial allowance allocation, embedding compensatory mechanisms for historically marginalized regions. Second, industrial relocation partnerships should be formalized, requiring high-carbon-capacity regions to provide technology transfer and green industrial spillovers to less-developed counterparts. The framework could be reinforced through fiscal equalization policies that redirect carbon tax revenues toward renewable energy infrastructure in carbon-vulnerable regions.

Author Contributions: Conceptualization, Z.Y. and H.Y.; methodology, Z.Y.; software, Z.Y.; validation, Z.Y., H.Y. and J.Z.; formal analysis, Z.Y.; investigation, H.Y.; resources, J.Z.; data curation, H.Y.; writing—original draft preparation, Z.Y.; writing—review and editing, J.Z.; visualization, H.Y.; supervision, J.Z. and Z.Y.; project administration, J.Z.; funding acquisition, J.Z. All authors have read and agreed to the published version of the manuscript.

Funding: This research was funded by the National Natural Science Foundation of China [grant number 41771187] and the Postgraduate Innovative Research Fund of the University of International Business and Economics.

Data Availability Statement: The original contributions presented in the study are included in the article, further inquiries can be directed to the corresponding author.

Conflicts of Interest: The authors declare no conflicts of interest.

Abbreviations

The following abbreviations are used in this manuscript:

NEDC	New Energy Demonstration Cities;
CDM	Clean Development Mechanism.

References

1. Bieszk-Stolorz, B.; Dmytrów, K.; Pietrzak, M.B. Quality of Consumed Energy as a Key Element in the Development of Processes of Energy Transformation in the European Union Countries. *Energies* **2025**, *18*, 460. [CrossRef]
2. Koufi, A.; Ziat, Y.; Belkhanchi, H. Study of the Gravimetric, Electronic and Thermoelectric Properties of XAlH3 (X= Be, Na, K) as hydrogen storage perovskite using DFT and the BoltzTrap Software Package. *Sol. Energy Sustain. Dev. J.* **2024**, *2024*, 53–66. [CrossRef]
3. Laghlimi, C.; Moutcine, A.; Ziat, Y.; Belkhanchi, H.; Koufi, A.; Bouyassan, S. Hydrogen, Chronology and Electrochemical Production. *Sol. Energy Sustain. Dev. J.* **2024**, *2024*, 22–37. [CrossRef]
4. He, Y.; Zhang, X.; Zhang, Y. Can new energy policy promote corporate total factor energy efficiency? Evidence from China's new energy demonstration city pilot policy. *Energy* **2025**, *318*, 134782, ISSN 0360-5442. [CrossRef]
5. Wang, N.; Qu, Z.; Li, J.; Zhang, Y.; Wang, H.; Xi, H.; Gu, Z. Spatial-temporal patterns and influencing factors of carbon emissions in different regions of China. *Environ. Res.* **2025**, *276*, 121447, ISSN 0013-9351. [CrossRef]
6. Alvarez, D.I.; González-Ladrón-de-Guevara, F.; Rojas Espinoza, J.; Borge-Diez, D.; Galindo, S.P.; Flores-Vázquez, C. The Evolution of AI Applications in the Energy System Transition: A Bibliometric Analysis of Research Development, the Current State and Future Challenges. *Energies* **2025**, *18*, 1523. [CrossRef]
7. Boyacı, S.; Kocięcka, J.; Jagosz, B.; Atılgan, A. Energy Efficiency in Greenhouses and Comparison of Energy Sources Used for Heating. *Energies* **2025**, *18*, 724. [CrossRef]
8. Wei, C.; Ni, J.; Li, C. CO_2 allocation and equity issues under China's carbon neutrality targets: Recent advances and a review. *China Econ. Rev.* **2024**, *83*, 102108, ISSN 1043-951X. [CrossRef]
9. Schmidt, V.A. Taking Ideas and Discourse Seriously: Theorizing Ideas and Their Use in the Study of Institutions. *Ann. Am. Acad. Political Soc. Sci.* **2010**, *634*, 126–148. [CrossRef]
10. Ning, X.; Ye, X.; Pan, Y.; Li, Y.; Sun, H. Towards urban-rural development and equity in Western China: Insights gained from residential buildings' carbon emissions and per capita income. *Energy Sustain. Dev.* **2025**, *85*, 101633, ISSN 0973-0826. [CrossRef]
11. Pan, X.; Teng, F.; Du Pont, Y.R.; Wang, H. Understanding equity–efficiency interaction in the distribution of global carbon budgets. *Adv. Clim. Change Res.* **2023**, *14*, 13–22, ISSN 1674-9278. [CrossRef]
12. Yang, M.; Liang, X.; Chen, H.; Ma, Y.; Aihemaiti, G. Regional carbon inequality prediction in China based on shared socioeconomic pathways: A human well-being equity perspective. *Sustain. Prod. Consum.* **2024**, *50*, 486–498, ISSN 2352-5509. [CrossRef]
13. Tian, M.; Hu, Y.; Wang, H.; Li, C. Regional allowance allocation in China based on equity and efficiency towards achieving the carbon neutrality target: A composite indicator approach. *J. Clean. Prod.* **2022**, *342*, 130914, ISSN 0959-6526. [CrossRef]
14. Zhou, R.; Guan, S.; He, B. The Impact of Trade Openness on Carbon Emissions: Empirical Evidence from Emerging Countries. *Energies* **2025**, *18*, 697. [CrossRef]
15. Rao, A.; Kumar, S.; Gupta, P. Decarbonization dilemmas: Evaluating the impact on energy security and equity in emerging economies. *Energy Econ.* **2025**, *145*, 108448, ISSN 0140-9883. [CrossRef]
16. Li, Y.; Wei, Y.; Zhang, X.; Tao, Y. Regional and provincial CO_2 emission reduction task decomposition of China's 2030 carbon emission peak based on the efficiency, equity and synthesizing principles. *Struct. Change Econ. Dyn.* **2020**, *53*, 237–256, ISSN 0954-349X. [CrossRef]
17. Wang, Q.; Gao, Z.; Ning, J. Model-based assessment of the pattern differences and the equity of national carbon emissions in China during 2000–2010. *J. Clean. Prod.* **2015**, *103*, 696–704, ISSN 0959-6526. [CrossRef]
18. Chen, F.; Lei, J.; Liu, Z.; Xiong, X. A Comparative Study on the Average CO_2 Emission Factors of Electricity of China. *Energies* **2025**, *18*, 654. [CrossRef]
19. Chaudary, M.S.A. Lithium dreams, local struggles: Navigating the geopolitics and socio-ecological costs of a low-carbon future. *Energy Res. Soc. Sci.* **2025**, *121*, 103952. [CrossRef]
20. Wang, L.; Chen, S.; Wang, Y.; Wu, Y. From ancient philosophy to modern governance: Confucianism shaping low-carbon corporate transformation in China's listed companies. *J. Asian Econ.* **2024**, *95*, 101835. [CrossRef]
21. Dong, Z.; Li, H. The impact of Confucianism on the efficiency of enterprises green innovation. *Financ. Res. Lett.* **2023**, *58*, 104233. [CrossRef]
22. Wang, X.; Lo, K. Civil society, environmental litigation, and Confucian energy justice: A case study of an environmental NGO in China. *Energy Res. Soc. Sci.* **2022**, *93*, 102831. [CrossRef]
23. Yang, X.; Weber, A. Who can improve the environment—Me or the powerful others? An integrative approach to locus of control and pro-environmental behavior in China. *Resour. Conserv. Recycl.* **2019**, *146*, 55–67. [CrossRef]
24. Birkin, F.; Margerison, J.; Monkhouse, L. Chinese environmental accountability: Ancient beliefs, science and sustainability. *Resour. Environ. Sustain.* **2021**, *3*, 100017. [CrossRef]
25. Xu, Y.; Peng, J.; Tian, J.; Fu, S.; Hu, C.; Fu, S.; Feng, Y. The impact and mechanism analysis of Clean Development Mechanism on the synergistic effects of pollution mitigation and carbon reduction. *Environ. Res.* **2024**, *260*, 119659. [CrossRef]

26. Hou, Y.; Yang, M.; Ma, Y.; Zhang, H. Study on city's energy transition: Evidence from the establishment of the new energy demonstration cities in China. *Energy* **2024**, *292*, 130549. [CrossRef]
27. Zhang, Q.; Huang, X.; Xu, Y.; Bhuiyan, M.A.; Liu, P. New energy demonstration city pilot and green energy consumption: Evidences from China. *Energy Rep.* **2022**, *8*, 7735–7750. [CrossRef]
28. Che, S.; Wang, J.; Chen, H. Can China's decentralized energy governance reduce carbon emissions? Evidence from new energy demonstration cities. *Energy* **2023**, *284*, 128665. [CrossRef]
29. Wu, J.; Zuidema, C.; Gugerell, K. Experimenting with decentralized energy governance in China: The case of New Energy Demonstration City program. *J. Clean. Prod.* **2018**, *189*, 830–838. [CrossRef]
30. Zhang, Z.; Luo, C.; Zhang, G.; Shu, Y.; Shao, S. New energy policy and green technology innovation of new energy enterprises: Evidence from China. *Energy Econ.* **2024**, *136*, 107743. [CrossRef]
31. Wang, Q.; Yi, H. New energy demonstration program and China's urban green economic growth: Do regional characteristics make a difference? *Energy Policy* **2021**, *151*, 112161. [CrossRef]
32. Manta, A.G.; Gherțescu, C.; Bădîrcea, R.M.; Manta, L.F.; Popescu, J.; Lăpădat, C.V.M. How Does the Interplay Between Banking Performance, Digitalization, and Renewable Energy Consumption Shape Sustainable Development in European Union Countries? *Energies* **2025**, *18*, 571. [CrossRef]
33. Zhao, C.; Dong, K.; Lee, C.-C. Carbon lock-in endgame: Can energy trilemma eradication contribute to decarbonization? *Energy* **2024**, *293*, 130662. [CrossRef]
34. Kung, J.K.S.; Ma, C. Can cultural norms reduce conflicts? Confucianism and peasant rebellions in Qing China. *J. Dev. Econ.* **2014**, *111*, 132–149. [CrossRef]
35. Chen, S.; Chen, Y.; Jebran, K. Trust and corporate social responsibility: From expected utility and social normative perspective. *J. Bus. Res.* **2021**, *134*, 518–530. [CrossRef]
36. Wang, Y.; Chen, Y. Confucianism and farmers' pro-environmental behaviors: Self-interest or mutual benefit? *J. Environ. Psychol.* **2025**, *102*, 102548. [CrossRef]
37. Chen, L.; Duan, L. Can informal environmental regulation restrain air pollution? Evidence from media environmental coverage. *J. Environ. Manag.* **2025**, *377*, 124637. [CrossRef]
38. Ip, E.C. Planetary health ethics: A Confucian alternative. *J. Clim. Change Health* **2025**, *22*, 100404. [CrossRef]
39. Cheng, M.; Fan, Z.; Feng, C.; Tian, B. Cultural norms and tax compliance: Evidence from China. *J. Econ. Behav. Organ.* **2024**, *227*, 106720. [CrossRef]
40. Wan, J.; Wang, Y.; Wang, Y. Promoting or Hindering: The Impact of ESG Rating Differences on Energy Enterprises' Green Transformation—A Causal Test from Double Machine-Learning Algorithms. *Energies* **2025**, *18*, 464. [CrossRef]
41. Pfaff, A.; Velez, M.A.; Broad, K.; Hamoudi, A.; Taddei, R. Contracts versus trust for transfers of ecosystem services: Equity and efficiency in resource allocation and environmental provision. *Water Resour. Econ.* **2019**, *28*, 100118. [CrossRef]
42. Tian, Y. Will Informal Institutions Affect ESG Rating Divergence? Evidence from Chinese Confucian Culture. *Sustainability* **2024**, *16*, 9951. [CrossRef]
43. Chen, H.; Shen, Q.; Li, H. Settlement and the nature of the dawn: The influence of Confucianism on farmers' migrant labor. *Contemp. Econ. Sci.* **2021**, *43*, 124–132. [CrossRef]
44. Chen, T.; Kung, J.K.S.; Ma, C.C. Long live Keju! The persistent effects of China's civil examination system. *Econ. J.* **2020**, *130*, 2030–2064. [CrossRef]
45. Zhang, Y.; Huang, H. The impact of digital economy development on inequity of carbon emissions. *Resour. Sci.* **2023**, *45*, 1223–1238. [CrossRef]
46. Qian, Y.; Pu, Z. Can digital innovation reduce carbon inequality within cities? *China Popul. Resour. Environ.* **2019**, *34*, 64–74. [CrossRef]
47. Fraser, T.; Chapman, A.J.; Shigetomi, Y. Leapfrogging or lagging? Drivers of social equity from renewable energy transitions globally. *Energy Res. Soc. Sci.* **2023**, *98*, 103006. [CrossRef]
48. Lee, C.; Ogata, S. Every coin has two sides: Dual effects of energy transition on regional sustainable development—A quasi-natural experiment of the New Energy Demonstration City Pilot Policy. *Appl. Energy* **2025**, *390*, 125772. [CrossRef]
49. Hua, W.; Wu, Z.; Gao, C.; Luo, L. Sustaining resource-oriented sanitation behaviors through informal institutions: Evidence from China. *J. Clean. Prod.* **2023**, *425*, 138886. [CrossRef]
50. Yu, Z.; Wang, Z.; Ma, M.; Ma, L. The impact of carbon leakage from energy-saving targets: A moderating effect based on new-energy model cities. *Appl. Energy* **2024**, *375*, 124113. [CrossRef]
51. Park, M.; Chesla, C. Revisiting Confucianism as a conceptual framework for Asian family study. *J. Fam. Nurs.* **2007**, *13*, 293–311. [CrossRef] [PubMed]
52. Wu, L.; Wang, G. Fairness, Trust, and Well-Being Among Young Adults: Evidence from 2021 Chinese General Social Survey (CGSS). *Healthcare* **2024**, *12*, 2186. [CrossRef] [PubMed]

53. Dong, X.; Zhang, J.F. Heterogeneity of regional carbon emission markets in China: Evidence from multidimensional determinants. *Energy Econ.* **2024**, *138*, 107835. [CrossRef]
54. Xing, P.; Wang, Y.; Ye, T.; Sun, Y.; Li, Q.; Li, X.; Li, M.; Chen, W. Carbon emission efficiency of 284 cities in China based on machine learning approach: Driving factors and regional heterogeneity. *Energy Econ.* **2024**, *129*, 107222. [CrossRef]
55. Mao, Q.; Mao, X. The shaping of inter-regional industrial linkages by institutional and cultural division in China: Characteristics and differences. *Econ. Anal. Policy* **2024**, *81*, 1113–1132. [CrossRef]
56. Dong, S.; Wang, X.C.; Dong, X.; Kong, F. Unsustainable imbalances in urbanization and ecological quality in the old industrial base province of China. *Ecol. Indic.* **2024**, *158*, 111441. [CrossRef]
57. Chen, J.; Liu, S.; Zhang, X. Resources coupled with executive authority: Implications of relocating government administrative headquarters for local economic development. *World Dev.* **2025**, *185*, 106798. [CrossRef]
58. Xiong, Y.; Tian, X.; Kang, A. A fine-scale atlas reveals the systematic imbalance between carbon emissions and economic benefits along value chain within China. *Sci. Total Environ.* **2024**, *957*, 177311. [CrossRef]
59. Wu, W.; Zhang, N.; Li, A.; Chen, Y. The path to global climate justice: From the perspective of regional discrepancy in embodied carbon emissions. *Environ. Impact Assess. Rev.* **2024**, *105*, 107410. [CrossRef]
60. Goggins, G.; Rau, H.; Moran, P.; Fahy, F.; Goggins, J. The role of culture in advancing sustainable energy policy and practice. *Energy Policy* **2022**, *167*, 113055. [CrossRef]
61. Li, X. Local government decision-making competition and regional carbon emissions: Experience evidence and emission reduction measures. *Sustain. Energy Technol. Assess.* **2022**, *50*, 101800. [CrossRef]
62. Zhang, N.; Bo, L.; Wang, X. Confucianism and corporate default risk: Assessing the governance influence of traditional culture. *Int. Rev. Econ. Finance* **2024**, *94*, 103378. [CrossRef]

field-specific statistical yearbooks, and the annual statistical bulletins on national economic and social development published by individual cities. To further enhance data accuracy, the study leverages county-level data as the foundational unit for carbon equity calculation, which was then aggregated to the city level. This approach ensures a high degree of granularity and precision in the emissions estimates.

4. Results

4.1. Descriptive Statistics

Figure 2 shows the carbon Lorentz curve at the country level from 2000 to 2020. With the year 2000 as the starting year, the deviation degree of China's carbon Lorentz curve from the absolute fairness line (solid diagonal black line) continues to increase, indicating that the unfair distribution of carbon emissions is worsening. As shown in Figure 3, China's carbon emission Gini coefficient has remained within a relatively equitable range (Gini < 0.4). However, the fluctuating upward trend suggests growing disparities in carbon emissions across districts and counties. According to Figure 3, the Gini coefficient of China's carbon emissions falls within a relative average range (Gini < 0.4). However, the trend observed displays a fluctuating pattern with a discernible upward trajectory, suggesting an increasing disparity in the distribution of carbon emissions across districts and counties.

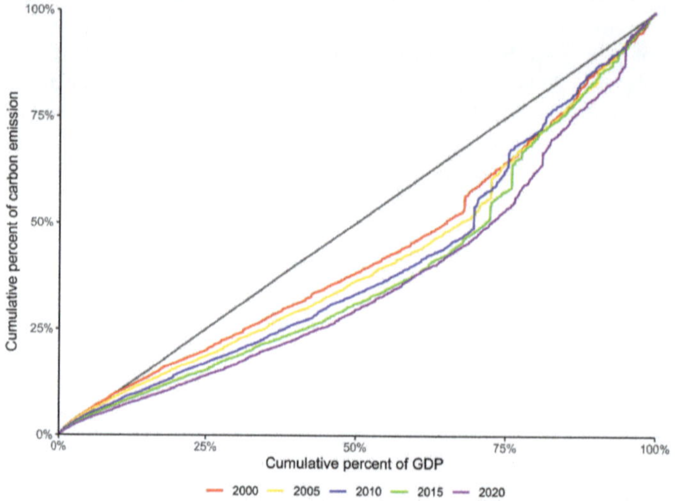

Figure 2. Lorenz curve: cumulative distribution of carbon emissions—cumulative distribution of GDP.

Against the background of global low-carbon development and energy transition in depth, the spatial coupling relationship between carbon equity and efficiency has become an important proposition in the study of sustainable urban development. We used the Gini index and the Super-SBM model with non-expected output to calculate the carbon emission equity and carbon emission efficiency levels in Chinese cities. The study selected the data results of the time node of 2021, and carried out spatial descriptive analysis and visual expression through ArcGIS 10.8 software to map out the spatial differentiation of carbon emission equity and efficiency in Chinese cities. As can be seen from Figure 4, the eastern coastal urban agglomerations (e.g., the Yangtze River Delta urban agglomerations) generally show higher carbon emission efficiency, but the fairness level is obviously differentiated; while the cities in central and western China, although the efficiency level needs to be improved, have excellent performance in the evaluation of the fairness of carbon

energies

MDPI

Article

The Impact of Environmental Policies on Renewable Energy Storage Decisions in the Power Supply Chain

Chunyi Ji *, Xinyue Wang, Wei Zhao, Xuan Wang and Wuyong Qian

School of Business, Jiangnan University, Wuxi 214122, China; 6220910015@stu.jiangnan.edu.cn (X.W.); 6220910025@stu.jiangnan.edu.cn (W.Z.); 6220911003@stu.jiangnan.edu.cn (X.W.); wuyongqian@jiangnan.edu.cn (W.Q.)

* Correspondence: 8333300181@jiangnan.edu.cn

Abstract: Energy storage is a proficient means of enhancing power supply reliability and facilitating the use of renewable energy. To study the impact of policies on energy storage decisions in the power supply chain, this paper constructs an electricity supply chain and compares the equilibrium results under four scenarios based on the Stackelberg game theory. The research reveals that both discharge subsidy and investment subsidy play a beneficial role in improving the level of energy storage technology, regardless of whether power producers choose to invest in or lease energy storage equipment. Furthermore, when combined with the implementation of a renewable portfolio standard, these subsidies can have beneficial outcomes. During the early stages of development in the energy storage industry, investment subsidy proves more advantageous for enhancing both technology levels and electricity demand. Conversely, at later stages of industry development, discharge subsidies become increasingly advantageous for enhancing technological advancements and fulfilling electricity demand. Furthermore, implementing a strategy in which power generators invest in energy storage can enhance their profitability while concurrently advancing technological standards and satisfying electricity demand.

Keywords: power supply chain; energy storage subsidy; renewable portfolio standard; energy storage investment

1. Introduction

Renewable energy plays a critical role in reducing emissions and addressing climate change. Ensuring a stable and efficient energy supply is essential for achieving sustainable economic growth [1]. Consequently, there has been an increasing focus on implementing policies to promote renewable energy in recent years [2]. The Renewable Portfolio Standard (RPS) has been implemented in various nations and is recognized as one of the most efficacious programs for promoting renewable energy [3]. However, with the continuous development and widespread use of renewable energy, its volatility and uncertainty have posed significant challenges to energy dispatch and balancing. In response to this issue, energy storage technology has emerged as a solution [4]. Energy storage serves the purpose of storing surplus energy and releasing it during periods of high or low demand. This addresses the necessity for a swift response from the power system to supply changes, optimizes the distribution and scheduling of electrical energy, and enhances the dependability and cost-effectiveness of energy delivery [5].

Energy storage systems are utilized in conjunction with wind and solar power generation to tackle the challenges of renewable energy consumption and to alleviate its variability [6]. According to industry statistics from the China Electricity Council, as of the

 https://doi.org/10.3390/en18092152

end of 2023, the majority of commissioned electrochemical energy storage power stations are situated on the power generation side, accounting for 49.11% of the total. Renewable energy dominates energy storage on the power generation side, accounting for 90.63%. However, significant initial costs and uncertain profits of future returns make renewable energy power generation enterprises exceedingly prudent when it comes to investing in energy storage [7]. Consequently, many power generation enterprises also consider collaborating with energy storage enterprises.

Provincial and municipal governments in China have enacted various energy storage subsidy plans [8]. The two most commonly implemented types of subsidies are discharge subsidies and investment subsidies. Table 1 compiles data on energy storage subsidy policies in selected regions of China for 2024. Despite ongoing efforts by the Chinese government to implement energy storage policies, there is still a lack of well-defined policy objectives and subsidy schemes. Additionally, significant variations in subsidy policies exist across different regions. In contrast, some developed countries have well-established policies regarding energy storage. For example, as early as 2009, California in the United States included energy storage in the subsidy coverage of its self-generation incentive program. In 2021, they issued more precise instructions about the subsidy amount for energy storage installation. The Australian Renewable Energy Agency was established in 2012 to offer financial support for the development and implementation of renewable energy technologies ranging from initial laboratory research to large-scale commercial deployment.

Table 1. Energy storage subsidy policies in selected regions of China in 2024.

Regions in China	Type of Subsidy	Subsidized Standards
Chengdu, Sichuan Province	Discharge Subsidy	0.3 CNY/kWh
Huangpu District, Guangzhou	Discharge Subsidy	0.2 CNY/kWh
Binhai New District, Tianjin	Discharge Subsidy	0.5 CNY/kWh
Haiyan County, Jiaxing	Discharge Subsidy	0.25 CNY/kWh
Pudong New District, Shanghai	Investment Subsidy	10%
Baiyun District, Guangzhou	Investment Subsidy	≤10%
Guangming District, Shenzhen	Investment Subsidy	20%
Beijing	Investment Subsidy	≤30%

Considering RPS and energy storage subsidy, this study categorizes the types of energy storage for power generators into two modes: investing in energy storage and leasing energy storage equipment. It utilizes the Stackelberg game model to address the following issues:

RQ1: What are the optimal energy storage decisions for different entities in the power supply chain under different environmental policies?

RQ2: How does the implementation of environmental policies affect different modes of energy storage? Which energy storage model is more likely to contribute to the development of renewable energy?

RQ3: How can the appropriate adjustment of various policies promote the advancement of energy storage technology and encourage the growth of the energy storage industry and renewable energy generation, considering the impacts of the renewable energy quota obligation, subsidy amount, subsidy ratio, and lease coefficient of energy storage equipment on supply chain energy storage decisions?

2. Materials and Methods

Currently, there are already numerous studies focused on renewable energy, RPS, and energy storage.

In the field of renewable energy studies, numerous scholars have focused on the impact in mitigating carbon emissions. Bird et al. found that cap-and-trade mechanisms had a significant effect on the quantity of electricity generated from renewable energy sources, thereby influencing overall carbon emission levels [9]. Dahal et al. emphasized the importance of implementing renewable energy policies to achieve carbon neutrality [10]. Hanif et al. confirmed that the use of renewable energy effectively regulates carbon emissions, while the reliance on non-renewable energy exacerbates them [11]. Yu et al. indicated that the development of renewable energy in China has a positive impact on reducing emissions [12]. Zhang et al. utilized the STIRPAT model to explore the relationship between investment in renewable energy and carbon emissions, revealing variations in influence across different stages of investment [13]. Meng et al. examined how the consumption of renewable energy affects carbon emissions across 30 provinces in China [14]. Yang et al. found that allocating a higher percentage of investment towards wind energy leads to decreased carbon emissions, whereas directing more investment towards solar and bioenergy results in increased emissions [15]. Siddik et al. proposes that the government should actively promote the use of renewable energy sources in order to cultivate a low-carbon economy [16]. Xu et al. recommended that the Chinese government continue to expand investment in renewable energy, improve financial efficiency, and provide an efficient financial market environment for renewable energy [17].

In the realm of research on Renewable Portfolio Standards (RPS), the effects of RPS implementation and potential policy improvements have been widely discussed. García-Álvarez et al. emphasized the necessity of a more risk-free framework to bolster investor confidence in RPS [18]. Young and Bistline highlighted the significant impact of future gas pricing on the ability of RPS to reduce carbon dioxide emissions [3]. Meanwhile, Bento et al. revealed that, while the implementation of RPS can lead to either a substantial increase in available resources or significant reductions in emissions, achieving both simultaneously is challenging [19]. Zhao et al. delved into the impact of incentives on the costs and effects of RPS, offering valuable insights for government development of dynamic incentive mechanisms [20]. Zhu et al. found that increasing green certificate prices encourages greater investment in renewable energy, but only when renewable energy quota obligations are excessively high; otherwise, it diminishes motivation for power companies to invest in renewable energy [21]. Yan et al. demonstrated that a combination of cap-and-trade mechanisms and RPS is most effective for promoting renewable energy [22], while Hu et al. discovered that an increase in renewable energy quota obligations can lead to a decline in green electricity consumption and higher carbon emissions [23]. Lee analyzed and examined the long-term impact of renewable portfolio standard on electricity prices and found that it depends on changes in long-term average costs [24].

In the field of energy storage studies, many scholars have focused on the technology of energy storage rather than considering its role in decision-making within the power supply chain. Rohit et al. highlighted the benefits of energy storage in the power supply chain and conducted a comprehensive comparison of various energy storage technologies [25]. Wang et al. examined two energy infrastructures in Chongming, China, from 2016 to 2040, demonstrating that the implementation of energy storage systems can enhance the efficiency of renewable energy utilization by 2040 [26]. Arbabzadeh et al. found that implementing energy storage technology has a positive impact on reducing carbon emissions [27]. Liu and Bao explored how energy storage could improve coordination and benefits within the wind power supply chain, finding that setting appropriate prices promotes wind energy utilization and enhances overall supply chain efficiency [28]. Zhang et al. demonstrated that energy storage can enhance power system reliability and increase renewable energy penetration [29]. Sayed et al. discussed issues related to managing various renewable energy sources and storage systems as well as recent advances in green hydrogen production and fuel cells, which pave the way for a wide range of renewable energy applications [30]. Zhao et al., through numerical results, showed that integrating with an effective strategy for utilizing stored energies enhances power system flexibility and increases utilization of renewable generation [31].

We thoroughly reviewed the existing research in the aforementioned three areas. In summary, the innovative points of this study can be summarized as follows:

1. The existing literature on renewable energy predominantly focuses investment, pricing, and the reduction in carbon emissions within the power supply chain. There is a greater emphasis on technical research regarding energy storage, while there is a scarcity of studies that examine the selection of energy storage models. Additionally, there is limited literature on improving energy storage technology from a supply chain perspective. This paper examines investment in energy storage equipment from a supply chain perspective and analyzes the effects of various stakeholders investing in such equipment, hence enhancing research on new power systems.
2. China has implemented various environmental policies for the power industry, such as RPS and cap-and-trade mechanisms, with relatively mature research in this area. However, China's policies on energy storage are still in an exploratory stage. While various provinces and cities in China have implemented different energy storage policies, the literature specifically focused on these policies remains scarce. The paper analyzes and contrasts two prevalent energy storage subsidy schemes in China, offering a foundation for governmental decision-making about the implementation of such policies.

3. Problem Definition

In the context of RPS, the power generator has both renewable energy generation units and traditional thermal power generation units. The electricity seller, who is responsible for RPS, is required to purchase a specified percentage of renewable energy and green certificates from the power generator. In this paper's power supply chain, the power generator stores renewable energy and makes decisions on the energy storage model to maximize profits. The power generator decides whether to invest in energy storage directly or to lease energy storage equipment from a professional energy storage provider. In this study, we used the Stackelberg game to determine the equilibrium solution, with a focus on the impact of various incentive policies and energy storage models on power generators' energy storage decisions. The Stackelberg game is able to reflect the structure of the electricity market well, and conforms to the decision-making sequence of the electricity supply chain. It is able to capture not only the impact of the policy on the subsidized (generators and storage service providers), but also the impact on the entire supply chain and other players in the supply chain.

In the case where the power generator invests in energy storage, this paper constructed a power supply chain with the power generator as the leader and the electricity seller as the follower, as shown in Figure 1, and the decision-making sequence of the power generator and the electricity seller is as follows: firstly, the generator decides on the wholesale electricity price and the level of energy storage technology, and secondly, the seller decides on the amount of electricity demand. In the case where the power generator leases the energy storage equipment, this paper constructed a power supply chain with the power generator and the energy storage provider as the leader and the electricity seller as the follower, as shown in Figure 2, and the decision-making sequence of the power generator, the electricity seller, and the energy storage provider is as follows: firstly, the generator decides on the wholesale electricity price, the energy storage provider decides on the level of the storage technology, and then the seller decides on the amount of electricity demand.

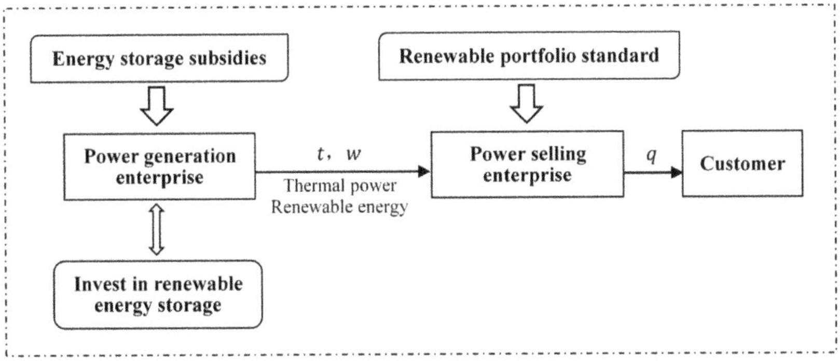

Figure 1. Supply chain structure in the generator investment scenario.

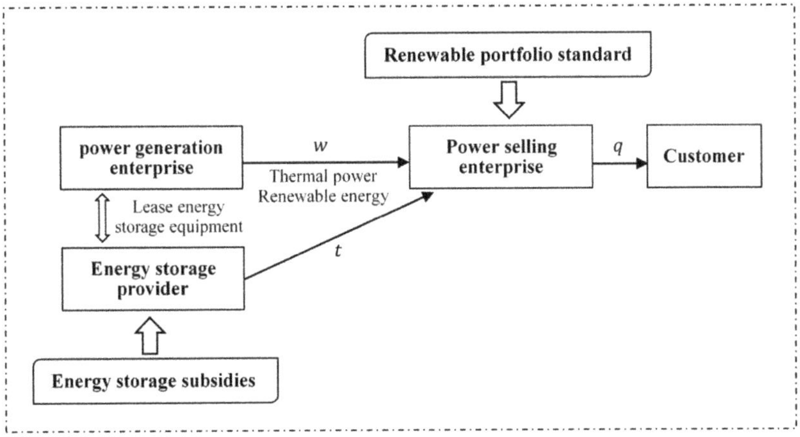

Figure 2. Supply chain structure in the generator leasing scenario.

To facilitate the modeling and solution, the following assumptions were made in this paper:

Assumption 1. *With energy storage devices, the power generator can deliver electricity with more stability. According to Chen et al., customers' electricity demand is determined not only by the price of power but also by its stability [32]. Consumers will prefer more stable electricity because it is safer and more reliable. This study assumed that the power demand function is connected to the level of energy storage technology, as follows:* $q = a - p + bt$, *where a is the basic electricity demand of the market; p is the market price of electricity, where the higher the p, the lower the electricity demand; b* $(0 < b < 1)$ *is the sensitivity coefficient of energy storage technology level, where the larger the b, the more importance consumers attach to the level of energy storage technology; and t is the level of energy storage technology of the energy storage equipment, where the higher the t, the higher the electricity demand.*

Assumption 2. *The electricity seller purchases thermal and renewable electricity from the power generator at the wholesale electricity price w and sells it to the consumption side at the market electricity price p. It is further assumed that both thermal power and renewable energy production facilities already exist. The unit production cost of thermal power is* c_t *and that of renewable energy is* c_r*, and the cost of transmission and distribution of electricity and the rate of energy loss are not taken into consideration. Currently, the cost of renewable energy is decreasing year by year, but it still lacks competitiveness relative to fossil energy sources such as coal [33]; so, in this study, it was* *assumed that the unit production cost of renewable energy is greater than the unit production cost of thermal power, i.e.,* $c_r > c_t$.

Assumption 3. *In this paper, only the storage of renewable energy was considered, and thermal power storage was not considered. In order to mitigate the problem of intermittency of renewable energy and thus improve the stability of power supply, an effective solution is to invest in energy storage technology and equipment to maximize the balance between power supply and demand [34]. It was assumed that the investment cost of energy storage is related to the level of energy storage technology. In order to pursue a higher level of energy storage technology, a higher investment cost of energy storage needs to be invested in. Therefore, it can be assumed that the energy storage investment cost function of the power generator or energy storage provider is* $\frac{1}{2}\beta t^2$ *[35], where* β *is the cost coefficient of energy storage investment. There exists a boundary condition for it, and when it is too high, it indicates that the current energy storage industry has not yet developed to the extent that it can support large-scale investment, and it needs to wait for further technological progress. This functional form is widely used in economic and engineering modeling to represent the case of increasing marginal costs, and can effectively represent the trend of rising costs during the advancement of energy storage technology.*

Assumption 4. *When a power generator leases energy storage equipment from an energy storage provider, the equipment, technology, and services are all leased together with the energy storage equipment. In this paper, it was assumed that the cost that an energy storage provider charges is correlated with the equipment's level of energy storage technology. Therefore, it was assumed that the cost function of energy storage equipment leasing for the power generator is* ft *[32], where* f *is the lease coefficient of the energy storage equipment.*

The lease coefficient of the energy storage equipment was impacted by market considerations in this study, rather than being used as a decision variable. In the pre-development period of the energy storage industry, due to the high cost of investing in energy storage, few companies invested in energy storage; so, few companies in the market can provide leasing services. Therefore, the cost of leasing energy storage equipment was higher at this time. When the energy storage industry develops into a mature stage, the cost of leasing energy storage equipment will be reduced.

Assumption 5. *Electricity sellers are responsible for RPS and are required to provide consumers with a certain percentage of the total electricity supplied from renewable energy sources. Assume that one green certificate is issued for each unit of renewable energy produced by the power generator, and the two are bundled and sold to the electricity seller. The price of the green certificate is* p_c*, the renewable energy quota obligation is* θ*, the amount of renewable energy purchased by the electricity seller from the power generator is* θq*, and the amount of thermal power is* $(1-\theta)q$ *[36].*

Assumption 6. *Governments can support energy storage through two key channels: investment subsidy and discharge subsidy. The discharge amount of energy storage equipment is influenced by its storage capacity, discharge efficiency, discharge depth, etc. To simplify the model and parameters, the storage capacity, discharge efficiency, and discharge depth were uniformly summarized in this paper as the level of energy storage technology. The discharge amount of the energy storage equipment is positively correlated with the level of energy storage technology, and the higher the level of energy storage technology, the higher the discharge amount [37]. To simplify the subsequent calculations, this paper introduced the discharge coefficient of energy storage equipment* γ*, assuming that the discharge volume is a primary function of the energy storage technology level. Under the discharge subsidy, the government provides subsidies to energy storage investment enterprises by the standard of RMB d/kWh, and the amount of the discharge subsidy is* $d\gamma q_e$*. There is a boundary condition for* d *such that, when it is too high, the marginal benefits of additional subsidies are reduced. This is because, at very high levels of subsidies, the costs of policy implementation and* *administration may outweigh the benefits. Under the investment subsidy, the government grants enterprises a certain percentage of the energy storage investment amount, and the amount of the investment subsidy is $\frac{1}{2}\alpha\beta t^2$, where α is the subsidy ratio of energy storage investment. There exists a boundary condition for α that, when too high, may lead to over-investment in energy storage technologies that are not yet cost-effective, thus reducing the overall return on investment.*

3.1. Problem Modeling

To investigate the impacts of different energy storage models and energy storage subsidies on the energy storage decision-making of the power supply chain under RPS, four power supply chain decision-making models were constructed in this section with two models of power generator investing in energy storage or leasing energy storage equipment, and two energy storage policies of discharge subsidy or investment subsidy. The four models are denoted by superscripts 1, 2, 3, and 4, respectively, and (*) represents the equilibrium solution. The symbols involved in this paper and their meanings are explained in Table 2.

Table 2. Notation and variable definitions.

Notation	Description
w	Wholesale electricity price
q	Electricity demand
t	Level of energy storage technology
a	Basic electricity demand of the market
b	Sensitivity coefficient of the energy storage technology level
p	Market price of electricity
c_t	Unit production cost of thermal power
c_r	Unit production cost of renewable energy
d	Subsidy amount per unit of discharge
α	Subsidy ratio of energy storage investment
β	Cost coefficient of energy storage investment
f	Lease coefficient of energy storage equipment
γ	Discharge coefficient of energy storage equipment
θ	Renewable energy quota obligation
p_c	Green certificate transaction price
π_G	Power generation's profit
π_S	Power seller's profit
π_E	Energy storage provider's profit

3.1.1. Scenario 1: Supply Chain Decision Model for the Power Generator Investing in Energy Storage Under Discharge Subsidy

The decision functions of the power generator and seller are, respectively:

$$\pi_G^1 = \theta q(w + p_c - c_r) + (1-\theta)q(w - c_t) - \frac{1}{2}\beta t^2 + d\gamma q_e \tag{1}$$

$$\pi_S^1 = \theta q(p - p_c - w) + (1-\theta)q(p - w) \tag{2}$$

According to the inverse induction method, it can be obtained:

$$q^1 = \frac{a + bt - w - p_c\theta}{2} \tag{3}$$

When $\beta - \frac{b^2}{4} > 0$ is satisfied, the Hessian matrix is negative definite and there are extreme values. Let $\frac{\partial \pi_G}{\partial t} = 0$ and $\frac{\partial \pi_G}{\partial w} = 0$, which yields:

$$t^{1*} = \frac{-4d\gamma + b(-a + c_t + c_r\theta - c_t\theta)}{b^2 - 4\beta} \tag{4}$$

$$w^{1*} = \frac{-2bd\gamma + b^2[c_t + c_r\theta - (c_t + p_c)\theta] - 2\beta[a + c_t + c_r\theta - (c_t + 2p_c)\theta]}{b^2 - 4\beta} \tag{5}$$

$$q^{1*} = \frac{-bd\gamma + \beta(-a + c_t + c_r\theta - c_t\theta)}{b^2 - 4\beta} \quad (6)$$

Substituting Equations (4)–(6) into Equations (1) and (2) yields the equilibrium profits of the power generation's profit and the power seller's profit:

$$\pi_G^{1*} = -\frac{1}{2(b^2-4\beta)}[a^2\beta - 2ac_t\beta + c_t{}^2\beta + 2abd\gamma - 2bc_td\gamma + 4d^2\gamma^2 + 2(-c_t + ct)(a\beta - c_t\beta + bd\gamma)\theta + (c_r - c_t)^2\beta\theta^2] \quad (7)$$

$$\pi_S^{1*} = \frac{\{bd\gamma + \beta[a + c_t(-1+\theta) - c_r\theta]\}^2}{(b^2 - 4\beta)^2} \quad (8)$$

Proposition 1. *The optimal energy storage technology level t^{1*}, the optimal wholesale electricity price w^{1*}, the optimal demand q^{1*}, as well as the generator's profit π_G^{1*} and the seller's profit π_S^{1*} for the power supply chain in Scenario 1 are shown above.*

3.1.2. Scenario 2: Supply Chain Decision Model for the Power Generator Investing in Energy Storage Under Investment Subsidy

The decision functions of the power generator and seller are, respectively:

$$\pi_G^2 = \theta q(w + p_c - c_r) + (1-\theta)q(w - c_t) - \frac{1}{2}\beta t^2 + \frac{1}{2}\alpha\beta t^2 \quad (9)$$

$$\pi_S^2 = \theta q(p - p_c - w) + (1-\theta)q(p - w) \quad (10)$$

According to the inverse induction method, it can be obtained:

$$q^2 = \frac{a + bt - w - p_c\theta}{2} \quad (11)$$

When $(1-\alpha)\beta - \frac{b^2}{4} > 0$ is satisfied, the Hessian matrix is negative definite and there are extreme values. Let $\frac{\partial \pi_G}{\partial t} = 0$ and $\frac{\partial \pi_G}{\partial w} = 0$, which yields:

$$t^{2*} = \frac{b(-a + c_t + c_r\theta - c_t\theta)}{b^2 + 4(-1+\alpha)\beta} \quad (12)$$

$$w^{2*} = \frac{b^2[c_t + c_r\theta - (c_t + pc)\theta] + 2(-1+\alpha)\beta[a + c_t + c_r\theta - (c_t + 2p_c)\theta]}{b^2 + 4(-1+\alpha)\beta} \quad (13)$$

$$q^{2*} = \frac{(-1+\alpha)\beta[a + c_t(-1+\theta) - c_r\theta]}{b^2 + 4(-1+\alpha)\beta} \quad (14)$$

Substituting Equations (12)–(14) into Equations (9) and (10) yields the equilibrium profits of the power generation's profit and the power seller's profit:

$$\pi_G^{2*} = \frac{(-1+\alpha)\beta[a + c_t(-1+\theta) - c_r\theta]^2}{2b^2 + 8(-1+\alpha)\beta} \quad (15)$$

$$\pi_S^{2*} = \frac{(-1+\alpha)^2\beta^2[a + c_t(-1+\theta) - c_r\theta]^2}{[b^2 + 4(-1+\alpha)\beta]^2} \quad (16)$$

Proposition 2. *The optimal energy storage technology level t^{2*}, the optimal wholesale electricity price w^{2*}, the optimal demand q^{2*}, as well as the generator's profit π_G^{2*} and the seller's profit π_S^{2*} for the power supply chain in Scenario 2 are shown above.*

3.1.3. Scenario 3: Supply Chain Decision Model for the Power Generator Leasing Energy Storage Equipment Under Discharge Subsidy

The decision functions of the power generator, seller, and energy storage provider are, respectively:

$$\pi_G^3 = \theta q(w + p_c - c_r) + (1-\theta)q(w - c_t) - fq_e \tag{17}$$

$$\pi_S^3 = \theta q(p - p_c - w) + (1-\theta)q(p - w) \tag{18}$$

$$\pi_E^3 = fq_e - \frac{1}{2}\beta t^2 + d\gamma q_e \tag{19}$$

According to the inverse induction method, it can be obtained:

$$q^3 = \frac{a + bq_e - w - p_c\theta}{2} \tag{20}$$

Next, the power generator decides on the wholesale electricity price w and the energy storage provider decides on the level of energy storage technology t simultaneously. It can be obtained that the wholesale electricity price w and the level of energy storage technology t are, respectively:

$$t^{3*} = \frac{f + d\gamma}{\beta} \tag{21}$$

$$w^{3*} = \frac{1}{2}[a + c_t + c_r\theta + \frac{b(f + d\gamma)}{\beta} - (c_r + 2p_c)\theta] \tag{22}$$

$$q^{3*} = \frac{b(f + d\gamma) + \beta[a + c_t(-1+\theta) - c_r\theta]}{4\beta} \tag{23}$$

Substituting Equations (21)–(23) into Equations (17) and (19) yields the equilibrium profits of the power generation's profit, the power seller's profit, and the energy storage provider's profit:

$$\pi_G^{3*} = \frac{1}{8\beta^2}\{b^2(f + d\gamma)^2 + 2b\beta(f + d\gamma)[a + c_t(-1+\theta) - c_r\theta] + \beta\{-8f(f + d\gamma) + \beta[a + c_t(-1+\theta) - c_r\theta]^2\}\} \tag{24}$$

$$\pi_S^{3*} = \frac{\{b(f + d\gamma) + \beta[a + c_t(-1+\theta) - c_r\theta]\}^2}{16\beta^2} \tag{25}$$

$$\pi_E^{3*} = \frac{(f + d\gamma)^2}{2\beta} \tag{26}$$

Proposition 3. *The optimal energy storage technology level t^{3*}, the optimal wholesale electricity price w^{3*}, the optimal demand q^{3*}, as well as the generator's profit π_G^{3*} and the seller's profit π_S^{3*} for the power supply chain in Scenario 1 are shown above.*

3.1.4. Scenario 4: Supply Chain Decision Model for the Power Generator Leasing Energy Storage Equipment Under Investment Subsidy

The decision functions of the power generator, seller, and energy storage provider are, respectively:

$$\pi_G^4 = \theta q(w + p_c - c_r) + (1-\theta)q(w - c_t) - fq_e \tag{27}$$

$$\pi_S^4 = \theta q(p - p_c - w) + (1-\theta)q(p - w) \tag{28}$$

$$\pi_E^4 = fq_e - \frac{1}{2}\beta t^2 + \frac{1}{2}\alpha\beta t^2 \tag{29}$$

According to the inverse induction method, it can be obtained:

$$q^4 = \frac{a + bq_e - w - p_c\theta}{2} \tag{30}$$

Next, the power generator decides on the wholesale electricity price w and the energy storage provider decides on the level of energy storage technology t simultaneously. It can be obtained that the wholesale electricity price w and the level of energy storage technology t are, respectively:

$$t^{4*} = -\frac{f}{(-1+\alpha)\beta} \tag{31}$$

$$w^{4*} = \frac{1}{2}[a + c_t + c_r\theta + \frac{bf}{\beta - \alpha\beta} - (c_t + 2p_c)\theta] \tag{32}$$

$$q^{4*} = \frac{1}{4}[a + \frac{bf}{\beta(1-\alpha)} + c_t(-1+\theta) - c_r\theta] \tag{33}$$

Substituting Equations (31)–(33) into Equations (27) and (29) yields the equilibrium profits of the power generation's profit, the power seller's profit, and the energy storage provider's profit:

$$\begin{aligned}\pi_G^{4*} = \frac{1}{8(1-\alpha)^2\beta^2}\{&f^2[b^2 + 8(-1+\alpha)\beta] - 2bf(-1+\alpha)\beta[a + c_t(-1+\theta) - c_r\theta] \\ &+(-1+\alpha)^2\beta^2[a + c_t(-1+\theta) - c_r\theta]^2\}\end{aligned} \tag{34}$$

$$\pi_S^{4*} = \frac{\{bf + (1-\alpha)\beta[a + c_t(-1+\theta) - c_r\theta]\}^2}{16(-1+\alpha)^2\beta^2} \tag{35}$$

$$\pi_E^{4*} = \frac{f^2}{2\beta(1-\alpha)} \tag{36}$$

Proposition 4. *The optimal energy storage technology level t^{4*}, the optimal wholesale electricity price w^{4*}, the optimal demand q^{4*}, as well as the generator's profit π_G^{4*} and the seller's profit π_S^{4*} for the power supply chain in Scenario 1 are shown above.*

3.2. Model Analysis

This section analyses the impacts of renewable energy quota obligation, subsidy amount per unit of discharge, and subsidy ratio of energy storage investment on the level of energy storage technology, electricity demand, power generator profits, and electricity seller profits. The equilibrium solutions for the four scenarios are also compared.

Inference 1. *As the renewable energy quota obligation increases, there is*

(1) $\frac{\partial t^{1*}}{\partial\theta} < 0,\ \frac{\partial t^{2*}}{\partial\theta} < 0,\ \frac{\partial t^{3*}}{\partial\theta} = 0,\ \frac{\partial t^{4*}}{\partial\theta} = 0;$

(2) $\frac{\partial q^{1*}}{\partial\theta} < 0,\ \frac{\partial q^{2*}}{\partial\theta} < 0,\ \frac{\partial q^{3*}}{\partial\theta} < 0,\ \frac{\partial q^{4*}}{\partial\theta} < 0;$

(3) $\frac{\partial \pi_G^{1*}}{\partial\theta} < 0,\ \frac{\partial \pi_G^{2*}}{\partial\theta} < 0,\ \frac{\partial \pi_G^{3*}}{\partial\theta} < 0,\ \frac{\partial \pi_G^{4*}}{\partial\theta} < 0;$

(4) $\frac{\partial \pi_S^{1*}}{\partial\theta} < 0,\ \frac{\partial \pi_S^{2*}}{\partial\theta} < 0,\ \frac{\partial \pi_S^{3*}}{\partial\theta} < 0,\ \frac{\partial \pi_S^{4*}}{\partial\theta} < 0.$

Inference 1(1) demonstrates that, as the renewable energy quota obligation increases, the investment of the generator in the level of energy storage technology decreases. The reason for this is that, when the renewable energy quota obligation increases, the power generator is required to create a greater amount of renewable energy. According to Zou et al., the cost of renewable energy can be competitive with fossil energy by 2030–2050 as the relevant technology matures [33]. Therefore, at present, this study continues to believe that the cost of producing renewable energy is higher than that of conventional thermal power. Due to resource constraints, the increase in the cost of power generation will cause power producers to consider reducing the cost of investing in energy storage, thus inhibiting the level of energy storage technology. Furthermore, the level of energy storage technology will remain unchanged in both generator leasing scenarios with the increase in the renewable energy quota obligation. Since the energy storage provider determines the level of energy storage technology in both scenarios, the RPS does not impose any limitations on the energy storage provider. Inference 1(2) demonstrates that the electricity demand decreases in all scenarios as the renewable energy quota obligation increases. This is because, in generator investment scenarios, a decrease in the level of energy storage technology results in a drop in the amount of electricity needed. In generator leasing scenarios, when the renewable energy quota obligation increases, the electricity seller's profit decreases. To compensate, the seller often raises the market price of electricity to improve their profit. However, this increase in price results in a fall in the electricity demand. Inference 1(3)–(4) show that, as the renewable energy quota obligation increases, the profits of both the power generator and seller decrease. The increase in the renewable energy quota obligation leads to an increase in the cost of electricity generation for the generator and the cost of obtaining green certificates for the electricity seller. Consequently, both profits decline.

Inference 2. *As the subsidy amount per unit of discharge increases, there is*

(1) $\frac{\partial t^{1*}}{\partial d} > 0,\ \frac{\partial t^{3*}}{\partial d} > 0;$

(2) $\frac{\partial q^{1*}}{\partial d} > 0,\ \frac{\partial q^{3*}}{\partial d} > 0;$

(3) $\frac{\partial \pi_G^{1*}}{\partial d} > 0,\ when\ 0 < f < f_1,\ \frac{\partial \pi_G^{3*}}{\partial d} > 0,\ when\ f > f_1,\ \frac{\partial \pi_G^{3*}}{\partial d} < 0;$

(4) $\frac{\partial \pi_S^{1*}}{\partial d} > 0,\ \frac{\partial \pi_S^{3*}}{\partial d} > 0.$

$f_1 = \frac{b\{bd\gamma+\beta[a+c_t(-1+\theta)-c_r\theta]\}}{4\beta-b^2}$

Inference 2(1)–(2) show that, as the subsidy amount per unit of discharge increases, both the level of energy storage technology and the electricity demand rise. A larger subsidy per unit of discharge reduces the financial burden of investing in energy storage, making investors more inclined to invest more in higher technology levels. Advanced energy storage technology results in a more reliable electricity supply, hence boosting electricity demand. Inference 2(3) demonstrates that, as the subsidy for energy storage per unit of discharge increases, the power generator's profit in the generator investment scenario increases. However, in the generator leasing scenario, the power generator's profit first grows and subsequently decreases. In the generator investment scenario, the increase in the subsidy amount per unit of discharge will provide the generator with more financial assistance. Additionally, an increase in electricity demand will result in higher profits from the sale of electricity. In the generator leasing scenario, when the energy storage subsidy per unit of discharge exceeds a certain threshold, the energy storage provider is inclined to make substantial investments in energy storage technology. When the energy storage technology reaches a high level, the generator's profit decreases since the cost of leasing storage equipment exceeds the revenue gained from selling additional electricity. Inference 2(4) shows that, as the subsidy amount per unit of discharge increases, the power seller's profit rises in all cases. This is because the power seller's profit is not affected by the energy storage factor and increases only due to the increase in electricity demand.

Inference 3. *As the subsidy ratio of energy storage investment increases, there is*

(1) $\frac{\partial t^{2*}}{\partial \alpha} > 0,\ \frac{\partial t^{4*}}{\partial \alpha} > 0;$
(2) $\frac{\partial q^{2*}}{\partial \alpha} > 0,\ \frac{\partial q^{4*}}{\partial \alpha} > 0;$
(3) $\frac{\partial \pi_G^{2*}}{\partial \alpha} > 0,\ when\ 0 < f < f_2,\ \frac{\partial \pi_G^{4*}}{\partial \alpha} > 0,\ when\ f > f_2,\ \frac{\partial \pi_G^{4*}}{\partial \alpha} > 0;$
(4) $\frac{\partial \pi_S^{2*}}{\partial \alpha} > 0,\ \frac{\partial \pi_S^{4*}}{\partial \alpha} > 0.$

$f_2 = \frac{b(-1+\alpha)\beta[a+c_t(-1+\theta)-c_r\theta]}{b^2+4(-1+\alpha)\beta}$

Inference 3(1)–(2) show that both the level of energy storage technology and electricity demand rise as the subsidy ratio of energy storage investment increases. The reason is similar to those assigned to Inferences 2(1)–(2) and will not be repeated here. Inference 3(3) shows that, as the subsidy ratio of energy storage investment increases, the power generator's profit in the generator investment scenario also increases, while the power generator's profit in the generator leasing scenario first grows and subsequently decreases. The reason is similar to that for Inference 2(3). Inference 3(4) shows that the power seller's profit increases in all scenarios as the subsidy ratio of energy storage investment increases. The reason is similar to that for Inference 2(4).

Inference 4. *Comparison of discharge subsidy and investment subsidy in the scenario where the power generator invests in energy storage*

(1) When $0 < \alpha < \alpha_1, t^{1*} > t^{2*}$; when $\alpha_1 < \alpha < 1, t^{1*} < t^{2*}$.
(2) When $0 < \alpha < \alpha_1, q^{1*} > q^{2*}$; when $\alpha_1 < \alpha < 1, q^{1*} < q^{2*}$.
(3) When $0 < \alpha < \alpha_2, \pi_G^{1*} > \pi_G^{2*}$; when $\alpha_2 < \alpha < 1, \pi_G^{1*} < \pi_G^{2*}$.
(4) When $0 < \alpha < \alpha_1, \pi_S^{1**} > \pi_S^{2*}$; when $\alpha_1 < \alpha < 1, \pi_S^{1*} < \pi_S^{2*}$.

$\alpha_1 = \frac{d\gamma(4\beta-b^2)}{\beta\{4d\gamma+b[a+c_t(-1+\theta)-c_r\theta]\}}, \alpha_2 = \frac{2d\left(4\beta-b^2\right)\gamma\{2d\gamma+b[a+c_t(-1+\theta)-c_r\theta]\}}{\beta\{4d\gamma+b[a+c_t(-1+\theta)-c_r\theta]\}^2}$

Inference 4(1)–(2) demonstrate that, when the subsidy ratio of energy storage investment is above the threshold α_1, the investment subsidy has a greater capacity to enhance the level of energy storage technology and stimulate electricity demand. This is because a higher subsidy ratio for energy storage investment allows the power generator to obtain more financial support through the investment subsidy. As a result, they will have more funds available to invest in energy storage, which in turn promotes the level of energy storage technology. Since electricity demand is positively correlated with the level of energy storage technology, an increase in the latter will result in a corresponding increase in the former. When the subsidy ratio of energy storage investment is low, the financial assistance received by the generator through the investment subsidy will be comparatively less than the discharge subsidy. This leads to a higher level of energy storage technology and electricity demand under the discharge subsidy. In addition, the threshold α_1 is negatively correlated with the sensitivity coefficient of energy storage technology level, which is lower as the consumer prefers a more stable electricity supply. Thus, if customers exhibit a stronger preference towards the level of energy storage technology, the government can implement more investment subsidies, which will be more favorable to the level of energy storage technology and electricity demand.

Inference 4(3) demonstrates that the investment subsidy benefits the power generator's profit more when the subsidy ratio of energy storage investment exceeds the threshold α_2. This is because when the government sets a low investment subsidy ratio, the generator receives less financial support through the investment subsidy compared to the discharge subsidy. In this case, the discharge subsidy is more favorable to the power generator's profit, and vice-versa for the investment subsidy. In addition, the threshold α_2 is positively correlated with the cost coefficient of energy storage investment. As the energy storage industry develops, the cost coefficient of energy storage investment will decrease due to the learning effect and scale effect. Therefore, when the energy storage industry reaches a mature stage, the investment subsidy becomes more advantageous for the power generator's profit. Inference 4(3) demonstrates that, when the subsidy ratio of energy storage investment is high, the investment subsidy has a more favorable impact on the power seller's profit improvement. This is because the level of storage technology, electricity demand, and market price of electricity are higher under investment subsidy, and therefore, the power seller's profit is higher.

Inference 5. *Comparison of discharge subsidy and investment subsidy in the scenario where the power generator leases energy storage equipment*

(1) *When* $0 < \alpha < \alpha_3, t^{3*} > t^{4*}$; *when* $\alpha_3 < \alpha < 1, t^{3*^*} < t^{4*}$.
(2) *When* $0 < \alpha < \alpha_3, q^{3*} > q^{IL^*}$; *when* $\alpha_3 < \alpha < 1, q^{3*} < q^{4*^*}$.
(3) *When* $0 < f < f_3$*or* $f > f_4, \pi_G^{3*} > \pi_G^{4*}$; *when* $f_3 < f < f_4, \pi_G^{3*} < \pi_G^{4*}$.
(4) *When* $0 < \alpha < \alpha_3, \pi_S^{3*} > \pi_S^{4*}$; *when* $\alpha_3 < \alpha < 1, \pi_S^{3*} < \pi_S^{4*}$.

$\alpha_3 = \frac{d\gamma}{f+d\gamma}, f_3 = \frac{b(1-\alpha)\{bd\gamma+2\beta[a+c_t(-1+\theta)-c_r\theta]\}}{b^2(-2+\alpha)-8(-1+\alpha)\beta}, f_4 = \frac{d\gamma(1-\alpha)}{\alpha}$

Inference 5(1)–(2) demonstrate that, when the subsidy ratio of energy storage investment is below the threshold α_3, the discharge subsidy has a stronger positive impact on the level of energy storage technology and electricity demand. This occurs because a fall in the investment subsidy ratio results in a reduction in the financial support received by the energy storage provider, who acts as the investor in energy storage. Consequently, their motivation to invest in energy storage diminishes. Consumer preference for energy storage technology directly affects electricity demand. When energy storage technology is at a lower level, electricity demand decreases accordingly. In the pre-development period of the energy storage industry, energy storage technologies are not yet fully mature, and the initial investment costs of the equipment are relatively high. Consequently, the lease coefficient of energy storage equipment f is also higher. Since the threshold α_3 is negatively correlated with f, α_3 is relatively low at this stage, and the subsidy ratio of energy storage investment is likely to exceed this threshold. It is recommended that the government implement more investment subsidies. The practical reason is that the cost of investing in energy storage technologies is relatively high at this stage, which reduces the willingness to invest. The government needs to provide a higher subsidy ratio of energy storage investment to encourage investment in energy storage technologies and thereby promote the improvement in energy storage technology levels. Therefore, in the early stages of energy storage industry development, it is recommended that the government implement more investment subsidy policies. However, as the energy storage industry gradually matures, technologies continue to advance, equipment costs gradually decrease, and the market scale expands, the lease coefficient will also decrease accordingly. Implementing discharge subsidies can more directly increase the utilization rate of energy storage equipment, increase discharge volumes, and thereby enhance energy storage technology levels and electricity demand.

Inference 5(3) demonstrates that, if the lease coefficient of energy storage equipment is either too low or too high, the discharge subsidy is more conducive to the power generator's profit. This is because the low lease coefficient of energy storage equipment leads to a high α_3, then the level of storage technology and electricity demand under the discharge subsidy is more likely to be higher than under the investment subsidy. As a result, the power generator under the discharge subsidy can generate more profit by selling electricity. When the lease coefficient of energy storage equipment is too high, the revenue generated by energy storage is insufficient to offset the cost of the lease. At this time, α_3 is low, indicating that the level of energy storage technology under the discharge subsidy is lower than that under the investment subsidy, and therefore, the generator's storage leasing cost is lower; so, the generator's profit under the discharge subsidy will be relatively higher. Inference 5(4) demonstrates that a lower lease coefficient of energy storage equipment leads to a more advantageous discharge subsidy for the power seller's profit. The reason is similar to that for Inference 4(4).

Inference 6. *Comparison of the scenarios where the power generator invests in energy storage or leases energy storage equipment under the discharge subsidy policy*

(1) *When* $0 < f < f_5, t^{1*} > t^{3*}$; *when* $f > f_5, t^{1*} < t^{3*}$.
(2) *When* $0 < f < f_5, q^{1*} > q^{3*}$; *when* $f > f_5, q^{1*} < q^{3*}$.
(3) $\pi_G^{1*} > \pi_G^{3*}$.
(4) *When* $0 < f < f_5, \pi_S^{1*} > \pi_S^{3*}$; *when* $f > f_5, \pi_S^{1*} < \pi_S^{3*}$.

$f_5 = \frac{b\{bd\gamma+\beta[a+c_t(-1+\theta)-c_r\theta]\}}{4\beta-b^2}$

Inference 6(1)–(2) demonstrate that, when the lease coefficient of energy storage equipment is below the threshold f_5, it is more advantageous for the power generator to make its investment in energy storage, considering the level of energy storage technology and electricity demand. When the lease coefficient of energy storage equipment is high, it indicates that the energy storage industry is in its early stage. Encouraging power generators to lease energy storage equipment can not only reduce their initial investment risks but also accelerate the dissemination of and improvement in technology through learning effects. As the industry matures, power generators directly invest in energy storage equipment, enabling them to fully exploit economies of scale, reduce unit costs, and thereby promote the improvement in energy storage technology levels and the increase in electricity demand. In addition, the threshold f_5 is positively correlated with the subsidy amount per unit of discharge. Therefore, when the energy storage industry is dominated by power generators, the government can increase this threshold f_5 by increasing the subsidy amount per unit of discharge. Whereas when the energy storage industry is dominated by energy storage providers, the government can increase the threshold f_5 by decreasing the subsidy amount per unit of discharge. This will help maximize the overall level of energy storage technology and electricity demand. In addition, f_5 is also positively correlated with the sensitivity coefficient of energy storage technology level. Thus, when consumers have a stronger preference towards a specific level of energy storage technology, investment in energy storage by generators is more favorable to the level of storage technology and electricity demand.

Inference 6(3) demonstrates that the power generator's profit in the generator investment scenario is higher than that in the generator leasing scenario. The discharge subsidy can efficiently alleviate the cost of the generator's investment in energy storage. Inference 6(4) demonstrates that, when the lease coefficient of energy storage equipment is below the threshold f_5, the generator investment scenario is more favorable to the power seller's profit. The reason is similar to that for Inference 4(4).

Combined with Corollary 6(1)–(4), it can be seen that, when the lease coefficient of energy storage equipment is below the threshold f_5, all equilibrium solutions in the generator investment scenario will exceed those in the generator leasing scenario. In turn, this threshold is positively correlated with the subsidy amount per unit of discharge. Therefore, if the government implements a discharge subsidy, it is advisable to incentivize power generators to invest in energy storage when the subsidy amount per unit of discharge is high. This will guarantee that the level of energy storage technology, electricity demand, power generators' profits, and power sellers' profits are all relatively better.

Inference 7. *Comparison of the scenarios where the power generator invests in energy storage or leases energy storage equipment under the investment subsidy policy*

(1) *When* $0 < f < f_6, t^{2*} > t^{4*}$; *when* $f > f_6, t^{2*} < t^{4*}$.
(2) *When* $0 < f < f_6, q^{2*} > q^{4*}$; *when* $f > f_6, q^{2*} < q^{4*}$.
(3) $\pi_G^{2*} > \pi_G^{4*}$.
(4) *When* $0 < f < f_6, \pi_S^{2*} > \pi_S^{4*}$; *when* $f > f_6, \pi_S^{2*} < \pi_S^{4*}$.

$f_6 = \frac{b(1-\alpha)\beta[a+c_t(-1+\theta)-c_r\theta]}{4\beta(1-\alpha)-b^2}$

Inference 7(1)–(2) demonstrate that, when the lease coefficient of energy storage equipment exceeds the threshold f_6, it is more favorable for the power generator to lease energy storage equipment to increase the level of energy storage technology and electricity demand, for the same reason as that for Inference 6. In addition, the threshold f_6 is negatively correlated with the sensitivity coefficient of energy storage technology level. When consumer preference for energy storage is stronger, the level of energy storage technology and electricity demand in the generator leasing scenario is more likely to be greater than in the generator investment scenario. As a result, generators are more inclined to lease energy storage equipment at this time.

Inference 7(3) demonstrates that the power generator's profit is higher in the generator investment scenario than in the generator leasing scenario. The reason is similar to that for Inference 6(3) and will not be repeated here. Inference 7(4) demonstrates that, when the lease coefficient of energy storage equipment is below the threshold f_6, the generator investment scenario is more favorable to the power seller's profit. The reason is similar to that for Inference 6(4).

Combined with Corollary 7(1)–(4), it can be seen that, when the lease coefficient of energy storage equipment is below the threshold f_6, all equilibrium solutions in the generator investment scenario exceed those in the generator leasing scenario. The threshold is positively correlated with the lease coefficient of energy storage equipment. Thus, when the government implements the investment subsidy, it is advisable to incentivize power generators to invest in energy storage if the subsidy ratio is large. This will guarantee that the level of energy storage technology, electricity demand, power generators' profits, and power sellers' profits are all relatively better.

4. Numerical Analysis

In this section, we employ numerical analysis methods to effectively show the impact of energy storage models and energy storage subsidies on the decision-making of the power supply chain under RPS. This section refers to Yan et al. and Chen et al. and energy storage subsidy policies implemented by various provinces and cities in China [22,32].

4.1. Impact of Renewable Energy Quota Obligation and Subsidy Amount per Unit of Discharge on Equilibrium Solutions

This section analyses the impact of renewable energy quota obligation and subsidy amount per unit of discharge on the equilibrium solution. The parameters were set as $a = 10$, $b = 0.6$, $f = 0.5$, $c_t = 0.3$, $c_r = 0.4$, $\alpha = 0.1$, $\gamma = 0.8$, and $\beta = 0.5$, and we took $\theta \in (0, 1)$ and $d \in (0, 1)$. Observing Figure 3a–d, we find:

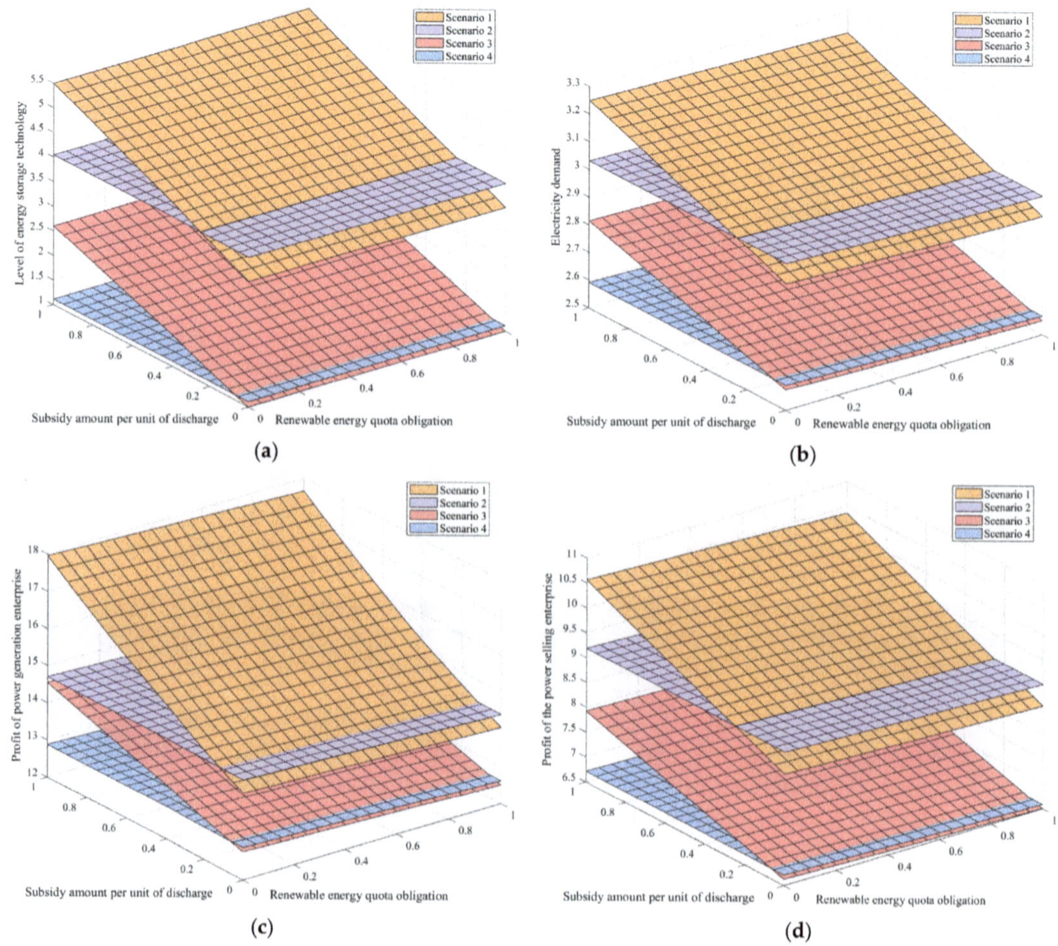

Figure 3. The impact of renewable energy quota obligation and subsidy amount per unit of discharge on the level of energy storage technology (**a**), electricity demand (**b**), power generator's profit (**c**), and power seller's profit (**d**).

Figure 3a shows that the level of energy storage technology in scenarios 1 and 2 decreases as the renewable energy quota obligation increases, and the level of energy storage technology in scenarios 1 and 3 increases as the subsidy amount per unit of discharge increases. Figure 3b shows that the electricity demand decreases with the increase in renewable energy quota obligation in all four scenarios, and the electricity demand increases with the increase in subsidy amount per unit of discharge in scenarios 1 and 3. Figure 3c,d shows that the power generator's profit and power seller's profit decreases with the increase in renewable energy quota obligation in all four scenarios, and the power generator's profit and power seller's profit increases with the increase in subsidy amount per unit of discharge in scenarios 1 and 3, which aligns with Inference 1. From Inference 1, it can be inferred that the renewable energy quota obligation has a negative impact on the equilibrium solution in all scenarios. However, when looking at the graph trend, it is evident that the inhibitory effect of the renewable energy quota obligation on the equilibrium solution is quite small compared to the effect of the subsidy amount per unit of discharge. Although RPS may reduce the level of energy storage technology and electricity demand, the implementation of a discharge subsidy can easily offset the suppressive effects of the RPS, even if the subsidy amount per unit of discharge is relatively small. Furthermore, RPS ensures the consumption of renewable energy; hence, the overall advantages surpass the drawbacks. In addition, the threshold of the subsidy amount per unit of discharge for each equilibrium solution in scenario 3 over scenario 4 is relatively small, between 0 and 0.1. It is worth noting that the storage subsidy policies implemented in China indicate that there is virtually no subsidy amount per unit of discharge within this range in practical terms. Consequently, the discharge subsidy in the generator leasing scenario exhibits a more favorable policy impact compared to the investment subsidy.

4.2. Impact of Renewable Energy Quota Obligation and Subsidy Ratio of Energy Storage Investment on the Equilibrium Solution

This section analyses the impact of renewable energy quota obligation and subsidy ratio of energy storage investment on the equilibrium solution. The parameters were set as $a = 10$, $b = 0.6$, $f = 0.5$, $c_t = 0.3$, $c_r = 0.4$, $\gamma = 0.8$, $\beta = 0.5$, and $d = 0.3$, and we took $\theta \in (0,1)$ and $\alpha \in (0,0.4)$. As can be seen from the implemented energy storage subsidy policies, no investment subsidy policy in practice sets the ratio at greater than 30%. The upper limit of a was set to 40% to facilitate a more intuitive comparison. Observing Figure 4a–d, we find:

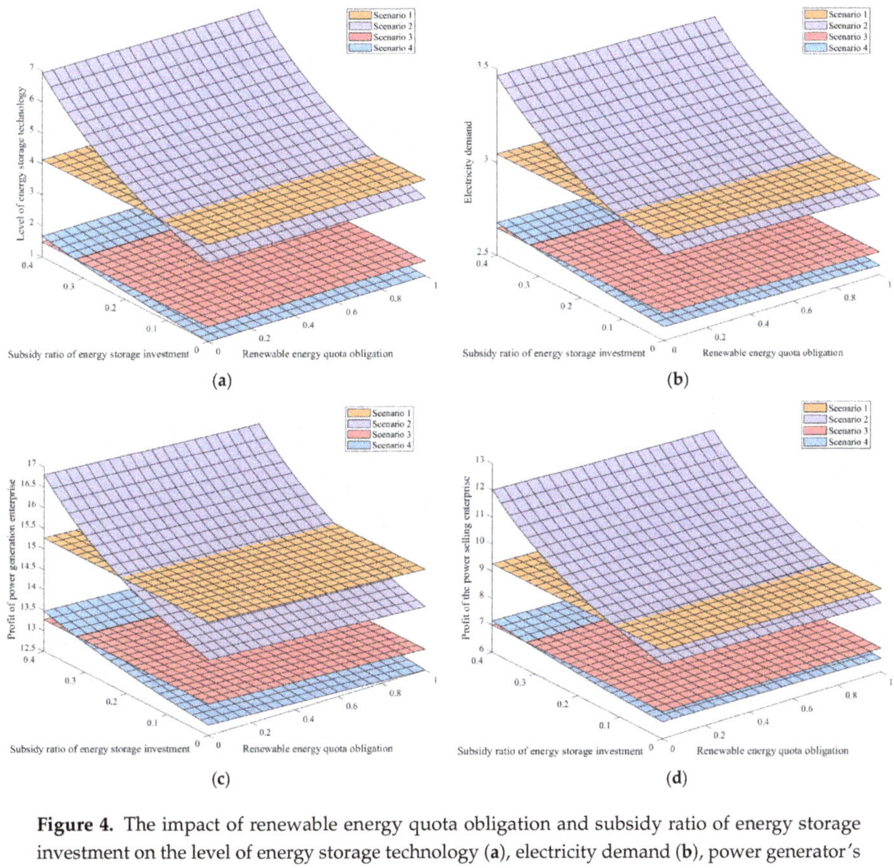

Figure 4. The impact of renewable energy quota obligation and subsidy ratio of energy storage investment on the level of energy storage technology (**a**), electricity demand (**b**), power generator's profit (**c**), and power seller's profit (**d**).

Figure 4a shows that the level of energy storage technology in scenarios 2 and 4 increases as the subsidy ratio of energy storage investment increases. This effect is particularly prominent in scenario 2, where a higher subsidy ratio leads to a more significant increase in the level of energy storage technology. Figure 4b shows that the electricity demand in scenarios 2 and 4 increases with the increase in the subsidy ratio of energy storage investment. Figure 4c,d show that the power generator's profit and the power seller's profit in scenarios 2 and 4 increase with the increase in the subsidy ratio of energy storage investment. Similar to the findings in Section 4.1, it is evident here that the reduction in the equilibrium solution caused by the renewable energy quota obligation is minimal compared to the influence of the subsidy ratio of energy storage investment. And the implementation of investment subsidies can easily compensate for this disincentive. Considering that RPS guarantees the consumption of renewable energy, it is beneficial to simultaneously apply RPS and investment subsidies. In addition, the threshold of the subsidy ratio of energy storage investment for each equilibrium solution in scenario 4 over scenario 3 is relatively large, ranging from 0.3 to 0.4. It is worth noting that the storage subsidy policies implemented in China indicate that there is virtually no subsidy ratio of energy storage investment within this range in practical terms. Therefore, the policy impact of the discharge subsidy in the generator leasing scenario is more favorable compared to the investment subsidy. Furthermore, the magnitude of the growth in each equilibrium solution in the generator investment scenario becomes more pronounced when the subsidy ratio of energy storage investment rises. This indicates that the impact of the policy is more noteworthy when power generators invest in energy storage under the investment subsidy policy.

The implementation of RPS can increase the consumption of renewable energy and enhance environmental benefits. However, it will also increase the production cost of renewable energy and the intermittency problem will affect the stability of power supply, which will affect the profit of power producers. However, combining Figures 3c and 4c, it can be observed that, under the combination of RPS and energy storage subsidy policy, the profit of power producers is enhanced. This is because, on the one hand, the electricity consumption side favors more stable electricity, and the improvement in the level of energy storage technology enhances the stability of electricity and raise the demand for electricity, thus boosting the profitability of power generators. On the other hand, the addition of the energy storage subsidy policy can very effectively reduce the pressure on power generators to invest in energy storage, thereby reducing the cost of energy storage investment.

4.3. Impact of Lease Coefficient of Energy Storage Equipment on the Equilibrium Solution

This section analyses the impact of the lease coefficient of energy storage equipment on the equilibrium solution. The parameters were set as $a = 10$, $b = 0.6$, $c_t = 0.3$, $c_r = 0.4$, $\alpha = 0.1$, $\gamma = 0.8$, $\beta = 0.5$, $d = 0.3$, and $\theta = 0.3$, and we took $f \in (0,1)$. Observing Figure 5a–d, we find:

Scenarios 1 and 2 are both the generator investment scenarios. The equilibrium solutions for these scenarios do not depend on the lease coefficient of energy storage equipment. To facilitate observation and comparison, these two scenarios are also depicted in the image. Figure 5a,b,d show that the level of energy storage technology, the electricity demand, and the power seller's profit in scenarios 3 and 4 all increase when the lease coefficient of energy storage equipment increases. When $f < 1$, the equilibrium solutions in scenarios 3 and 4 are smaller than those in scenarios 1 and 2, and they tend to surpass the equilibrium solutions in scenarios 1 and 2. Nevertheless, the lease coefficient of energy storage equipment in practice has already surpassed the range established in this study, making it challenging to obtain such a high number. Thus, it can be inferred that, overall, the energy storage technology level, electricity demand, and power seller's profit are higher in the generator investment scenario compared to the generator leasing scenario. It is worth noting that, in Figure 5c, the power generator's profit grows and subsequently decreases when the lease coefficient of energy storage equipment increases in scenarios 3 and 4, and it does not surpass the power generator's profits in scenarios 1 and 2. This is because, in our model, the profit of the electricity generator in the scenario of generator leasing is affected by both the cost of the energy storage lease and the revenue from electricity sales. The cost of leasing energy storage equipment is directly related to lease coefficient of energy storage equipment f. Revenue from electricity sales is affected by the level of energy storage technology, which in turn affects the stability and reliability of electricity supply. Higher levels of energy storage technology lead to a higher demand for electricity, which in turn leads to higher revenues. However, when f exceeds a certain threshold, the additional cost of leasing energy storage equipment can begin to outweigh the increased revenue from electricity sales. It is clear from Figure 5c that there exists a value of f that makes the generator's profit optimal, and the generator can flexibly adjust its storage strategy by observing the cost of storage leasing in the market.

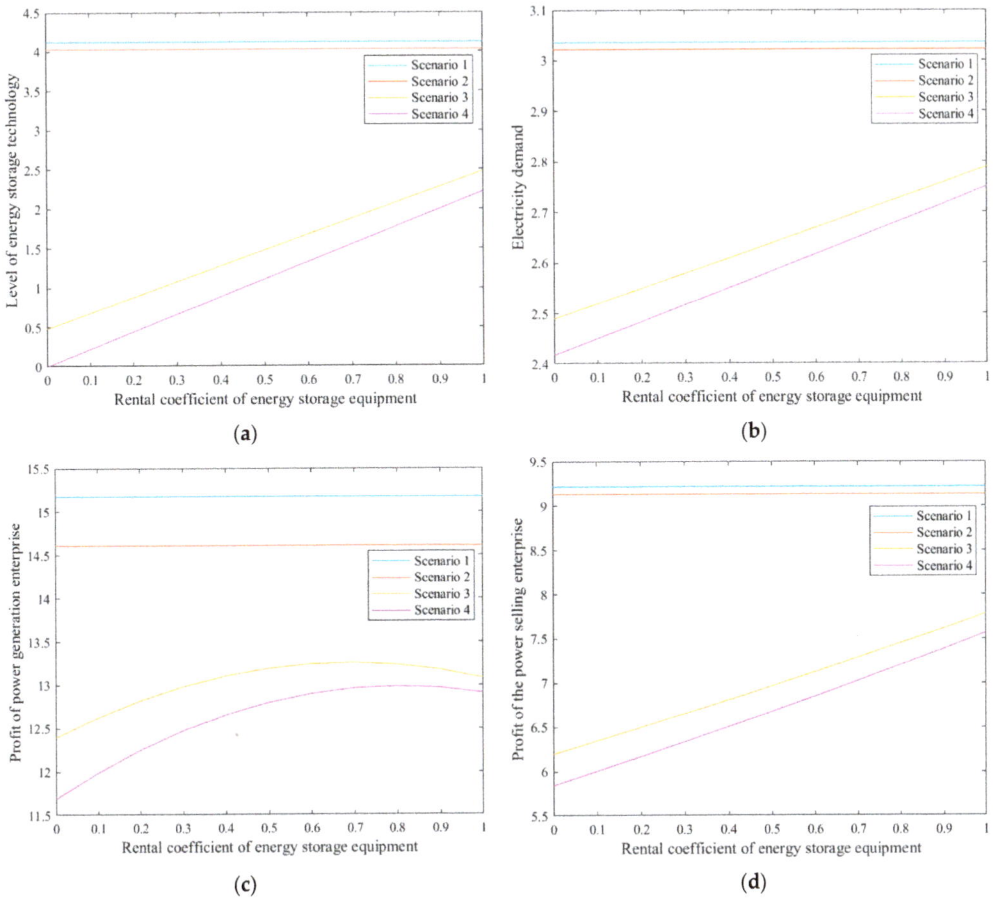

Figure 5. The impact of the lease coefficient of energy storage equipment on the level of energy storage technology (**a**), electricity demand (**b**), power generator's profit (**c**), and power seller's profit (**d**).

5. Conclusions and Policy Implications

Despite the promise of energy storage to enhance the proliferation of renewable energy, there is presently insufficient research regarding the influence of various energy storage policies and models on decision-making within the power supply chain. Therefore, this article focused on the impact of different energy storage subsidy policies combined with different energy storage models under RPS, which provides a basis for energy storage decision-making in the electricity supply chain. In this article, four scenarios were presented and two types of supply chain models were developed to illustrate this point. By analyzing the equilibrium solutions, we identified the following key findings:

1. In the scenario of generator investment or generator leasing, a higher subsidy ratio for energy storage investment is advantageous in increasing the level of energy storage technology and electricity demand. During the pre-development stage of the energy storage industry, stable power supply is crucial for the power consumption side. If power generators invest in energy storage, it will be more beneficial for the investment subsidy to enhance the level of energy storage technology and electricity demand. Additionally, during this time, the lease coefficient of energy storage equipment is also relatively high. Consequently, if power generators opt to lease energy storage equipment, it will be more beneficial for the investment subsidies to enhance the level of energy storage technology and electricity demand. Conversely, in the mature stage of the energy storage industry, regardless of the energy storage models, it is more beneficial for discharge subsidies to enhance both levels of technology and electricity demand.
2. The model of power generators investing in energy storage becomes more advantageous for increasing their profits when the discharge subsidy or investment subsidy is implemented. When the lease coefficient of energy storage equipment is low, this model also becomes more favorable for enhancing energy storage technology and meeting electricity demand. In the early stages of development in the energy storage industry, the lease coefficient of energy storage equipment tends to be high, leading power generators to invest in energy storage equipment primarily to maximize their profits. However, this approach does not contribute significantly to improving the level of energy storage technology or meeting electricity demand. Conversely, in the mature stage of the energy storage industry, power generators choose to invest in energy storage not only to maximize their profits but also to enhance the level of energy storage technology and meet electricity demand.
3. An increase in either the subsidy amount per unit of discharge or the subsidy ratio of energy storage investment has a positive impact on the level of energy storage technology, electricity demand, and the profits of electricity sellers. This policy effect is particularly significant when power generators invest in energy storage under the investment subsidy policy. In the mature stage of the energy storage industry, when the subsidy ratio of energy storage investment is low, power generators can improve their profits by relying on a higher subsidy amount per unit of discharge or a higher subsidy ratio of energy storage investment.
4. The increase in the renewable energy quota obligation imposes constraints on energy storage technology, electricity demand, and the revenues of power generators and sellers. However, whether it is implemented with a discharge subsidy or an investment subsidy, its constraining effect is relatively minor. The Renewable Portfolio Standard (RPS) significantly influences the consumption of renewable energy. As the quota obligation increases, so does the consumption of renewable energy. Therefore, RPS plays a beneficial role in advancing renewable energy.

The aforementioned findings may offer valuable managerial insights and policy implications.

Firstly, when the government considers providing subsidies for energy storage or when power generators choose energy storage models, it is important to consider the level of preference that consumers have for a reliable electricity supply. If consumers strongly prefer a certain level of energy storage technology, then government investment subsidies will be more beneficial in promoting the adoption of advanced energy storage technology and meeting electricity demand. Similarly, if consumers have a strong preference for specific energy storage technology when discharge subsidies are implemented by the government, power generators may choose to invest independently in order to maximize profits and meet consumer demands. In cases where consumers strongly favor particular levels of energy storage technology while the government implements investment subsidies, additional incentives should be provided to power generators who opt to lease energy storage equipment. This approach will encourage power generators to lease energy storage equipment and ultimately promote the advancement of energy storage technology and meet electricity demand.

Secondly, when the government considers providing subsidies for energy storage or when power generators are choosing energy storage models, it is important to consider the different development stages of the energy storage industry. During the pre-development stage of the energy storage industry, it is recommended that the government implement an investment subsidy policy and encourage power generators to lease energy storage equipment. As the energy storage industry matures, it is recommended that the subsidy ratio of energy storage investment be gradually reduced. This will help prompt energy storage investment companies to stop relying on subsidies and focus on improving their technological competitiveness, which will in turn promote sustainable development of the industry. In implementing this policy, the government may also consider providing additional incentives to power generators who choose to lease energy storage equipment. This approach will restructure the energy storage industry towards a focus on power generators leasing energy storage equipment, which can promote advancements in energy storage technology and meet electricity demand. When the energy storage industry develops to a mature stage, it is advisable for the government to implement discharge subsidies and encourage power generators to invest in their own energy storage systems. By increasing the subsidy amount per unit of discharge, more power generators can be encouraged to invest in this technology. This strategy will shift focus within the industry towards power generator investment in their own energy storage systems, promoting further improvements in technology and meeting electricity demand.

Finally, it is imperative for the government to persist in enforcing RPS. As renewable energy technology advances, it is essential to gradually raise the renewable energy quota obligation to encourage the consumption of renewable energy. Moreover, the government can simultaneously implement an energy storage subsidy policy alongside RPS. On the one hand, RPS holds the authority to establish quota obligations and compel enterprises to consume renewable energy. On the other hand, the purpose of the energy storage subsidy is to enhance energy storage technology and improve the stability of renewable energy. This will ultimately enable firms to consume more renewable energy and assist them in meeting their RPS consumption objectives.

Author Contributions: Conceptualization, C.J.; methodology, X.W. (Xinyue Wang); software, W.Z.; validation, X.W. (Xuan Wang) and W.Q.; formal analysis, C.J.; investigation, X.W. (Xinyue Wang); data curation, W.Z.; writing—original draft preparation, X.W. (Xinyue Wang); writing—review and editing, W.Q.; supervision, C.J.; project administration, C.J.; funding acquisition, C.J. All authors have read and agreed to the published version of the manuscript.

Funding: This research was funded by the General Program of the National Social Science Foundation of China (No. 24BGL218), Humanities and Social Science Fund of Ministry of Education of China (No. 22YJAZH033), Major Project of Philosophy and Social Science Research in Colleges and Universities of Jiangsu Province (No. 2021SJZDA134), and the Fundamental Research Funds for the Central Universities (No. JUSRP123137).

Data Availability Statement: Data will be made available upon request.

Conflicts of Interest: The authors declare no conflicts of interest.

References

1. Liang, C.; Umar, M.; Ma, F.; Huynh, T.L.D. Climate policy uncertainty and world renewable energy index volatility forecasting. *Technol. Forecast. Soc. Chang.* **2022**, *182*, 121810. [CrossRef]
2. Carley, S.; Davies, L.L.; Spence, D.B.; Zirogiannis, N. Empirical evaluation of the stringency and design of renewable portfolio standards. *Nat. Energy* **2018**, *3*, 754–763. [CrossRef]
3. Young, D.; Bistline, J. The costs and value of renewable portfolio standards in meeting decarbonization goals. *Energy Econ.* **2018**, *73*, 337–351. [CrossRef]
4. Wang, W.; Yuan, B.; Sun, Q.; Wennersten, R. Application of energy storage in integrated energy systems—A solution to fluctuation and uncertainty of renewable energy. *J. Energy Storage* **2022**, *52*, 104812. [CrossRef]
5. Liu, C.; Zhuo, J.; Zhao, D.; Li, S.; Chen, J.; Wang, J.; Yao, Q. A Review on the Utilization of Energy Storage System for the Flexible and Safe Operation of Renewable Energy Microgrids. *Proc. Chin. Soc. Electr. Eng.* **2020**, *40*, 1–18.
6. Cai, S.; Laijun, C.; Xinjie, Q.; Tianwen, Z.; Shengwei, M. A Generation-side Shared Energy Storage Planning Model Based on Cooperative Game. *J. Glob. Energy Interconnect.* **2019**, *2*, 360–366. [CrossRef]
7. Wang, Q.; Yi, C.; Zhang, X.; Xu, D. Optimization of Energy Storage Capacity Sizing Considering Carbon Capture Intensity and Renewable Energy Consumption. *Proc. Chin. Soc. Electr. Eng.* **2023**, *43*, 8295–8308.
8. Zhao, W.; Li, R.; Zhu, S. Subsidy Policies and Economic Analysis of Photovoltaic Energy Storage Integration in China. *Energies* **2024**, *17*, 2372. [CrossRef]
9. Bird, L.A.; Holt, E.; Carroll, G.L. Implications of carbon cap-and-trade for US voluntary renewable energy markets. *Energy Policy* **2008**, *36*, 2063–2073. [CrossRef]
10. Dahal, K.; Juhola, S.; Niemela, J. The role of renewable energy policies for carbon neutrality in Helsinki Metropolitan area. *Sustain. Cities Soc.* **2018**, *40*, 222–232. [CrossRef]
11. Hanif, I.; Aziz, B.; Chaudhry, I.S. Carbon emissions across the spectrum of renewable and nonrenewable energy use in developing economies of Asia. *Renew. Energy* **2019**, *143*, 586–595. [CrossRef]
12. Yu, S.; Hu, X.; Li, L.; Chen, H. Does the development of renewable energy promote carbon reduction? Evidence from Chinese provinces. *J. Environ. Manag.* **2020**, *268*, 110634. [CrossRef]
13. Zhang, M.; Yang, Z.; Liu, L.; Zhou, D. Impact of renewable energy investment on carbon emissions in China-An empirical study using a nonparametric additive regression model. *Sci. Total Environ.* **2021**, *785*, 147109. [CrossRef] [PubMed]
14. Meng, S.; Sun, R.; Guo, F. Does the use of renewable energy increase carbon productivity? --An empirical analysis based on data from 30 provinces in China. *J. Clean. Prod.* **2022**, *365*, 132647. [CrossRef]
15. Yang, Z.; Zhang, M.; Liu, L.; Zhou, D. Can renewable energy investment reduce carbon dioxide emissions? Evidence from scale and structure. *Energy Econ.* **2022**, *112*, 106181. [CrossRef]
16. Siddik, A.B.; Khan, S.; Khan, U.; Yong, L.; Murshed, M. The role of renewable energy finance in achieving low-carbon growth: Contextual evidence from leading renewable energy-investing countries. *Energy* **2023**, *270*, 126864. [CrossRef]
17. Xu, G.; Yang, M.; Li, S.; Jiang, M.; Rehman, H. Evaluating the effect of renewable energy investment on renewable energy development in China with panel threshold model. *Energy Policy* **2024**, *187*, 114029. [CrossRef]
18. Teresa Garcia-Alvarez, M.; Cabeza-Garcia, L.; Soares, I. Analysis of the promotion of onshore wind energy in the EU: Feed-in tariff or renewable portfolio standard? *Renew. Energy* **2017**, *111*, 256–264. [CrossRef]
19. Bento, A.M.; Garg, T.; Kaffine, D. Emissions reductions or green booms? General equilibrium effects of a renewable portfolio standard. *J. Environ. Econ. Manag.* **2018**, *90*, 78–100. [CrossRef]
20. Zhao, X.-G.; Zuo, Y.; Wang, H.; Wang, Z. How can the cost and effectiveness of renewable portfolio standards be coordinated? Incentive mechanism design from the coevolution perspective. *Renew. Sustain. Energy Rev.* **2022**, *158*, 112096. [CrossRef]
21. Zhu, Q.; Chen, X.; Song, M.; Li, X.; Shen, Z. Impacts of renewable electricity standard and Renewable Energy Certificates on renewable energy investments and carbon emissions. *J. Environ. Manag.* **2022**, *306*, 114495. [CrossRef] [PubMed]
22. Yan, Y.; Sun, M.; Guo, Z. How do carbon cap-and-trade mechanisms and renewable portfolio standards affect renewable energy investment? *Energy Policy* **2022**, *165*, 112938. [CrossRef]

23. Hu, F.; Zhou, D.; Zhu, Q.; Wang, Q. Renewable portfolio standard: When does it play a positive role? *J. Environ. Manag.* **2022**, *322*, 116122. [CrossRef]
24. Lee, K. Renewable portfolio standards and electricity prices. *Energy Econ.* **2023**, *126*, 106959. [CrossRef]
25. Rohit, A.K.; Devi, K.P.; Rangnekar, S. An overview of energy storage and its importance in Indian renewable energy sector Part I—Technologies and Comparison. *J. Energy Storage* **2017**, *13*, 10–23. [CrossRef]
26. Wang, K.; Chen, S.; Liu, L.; Zhu, T.; Gan, Z. Enhancement of renewable energy penetration through energy storage technologies in a CHP-based energy system for Chongming, China. *Energy* **2018**, *162*, 988–1002. [CrossRef]
27. Arbabzadeh, M.; Sioshansi, R.; Johnson, J.X.; Keoleian, G.A. The role of energy storage in deep decarbonization of electricity production. *Nat. Commun.* **2019**, *10*, 3413. [CrossRef] [PubMed]
28. Liu, J.; Bao, H. Research on interest coordination model of wind power supply chain with energy storage participation. *J. Energy Storage* **2022**, *49*, 104107. [CrossRef]
29. Zhang, H.; Fu, S.; Jiang, J. Multi-period network equilibrium in power system with energy storage on generation side. *Electr. Power Syst. Res.* **2023**, *223*, 109533. [CrossRef]
30. Sayed, E.T.; Olabi, A.G.; Alami, A.H.; Radwan, A.; Mdallal, A.; Rezk, A.; Abdelkareem, M.A. Renewable Energy and Energy Storage Systems. *Energies* **2023**, *16*, 1415. [CrossRef]
31. Zhao, Y.; Zhang, Y.; Li, Y.; Chen, Y.; Huo, W.; Zhao, H. Optimal configuration of energy storage for alleviating transmission congestion in renewable energy enrichment region. *J. Energy Storage* **2024**, *82*, 110398. [CrossRef]
32. Chen, W.; Tian, Y.; Ma, Y.; Bai, C. Impact of Leasing Mode with Renewable Energy Storage Equipment on Electricity Quality and Price. *Chin. J. Manag. Sci.* **2024**, *32*, 309–318.
33. Zou, C.; Xiong, B.; Xue, H.; Zheng, D.; Ge, Z.; Wang, Y.; Jiang, L.; Pan, S.; Wu, S. The role of new energy in carbon neutral. *Pet. Explor. Dev.* **2021**, *48*, 480–491. [CrossRef]
34. Kifle, Y.; Khan, B.; Singh, P. Assessment and Enhancementof Distribution System Reliabilityby Renewable Energy Sourcesand Energy Storage. *J. Green Eng.* **2018**, *8*, 219–262. [CrossRef]
35. Jamali, M.-B.; Rasti-Barzoki, M. A game-theoretic approach for examining government support strategies and licensing contracts in an electricity supply chain with technology spillover: A case study of Iran. *Energy* **2022**, *242*, 122919. [CrossRef]
36. Bao, X.; Zhao, W.; Song, Y.; Wang, X. Policy Effects Concerning the Coexistence of Renewable Portfolio Standards and Emissions Trading. *Oper. Res. Manag. Sci.* **2022**, *31*, 129–135.
37. Jian, L.; Hu, Y.; Gaocai, Y.; Ye, Z. Application of Battery Energy Storage Technology in Peak Regulation Optimization of Wind Power System. *Electr. Power Inf. Commun. Technol.* **2020**, *18*, 67–73. [CrossRef]

energies

MDPI

Article

Promoting or Hindering: The Impact of ESG Rating Differences on Energy Enterprises' Green Transformation—A Causal Test from Double Machine-Learning Algorithms

Jun Wan [1], Yuejia Wang [2] and Yuan Wang [3,*]

1 School of Management, Wuhan Textile University, Wuhan 430200, China; wanjun1230@wtu.edu.cn
2 School of International Tourism and Public Management, Hainan University, Haikou 570228, China; 20223008134@hainanu.edu.cn
3 School of Business Administration, South China University of Technology, Guangzhou 510642, China
* Correspondence: bmwangyuan@mail.scut.edu.cn

Abstract: There is a lack of comprehensive evaluation on the impact of ESG rating differences on the green transformation of energy enterprises in the transition era. This study leverages data from companies listed on the Shanghai Stock Exchange in China, applying double machine-learning algorithms to precisely estimate the causal relationship between variations in ESG ratings and the green transition efficiency of energy companies. The research shows that the difference in ESG ratings of third-party rating agencies significantly promotes the efficiency of green transformation of energy enterprises. This paper also studies the influencing factors of this effect: First, ESG rating differences significantly promote the improvement of green transition efficiency of energy enterprises; Second, the positive effect is more pronounced in energy companies with more balanced board structures. Finally, energy companies with high capital market attention can also contribute to this positive impact. Through the mechanism test, this paper finds that enterprise green innovation is an important mechanism for ESG rating divergence to positively promote the efficiency of energy enterprises' green transformation. Furthermore, this paper analyzes the impact of ESG rating on enterprises from the perspective of market cognition and short-term behavior, which provides a new perspective for analyzing the practice of enterprises pursuing long-term transformation. The study also calls for a more sober reflection on the trend toward ESG in society.

Keywords: ESG rating discrepancies; green transformation efficiency; internal governance; double machine learning

1. Introduction

Against the backdrop of global environmental deterioration, increasingly severe environmental issues have drawn widespread attention from various sectors around the world [1]. Currently, the pursuit of green and sustainable development has become a core issue of our time [2]. In China, all sectors of society have deeply realized the necessity of abandoning the previous development model characterized by "high energy consumption and high emissions" [3], shifting instead towards promoting a green transformation of the economy to achieve higher-quality growth [4]. In September 2020, China set clear goals for reaching carbon peaking by 2030 and achieving carbon neutrality by 2060 [5]. Under the new normal, green innovation transformation plays a more significant role than comprehensive energy innovation in driving economic growth and reducing energy consumption [4]. As the basic unit of economic development, the green transformation of enterprises is particularly critical, especially for energy companies [6]. Researchers have noted in their scientific reflections that energy companies hold a strategic and undeniable role in the operation and further development of the modern economy [7]. Therefore, the ability of energy companies to adopt green development practices and successfully undergo transformation is crucial for reshaping economic growth patterns, achieving the "dual carbon" targets, and fostering high-quality economic development [8].

In the current environment, environmental, social, and governance (ESG) issues have become a focal point in recent years [9]. The term ESG represents an emerging investment and business concept that refers to the incorporation of environmental protection, social responsibility, and corporate governance into investment decisions and business practices [10]. It has become a key indicator for measuring a company's sustainability performance [11]. With the increasing demand for ESG-related information from stakeholders, rating agencies have become essential players in this domain [12]. However, accurately assessing corporate ESG performance presents considerable challenges [13]. In practice, discrepancies in ESG ratings often arise due to conflicts of interest or differences in evaluation methodologies among rating agencies [14]. These rating discrepancies leave stakeholders confused when making decisions based on ESG scores and have attracted widespread attention from the academic community [15]. It is noteworthy that discrepancies in ESG ratings are a factor that stakeholders must consider carefully when making decisions based on ESG scores [16]. Particularly in corporate decision-making processes, ESG ratings directly impact stock prices, earnings, and market performance [17,18]. Therefore, whether companies will adopt strategic behaviors to reduce ESG rating discrepancies to cater to rating agencies' preferences and the potential consequences of these actions are crucial topics for further research in both academia and business practice.

In this study, we analyze whether discrepancies in ESG ratings affect the efficiency of green transformation in energy companies and explore the specific mechanisms of this impact. The starting point of the research is the insufficient attention paid to the ESG ratings of energy companies in the existing literature, limiting the possibility of an in-depth understanding of the ESG performance of energy companies. Most existing studies focus on the association between ESG ratings and corporate financial performance and their importance in measuring a company's ability to achieve sustainable development [19–24], but they neglect the ratings themselves and their potential to influence carbon reduction efforts in energy companies. Some scholars have begun to focus on the inherent issues of ESG ratings, expressing concerns about their lack of transparency [25], unclear standards [26], lack of consistency [14], and scale bias [27], pointing out that these issues may lead to stakeholder misunderstandings of companies' carbon reduction commitments [28]. The scale bias of ESG ratings marginalizes small and medium-sized energy companies in their carbon reduction efforts [29], despite these companies playing an equally indispensable role in the overall energy transition [30–32]. Moreover, in the limited studies on ESG rating discrepancies, scholars primarily focus on exploring their causes, such as differences in the selection of evaluation indicators [33], measurement methods [34], weight allocation [35], and subjective judgments [36].

While existing research provides valuable insights into the causes of ESG rating discrepancies [36,37], the effects of these discrepancies on corporate value remain relatively underexplored. Only a few studies have examined the impact of ESG rating discrepancies from the perspectives of stock market reactions [38], financial constraints [36,39], and greenwashing [40]. Regarding ESG discrepancies and energy company green transformation efficiency, existing research indicates that ESG investments can effectively secure the future economic safety of energy companies in the context of green transformation [6]. Ref. [41]

pointed out that the growing interest of investors in ESG encourages companies to seek more sustainable development paths to meet ESG standards and investor demands [41]. Ref. [36] also noted that significant differences among ESG rating providers indicate that investors and stakeholders must exercise caution when relying on these ratings for decision-making, particularly in energy efficiency projects that require consistent evaluation [36]. Ref. [42] mentioned that ESG ratings may not significantly improve companies' sustainable development behavior and could mislead stakeholders [42]. Clément et al. (2022) [25] further argued that ESG scores lack design features to measure sustainability concepts such as temporality, impact, resource management, and interconnectedness, failing to adequately capture sustainability [25].

In conclusion, the literature lacks sufficient evidence on the consequences of ESG rating discrepancies, particularly regarding their impact on the sustainability of energy companies. However, more scholars believe that ESG has a positive role in the green transformation of energy companies. Therefore, linking ESG performance and its discrepancies with the green transformation efficiency of energy companies provides a new perspective for analyzing the role of ESG performance in the sustainable development of energy companies. Ref. [43] found that differences in ESG practices across industries highlight the need for tailored approaches to corporate green transformation [43]. Ref. [44] also argued that investors' growing focus on ESG factors has led to increased scrutiny of ESG differences between companies, prompting them to strengthen their sustainability disclosures and accelerate green transformation initiatives to meet stakeholder expectations [44]. Energy companies play a crucial role in society's overall carbon reduction process, and ESG ratings are a market-based supervisory measure for the sustainable transformation of companies in the energy sector. However, there is still a gap in research on how ESG ratings affect the green transformation of companies in the energy industry, and no consensus has been reached on the relationship between the two. Whether energy companies can address ESG rating discrepancies by improving green transformation efficiency is still an unresolved question in academia. Therefore, under the guidance of the concept of sustainable development, our research aims to fill this gap by examining the role of ESG rating discrepancies in the sustainable development of energy companies from the perspective of green transformation efficiency, providing new insights into the economic consequences of ESG rating discrepancies.

To this end, we empirically examine whether and how ESG rating discrepancies affect the green transformation efficiency of energy companies. Using a sample of listed companies in China, we find strong evidence that ESG rating discrepancies significantly promote the improvement of energy companies' green transformation efficiency. From the internal and external perspectives of corporate governance, we further find that the positive impact of ESG rating discrepancies on energy companies' green transformation efficiency is less pronounced in companies with more balanced board structures and higher capital market attention. In addition, by exploring the role of green innovation in this supervisory process, this paper also provides an in-depth analysis of how ESG rating discrepancies influence energy companies' green transformation efficiency [45]. Finally, by examining the impact of ESG rating discrepancies on the uncertainty and financial performance of energy companies, we reveal their overall impact on the performance of energy companies. Overall, our research reveals the significant impact of ESG ratings on the sustainability outcomes of energy companies.

2. Theoretical Analysis and Hypothesis Formulation

To achieve sustainable development, energy companies have come under widespread ESG evaluations [46]. Whether ESG discrepancies affect the efficiency of energy companies' green transitions and contribute to their high-quality development is a question worth analyzing. Stakeholder theory suggests that companies should not only meet shareholder interests but also consider the needs of a broad range of stakeholders [47]. ESG ratings, as a key indicator of corporate sustainability performance, reflect how well companies perform in environmental protection, social responsibility, and good governance [48]. However, the efficiency of green transitions in energy companies, which is a more implicit assessment metric, is often not fully incorporated into ESG rating systems. Improving green transition efficiency means energy companies can use energy more effectively in their production processes, reducing waste and carbon emissions, thus reflecting their capacity to respond to climate change and dual carbon policies. Nevertheless, this efficiency improvement is essential for energy companies to progress efficiently towards a carbon-neutral future. For most energy companies, zero carbon emissions in production are impossible [49]. Only by improving green transition efficiency can companies advance their carbon neutrality process.

In the context of pursuing sustainable development, ESG ratings, as a vital measure of corporate sustainability performance, have been widely applied [50]. ESG ratings cover environmental, social, and governance dimensions and aim to comprehensively reflect corporate sustainability [51]. Prior research has demonstrated a significant positive correlation between favorable ESG ratings and stock price premiums [52]. This indicates that managers are strongly motivated to improve their companies' ESG ratings to gain positive recognition from the capital market and attract investors [53]. However, discrepancies in ESG ratings can lead to mistrust and skepticism from the capital market, prompting managers to take steps to reduce rating discrepancies to gain broader investor recognition [38]. Therefore, energy companies may aim to improve their sustainability performance by narrowing ESG rating discrepancies, with green transitions being a key pathway to achieving corporate sustainability. Through more efficient green transitions, companies can more effectively enhance resource utilization efficiency, reduce energy consumption, and promote sustainable development.

In summary, driven by ESG rating discrepancies, energy companies will improve their green transition efficiency to use energy more effectively, thereby pursuing enhanced sustainability performance. Therefore, we propose Hypothesis 1:

H1. *ESG rating discrepancies significantly promote the improvement of green transition efficiency in energy companies.*

Based on the above analysis, we believe that the extent of this positive effect varies across energy companies with different board structures, thus affecting the improvement of green transition efficiency. We will discuss this from three aspects.

First, we consider the impact of the proportion of independent directors on decision-making in energy companies. Independent directors, as a key component of corporate governance, bring more objective perspectives and higher professional expertise. They can objectively assess strategic decisions, such as formulating and implementing plans to improve green transition efficiency, thus improving the supervision and decision-making quality of the board [54]. In companies with a higher proportion of independent directors, ESG factors tend to carry more weight [55], as independent directors are more inclined to consider long-term interests, corporate image, and social responsibility [56], which leads to greater attention to ESG issues. Consequently, they more actively supervise environmental policies, social responsibilities, and governance practices. Moreover, independent directors often demand that energy companies take action to improve green transition efficiency to meet social and environmental expectations [57]. Therefore, in cases of significant ESG rating discrepancies, independent boards push companies to focus more on substantial sustainability metrics such as green transition efficiency [58], directing resources toward these areas rather than merely striving for short-term improvements in ESG scores. As a result, the positive impact of ESG discrepancies on improving green transition efficiency should be more evident in energy companies with a higher proportion of independent directors.

Second, gender diversity on boards is another important factor. Research shows that increasing the proportion of women on boards often leads to more cautious and forward-thinking decision-making, with greater attention to long-term sustainability rather than short-term gains [59]. Therefore, gender-diverse boards tend to be more innovative and inclusive, capable of considering a broader range of stakeholder needs and expectations and bringing different perspectives and approaches to decision-making [60]. This diversity helps avoid groupthink and improve decision quality [61]. In this context, boards with a higher gender ratio are more inclined to take actions that align with ESG standards rather than merely pursuing short-term profit maximization. They focus more on substantive sustainability metrics like green transition efficiency, as this aligns with environmental and social responsibilities and long-term economic interests [62]. However, in companies with a lower proportion of female directors, the lack of diversity may lead to boards being overly focused on short-term performance evaluations, with less emphasis on long-term projects. Consequently, resources may be allocated to issues yielding immediate returns, reducing investment in green transition efficiency. Therefore, the positive impact of ESG discrepancies on improving green transition efficiency should be more pronounced in energy companies with higher gender diversity on their boards.

Lastly, the influence of the largest director on corporate strategy and operations is crucial. The largest director typically refers to the board member with the most voting power or influence, holding dominance in decision-making. If the largest director's influence is too great, it can lead to a dictatorship-like governance structure where decisions are overly concentrated, lacking diversity and balance. In this case, companies may not fully weigh various interests or engage in collective discussions. Individual directors, being irrational, tend to pursue immediate returns and short-term interests, often prioritizing their own interests over those of other stakeholders [63]. In these companies, when faced with ESG rating discrepancies, managers may allocate resources to actions that reduce discrepancies to gain market recognition, boost stock prices, and enhance personal interests. As a result, green transition efficiency, which requires long-term investment, might be overlooked. According to this analysis, for energy companies with a lower proportion of shares held by the largest director, the positive impact of ESG discrepancies on improving green transition efficiency should be more evident.

Thus, by examining the proportion of independent directors, gender diversity, and the shareholding ratio of the largest director in governance structures, we propose the second hypothesis:

H2. *In energy companies with more balanced governance structures (higher proportions of independent and female directors and lower largest director shareholding), the positive impact of ESG rating discrepancies on green transition efficiency is more pronounced.*

The capital market's attention refers to the level of interest and responsiveness that the market shows toward a particular company or industry. For the capital market, companies with higher visibility tend to attract more investor attention, thereby securing more funding and lower financing costs [64]. Therefore, the extent of this positive effect varies across energy companies depending on their visibility in the capital market.

When capital market attention is high, companies' actions will come under broader public and investor scrutiny, and investors generally have higher expectations for companies' environmental actions and climate change responses [65]. This creates pressure for companies to not only focus on their ESG performance but also take substantive emission reduction measures. If a company's actions do not match its ESG performance claims, the company may face public skepticism and lose investor trust, risking stock price declines and divestment [66]. In this situation, high market attention acts as a soft supervisory mechanism [67], making companies more cautious about balancing ESG rating discrepancies and green transition efficiency improvements. Aware of the high level of market attention, companies understand that besides improving ESG performance, substantive emission reductions are equally important. Thus, energy companies will focus more on improving green transition efficiency to reduce carbon emissions and minimize negative environmental impacts, thereby avoiding public and investor scrutiny.

In summary, we propose the third hypothesis:

H3. *The positive impact of ESG rating discrepancies is more significant in energy companies with higher capital market attention.*

3. Research Design

3.1. Sample Selection and Data Sources

The sample in this paper consists of A-share companies listed on the Shanghai and Shenzhen stock exchanges from 2007 to 2022. The ESG data of the companies comes from the ratings of Wind, China Securities Index (CSI), FTSE Russell, SynTao Green Finance, Minglang, and Bloomberg. All other data are obtained from the CSMAR database. For the selection of energy sector companies, this paper follows the 2012 industry classification standards of the China Securities Regulatory Commission (CSRC), with sub-sector categories including coal mining and washing, oil and gas extraction, petroleum refining, coking and nuclear fuel processing, electricity and thermal power production and supply, and gas production and supply. Referring to existing literature, the following steps were taken to process the raw data: (1) Financial companies were excluded since they typically do not engage in industrial activities. (2) Companies labeled as ST or PT during the sample period were excluded for the corresponding year. (3) Due to the contagious nature of missing values in R Studio R 4.0.2, samples with missing data were removed.

3.2. Variable Quantification

3.2.1. ESG Rating Divergence

The explanatory variable ESG_dif represents ESG rating divergence. As ESG concepts have continued to develop and deepen, numerous ESG rating agencies have emerged, each with its unique calculation indicators and measurement methods. This paper selects six representative ESG rating agencies: CSI, FTSE Russell, Minglang, Bloomberg, SynTao Green Finance, and Wind, and calculates the standard deviation of their ratings to obtain ESG rating divergence data. CSI and Wind ratings are divided into 9 levels, and SynTao Green Finance ratings are divided into 10 levels, so these were assigned values ranging from 1 to 9 and 0 to 9, respectively. To make the scores of the other four agencies comparable, this paper processed them accordingly. Referring to Sustainable Investing with ESG Rating Uncertainty for data processing [38], Minglang ratings, for example, are divided into 19 enhanced levels from low to high, with the lowest level assigned a value of 0 points, the second-lowest level assigned 1 point, and so on, with the highest level assigned 18 points. The values were then multiplied by 9/18 to scale the scores down to 0 to 9. For Runling Global, which has 7 levels and lacks C and CC levels compared to CSI and Wind, the ratings would be 2 points lower. In other words, when a company is rated CCC, CSI and Wind would assign a value of 3. For FTSE Russell and Bloomberg, the specific ESG rating scores were processed by taking 10% of Bloomberg's ESG score and 200% of FTSE Russell's score as sample data. If only one ESG rating is available, the divergence value is 0.

After processing, the ESG ratings of the seven agencies all range from 0 to 9, meeting the comparability requirements.

3.2.2. Corporate Green Transformation

This paper draws on the research method from How does digitalization affect the green transformation of enterprises registered in China's resource-based cities? Further analysis of the mechanism and heterogeneity [68]. The measurement dimensions of environmental performance (efflnincom) are expanded into four levels: public welfare, work rewards and penalties, information disclosure, and pollution control. The specific scores of eight detailed variables are used to describe the environmental performance of companies. Data related to environmental performance from the CSMAR database are used as proxy variables, and the corporate green transformation index is obtained by summing these scores.

3.2.3. Board Structure

This paper measures corporate governance structure from three aspects. As mentioned earlier, companies with a higher proportion of independent directors, a higher proportion of female directors, and a lower shareholding ratio of the largest shareholder are considered to have a governance structure with a longer-term vision. Specifically, this paper uses the proportion of independent directors to the total number of directors as the proportion of independent directors (Indrcrat), the proportion of female directors to the total number of directors as the proportion of female directors (Feldrcrat), and the shareholding ratio of the largest shareholder to the total number of shares as the largest shareholder's ownership ratio (LrgHldRt).

3.2.4. Capital Market Attention

This paper uses analyst attention to companies as a proxy variable for capital market attention. The number of research reports published by securities analysts on a company is used as a measure to quantify analyst attention (gaze).

3.3. Control Variables

Referring to existing research, the following control variables are selected: firm size (natural logarithm of total assets), cash ratio (CR: ending balance of cash and cash equivalents/current liabilities), tangible asset ratio (TAR: total tangible assets/total assets), intangible asset ratio (RIA: net intangible assets/total assets), fixed asset ratio (FAR: net fixed assets/total assets), current asset ratio (CAR: total current assets/total assets), working capital ratio (WCR: (current assets − current liabilities)/total assets), inventory turnover ratio (RST: cost of sales/average net inventory balance), fixed asset turnover ratio (TFR: revenue/ending balance of net fixed assets), and effective tax rate (ERT: income tax expense/pre-tax profit) [69].

3.4. Construction of the Econometric Model

Double/Debiased Machine Learning (DML), proposed by [70], is a method that improves the estimation of the non-parametric parts of models such as Partially Linear Regression (PLR) by leveraging the predictive advantages of machine learning in high-dimensional scenarios [71]. In this paper, DML not only effectively addresses the endogeneity problem caused by confounding factors in causal inference by analyzing the variables that influence policy distortions but also eliminates regularization bias and overfitting bias when estimating treatment effects using complex machine-learning methods in high-dimensional environments through orthogonalization and cross-fitting. As a result, consistent estimates of the effect of ESG rating divergence on corporate green transformation are obtained [72].

This paper first constructs a simple, partially linear machine-learning model as follows:

$$Y_{i,t} = \theta_0 ESG\_dif_{i,t} + g(X_{i,t}) + U_{i,t} \quad (1)$$

$$E(U_{i,t}|Policy_{i,t}, X_{i,t}) = 0 \quad (2)$$

Let i represent the firm, and t represent the year. $Y_{i,t}$ denotes the dependent variable, which is the efficiency of green transformation in energy companies; $ESG\_dif_{i,t}$ is the policy variable, and θ_0 is the coefficient of interest. $X_{i,t}$ represents the set of high-dimensional variables to be estimated, specifically comprising a collection of control variables. $U_{i,t}$ is the error term with a conditional mean of zero. We estimate Equations (1) and (2), resulting in the following parameter estimates:

$$\hat{\theta} = (\frac{1}{n}\sum_{i \in I, t \in T} ESG\_dif_{i,t}{}^2)^{-1}\frac{1}{n}\sum_{i \in I, t \in T} ESG\_dif_{i,t}(Y_{i,t} - \hat{g}(X_{i,t})) \quad (3)$$

where n is the sample size. In general machine-learning models, after regularization, the given half of the sample information can be used to estimate the other half using the above formula.

However, the above expression suffers from regularization estimation bias. The random effects bias and factor influence bias are as follows:

$$\sqrt{n}(\hat{\theta_0} - \theta_0) = (\frac{1}{n}\sum_{i \in I, t \in T} ESG\_dif_{i,t}{}^2)^{-1}\frac{1}{n}\sum_{i \in I, t \in T} ESG\_dif_{i,t}\, U_{i,t} + (\frac{1}{n}\sum_{i \in I, t \in T} ESG\_dif_{i,t}{}^2)^{-1}\frac{1}{n}\sum_{i \in I, t \in T} ESG\_dif_{i,t}\, [g(X_{i,t}) - \hat{g}(X_{i,t})] \quad (4)$$

It should be noted that a = $\left(\frac{1}{n}\sum_{i \in I, t \in T} ESG\_dif_{i,t}{}^2\right)^{-1}\frac{1}{n}\sum_{i \in I, t \in T} ESG\_dif_{i,t}\, U_{i,t}$ represents the bias under the influence of random factors, and it follows a Gaussian distribution.

On the other hand, b = $(\frac{1}{n}\sum_{i \in I, t \in T} ESG\_dif_{i,t}{}^2)^{-1}\frac{1}{n}\sum_{i \in I, t \in T} ESG\_dif_{i,t}[g(X_{i,t}) - \hat{g}(X_{i,t})]$ represents the bias that exists after regularization, which is typically divergent [70]. To obtain a convergent unbiased estimate of the parameter, we employ cross-fitting estimation methods.

We construct a set of regressions for auxiliary cross-fitting:

$$ESG\_dif_{i,t} = m(X_{i,t}) + V_{i,t} \quad (5)$$

$$E(V_{i,t}|X_{i,t}) = 0 \quad (6)$$

Here, m denotes the coefficients of the high-dimensional parameters to be estimated, and we express its estimate as $\hat{m}(X_{i,t})$. $V_{i,t}$ also represents a random error term, following a normal distribution with a mean of zero.

We use the first machine-learning process to fit m, yielding the residual term: $\hat{V}_{i,t} = ESG\_dif_{i,t} - \hat{m}(X_{i,t})$; In this regression process, $\hat{V}_{i,t}$ reflects the influence of $X_{i,t}$ deviating from $Policy_{i,t}$.

Next, we utilize the second machine-learning process to fit the final parameter to be estimated g:

$$\hat{\theta} = (\frac{1}{n}\sum_{i \in I, t \in T} V_{i,t} ESG\_dif_{i,t}{}^2)^{-1}\frac{1}{n}\sum_{i \in I, t \in T} V_{i,t}(Y_{i,t} - \hat{g}(X_{i,t})) \quad (7)$$

Similar to Equation (4), we can approximate Equation (7) as:

$$\sqrt{n}(\hat{\theta_0} - \theta_0) = [E(V_{i,t}{}^2)]^{-1}\frac{1}{\sqrt{n}}\sum_{i \in I, t \in T} V_{i,t}\, U_{i,t} + [E(V_{i,t}{}^2)]^{-1}\frac{1}{\sqrt{n}}\sum_{i \in I, t \in T} [m(X_{i,t}) - \hat{m}(X_{i,t})]\, [g(X_{i,t}) - \hat{g}(X_{i,t})] \quad (8)$$

In this context, $[E(V_{i,t}{}^2)]^{-1}\frac{1}{\sqrt{n}}\sum_{i \in I, t \in T} V_{i,t}U_{i,t}$ follows a normal distribution with a mean of zero. After two rounds of estimation, $[E(V_{i,t}{}^2)]^{-1}\frac{1}{\sqrt{n}}\sum_{i \in I, t \in T} [m(X_{i,t}) - \hat{m}(X_{i,t})][g(X_{i,t})$ does not exhibit a slowing convergence rate, allowing us to obtain a consistent unbiased estimate.

4. Empirical Analysis

4.1. Analysis of Main Effects and Moderating Effects

In Section 4.1, the conclusions derived from Tables 1 and 2 are discussed separately. Table 1 presents the descriptive statistics and correlation analysis, which provide a foundational understanding of the data characteristics and relationships between key variables. The discussion here focuses on the general trends observed from the data in Table 1, offering insights into the underlying patterns.

Table 1. Main Effects and Moderating Effects.

	(1) Z1LM	(2) Z1ML	(3) T1LM	(4) T1ML	(5) T2LM	(6) T2ML	(7) T3LM	(8) T3ML	(9) T4LM	(10) T4ML
(Intercept)	0.359 **		0.384 **		−0.337 **		−0.351 **		−0.340 **	
	(0.113)		(0.119)		(0.114)		(0.118)		(0.113)	
ESGdif6	0.015 *	0.021 ***	0.017 ***	0.001 **	0.027 ***	0.001 ***	0.019 ***	0.002 ***	0.016 ***	0.011 ***
	(0.006)	(0.001)	(0.002)	(0)	(0.001)	(0.000)	(0.005)	(0.000)	(0.002)	(0.001)
size	0.333 ***		0.333 ***		0.331 ***		0.332 ***		0.328 ***	
	(0.006)		(0.006)		(0.007)		(0.007)		(0.007)	
CR	−0.158 ***		−0.155 ***		−0.155 ***		−0.156 ***		−0.146 ***	
	(0.044)		(0.044)		(0.044)		(0.045)		(0.044)	
TAR	−0.013 *		−0.013		−0.012		−0.013 *		−0.011	
	(0.007)		(0.007)		(0.007)		(0.007)		(0.007)	
RIA	−0.031		−0.030		−0.042		−0.037		−0.010	
	(0.117)		(0.117)		(0.118)		(0.119)		(0.117)	
FAR	0.102		0.100		0.083		0.096		0.112	
	(0.121)		(0.121)		(0.122)		(0.123)		(0.121)	
CAR	0.196 ***		0.199 ***		0.199 ***		0.195 ***		0.182 ***	
	(0.038)		(0.038)		(0.038)		(0.038)		(0.039)	
WCR	0.301 ***		0.298 ***		0.299 ***		0.299 ***		0.299 ***	
	(0.050)		(0.051)		(0.051)		(0.051)		(0.050)	
RST	−0.177 ***		−0.176 ***		−0.178 ***		−0.175 ***		−0.186 ***	
	(0.048)		(0.048)		(0.048)		(0.049)		(0.048)	
TFR	0.010 ***		0.010 ***		0.010 ***		0.010 ***		0.010 ***	
	(0.001)		(0.001)		(0.001)		(0.001)		(0.001)	
ERT	0.000		0.000		0.000		0.000		0.000	
	(0.000)		(0.000)		(0.000)		(0.000)		(0.000)	
Etaxrt	−0.000		−0.000		−0.000		−0.000		0.000	
	(0.003)		(0.003)		(0.003)		(0.003)		(0.003)	
Concurrent Position	−0.010		−0.010		−0.009		−0.010		−0.009	
	(0.011)		(0.011)		(0.011)		(0.011)		(0.011)	
Indrcrat2			0.059 ***							
			(0.009)							
ESGdif6:Indrcrat2			0.067 ***							
			(0.006)							
Feldrcrat					0.003 ***					
					(0.001)					
ESGdif6:Feldrcrat					0.027 ***					
					(0.000)					

Table 1. *Cont.*

	(1) Z1LM	(2) Z1ML	(3) T1LM	(4) T1ML	(5) T2LM	(6) T2ML	(7) T3LM	(8) T3ML	(9) T4LM	(10) T4ML
LrgHldRt							0.009 ***			
							(0.000)			
ESGdif6:LrgHldRt							0.0227 ***			
							(0.000)			
gaze									0.001 ***	
									(0.000)	
ESGdif6:gaze									0.039 ***	
									(0.001)	
R^2	0.903		0.903		0.904		0.903		0.905	
Adj. R^2	0.901		0.900		0.901		0.900		0.901	
Num. obs.	1081		1081		1081		1081		1081	

Note. Unstandardized regression coefficients are displayed, with standard errors in parentheses. * $p < 0.05$. ** $p < 0.01$. *** $p < 0.001$.

Table 2. Mechanism Test of Green Innovation.

	(1) Efflnincom	(2) Greinn	(3) Efflnincom
(Intercept)	3.168 ***	24.216 ***	3.168 ***
	(0.003)	(4.249)	(0.003)
size	0.333 ***	86.750 ***	0.321 ***
	(0.006)	(7.964)	(0.007)
CR	−0.158 ***	−208.631 ***	−0.130 **
	(0.044)	(54.126)	(0.044)
TAR	−0.013 *	−10.235	−0.012
	(0.007)	(8.297)	(0.007)
RIA	−0.031	−189.589	−0.005
	(0.117)	(145.128)	(0.116)
FAR	0.102	−354.431 *	0.149
	(0.121)	(149.670)	(0.120)
CAR	0.196 ***	−32.628	0.201 ***
	(0.038)	(47.232)	(0.038)
WCR	0.301 ***	47.771	0.295 ***
	(0.050)	(62.407)	(0.050)
RST	−0.177 ***	−98.288	−0.164 ***
	(0.048)	(59.335)	(0.047)
TFR	0.010 ***	1.351	0.009 ***
	(0.001)	(1.092)	(0.001)
ERT	0.000	−0.088	0.000
	(0.000)	(0.343)	(0.000)
Etaxrt	−0.000	0.443	−0.000
	(0.003)	(3.463)	(0.003)
ConcurrentPosition	−0.010	26.841 *	−0.014
	(0.011)	(13.219)	(0.011)
ESGdif6	0.015 *	4.662 ***	0.015 *
	(0.006)	(0.483)	(0.006)
greinn			0.002 ***
			(0.000)
R^2	0.903	0.256	0.906
Adj. R^2	0.901	0.236	0.903
Num. obs.	1081	1081	1081

Note. Unstandardized regression coefficients are displayed, with standard errors in parentheses. * $p < 0.05$. ** $p < 0.01$. *** $p < 0.001$.

Following that, Table 2 presents the results from the regression analysis, which tests the hypotheses and provides deeper insights into the relationships between ESG rating discrepancies and the green transformation efficiency of energy enterprises. It is important to note that the conclusions drawn from Table 1 are based on descriptive and correlational analyses, whereas the conclusions from Table 2 are specifically tied to the regression models and statistical tests.

Columns 1 and 2 of Table 1 present the impacts of ESG rating discrepancies on firms' green transformation using linear regression and double machine-learning algorithms, respectively. Columns 3 and 4 display the results of the moderating effects of independent director ratios on the main effects obtained from the two methods. Columns 5 and 6 show the moderating effects of board gender ratios on the main effects derived from both approaches. Columns 7 and 8 illustrate the moderating effects of ownership equilibrium on the main effects using the two methodologies. Finally, columns 9 and 10 present the moderating effects of analyst attention on the main effects as determined by both methods.

From Table 2, it can be observed that ESG rating discrepancies positively influence the efficiency of green transformation in energy companies, validating Hypothesis 1. Specifically, a one-unit increase in ESG rating discrepancy corresponds to an average 15.8% improvement in green transformation efficiency, as demonstrated by the results in Table 2, Column 2. Additionally, in energy companies with more balanced governance structures, such as those with higher proportions of independent directors, this effect is magnified to a 21.3% improvement. Additionally, each moderating variable positively enhances the main effects, indicating that internal governance within firms can increase their responsiveness to market supervision, thereby improving substantial environmental protection measures undertaken by the firms.

4.2. Robustness Checks

This study conducted a series of robustness checks to validate the consistency of our findings. Due to space constraints, these results are not presented in the main text but are shown in Appendix A. First, we replaced the neural network algorithm with a random forest model, as specified in the DMLLZU package 1.0.0 on CRAN, and the results, presented in Appendix A Table A1, indicated that the conclusions remained unchanged in terms of significance, though the coefficient magnitudes differed. Additionally, we employed a stacking bagging model, as shown in Appendix A Table A2, which confirmed the robustness of the results.

To address potential omitted variable bias, we included two control variables: economic policy uncertainty and the number of employees [73]. Economic policy uncertainty affects corporate decision-making, particularly under economic pressure, influencing the efficiency of green transformation [74,75]. The inclusion of these control variables did not alter the original conclusions, as shown in Appendix A Table A3.

Furthermore, we modified the quantification method for green transformation efficiency by replacing total revenue with sales revenue in the calculations. The findings, presented in Appendix A Table A4, remained consistent with the benchmark regression, further supporting the robustness of the results. Lastly, we adjusted the quantification method for ESG rating discrepancies, as shown in Appendix A Table A5, and observed consistent conclusions, confirming the robustness of our analysis.

4.3. Mechanism Analysis

The preceding analysis has thoroughly examined the causal relationship between ESG discrepancies and firms' green transformation [76]. A key question remains: How do ESG discrepancies influence firms' green transformation? A review of existing literature reveals that a crucial factor affecting corporate transformation, which requires additional attention, is the role of green innovation. As firms enhance production efficiency and reduce carbon emissions through green innovation, many innovative outcomes can improve existing emission equipment, thereby advancing firms' green transformation. Since ESG discrepancies motivate firms to cater to ESG rating agencies, this results in green innovation and clean production transformation. Therefore, this paper posits that corporate green innovation can serve as a positive mediating factor influencing the relationship between ESG discrepancies and firms' green transformation. We utilized the CSMAR database to obtain data on corporate green innovation and constructed a mediation model to analyze its mediating role. As shown in the table below, corporate green innovation plays a positive mediating role in the positive relationship between ESG rating discrepancies and the efficiency of green transformation in energy companies. In other words, corporate green innovation enhances market oversight of firms, thus strengthening the positive impact of ESG discrepancies on the efficiency of green transformation in energy companies to some extent. The following conclusion was also validated through 1000 bootstrap sampling tests, yielding the same results.

4.4. Further Research

In addition to the impact of ESG rating discrepancies on corporate green transformation (see Table 3), do they also have subsequent effects on corporate performance due to green transformation? Specifically, as shown in Table 4, ESG rating discrepancies positively influence corporate growth potential while reducing stock price volatility, suggesting that ESG discrepancies serve as important market oversight signals. To explore this, this paper considers the effects of ESG rating discrepancies on both corporate development uncertainty and corporate development returns. Specifically, we use stock price volatility as a proxy for corporate development uncertainty and growth indicators as a proxy for corporate development returns for analysis. Furthermore, we examine whether corporate green transformation can serve as a mediating variable that allows firms to mitigate the negative impact of market rating discrepancies by enhancing their green transformation, thereby improving their growth potential and reducing uncertainty.

Table 3. Further Research: Analysis of Factors Affecting Corporate ESG Rating Discrepancies and Stock Price Volatility.

	(1) MeanAmret	(2) Efflnincom	(3) MeanAmret
(Intercept)	1.026 ***	3.168 ***	1.026 ***
	(0.150)	(0.003)	(0.150)
size	−0.384	0.333 ***	−1.686 *
	(0.281)	(0.006)	(0.727)
CR	2.586	−0.158 ***	3.203
	(1.911)	(0.044)	(1.932)
TAR	−0.196	−0.013 *	−0.144
	(0.293)	(0.007)	(0.293)
RIA	7.126	−0.031	7.246
	(5.124)	(0.117)	(5.109)
FAR	12.185 *	0.102	11.787 *
	(5.284)	(0.121)	(5.272)
CAR	2.052	0.196 ***	1.283
	(1.668)	(0.038)	(1.709)
WCR	−0.164	0.301 ***	−1.343
	(2.203)	(0.050)	(2.279)
RST	3.868	−0.177 ***	4.562 *
	(2.095)	(0.048)	(2.119)
TFR	0.012	0.010 ***	−0.025
	(0.039)	(0.001)	(0.043)
ERT	0.000	0.000	−0.001
	(0.012)	(0.000)	(0.012)
Etaxrt	0.089	−0.000	0.090
	(0.122)	(0.003)	(0.122)

Table 3. *Cont.*

	(1) MeanAmret	(2) Efflnincom	(3) MeanAmret
ConcurrentPosition	0.288	−0.010	0.327
	(0.467)	(0.011)	(0.466)
ESGdif6	−0.116 ***	0.015 *	−0.106 ***
	(0.064)	(0.006)	(0.065)
efflnincom			3.915 ***
			(1.015)
R^2	0.037	0.903	0.045
Adj. R^2	0.011	0.901	0.017
Num. obs.	1081	1081	1081

Note. Unstandardized regression coefficients are displayed, with standard errors in parentheses. * $p < 0.05$. *** $p < 0.001$.

Table 4. Further Research: Analysis of Factors Influencing the Relationship Between Corporate ESG Rating Discrepancies and Growth Potential.

	(1) GRO	(2) Efflnincom	(3) GRO
(Intercept)	0.054 **	3.168 ***	0.054 **
	(0.021)	(0.003)	(0.021)
size	−0.068	0.333 ***	−0.258 *
	(0.039)	(0.006)	(0.101)
F011201A	0.091	−0.158 ***	0.181
	(0.265)	(0.044)	(0.268)
F010401A	−0.049	−0.013 *	−0.041
	(0.041)	(0.007)	(0.041)
F031001A	0.270	−0.031	0.287
	(0.710)	(0.117)	(0.708)
F030901A	1.367	0.102	1.309
	(0.732)	(0.121)	(0.730)
F030801A	−0.293	0.196 ***	−0.404
	(0.231)	(0.038)	(0.237)
F030101A	−0.543	0.301 ***	−0.714 *
	(0.305)	(0.050)	(0.316)
F030501A	0.363	−0.177 ***	0.464
	(0.290)	(0.048)	(0.293)
F041401B	0.010	0.010 ***	0.004
	(0.005)	(0.001)	(0.006)
F040505C	0.001	0.000	0.000
	(0.002)	(0.000)	(0.002)
Etaxrt	−0.002	−0.000	−0.002
	(0.017)	(0.003)	(0.017)
ConcurrentPosition	0.026	−0.010	0.031
	(0.065)	(0.011)	(0.065)
ESGdif6	0.023 ***	0.015 *	0.024 ***
	(0.007)	(0.006)	(0.037)
efflnincom			0.569 *
			(0.279)
R^2	0.075	0.903	0.084
Adj. R^2	0.050	0.901	0.056
Num. obs.	1081	1081	1081

Note. Unstandardized regression coefficients are displayed, with standard errors in parentheses. * $p < 0.05$. ** $p < 0.01$. *** $p < 0.001$.

As shown in the table below, ESG rating discrepancies negatively impact corporate stock price volatility, indicating that ESG rating discrepancies serve as an important market oversight signal for firms, reducing stock price uncertainty and increasing the market's generally cautious outlook on firms. Corporate green transformation acts as a positive mediating factor, suggesting that green transformation enhances business stability, thereby increasing the influence of market perceptions on stock price volatility.

As shown in the table below, ESG rating discrepancies can positively influence corporate growth potential. This paper analyzes this situation, and relevant research suggests that ESG rating discrepancies, to some extent, represent disparities in market attention. Therefore, they symbolize corporate development potential and controversy, with increased ESG discrepancies contributing to enhanced corporate growth. Corporate green transformation serves as a positive mediating factor, indicating that green transformation can increase corporate growth potential by reducing institutional costs faced by firms in the context of environmental regulations and enhancing opportunities for transformational development.

5. Discussion

In the context of current climate change and dual carbon policies, the efficiency of green transformation in energy enterprises has become a crucial indicator of their sustainable development capabilities. For instance, energy companies like those in the oil and gas sector can become advocates for energy transition by embracing climate change and taking proactive measures to address its impacts [77]. Ref. [78] also noted that oil and gas companies must develop strategies to adapt to climate change and renewable energy to ensure sustainability in the global energy transition context [78]. Although this indicator is not commonly included in ESG ratings, it reflects how energy companies respond to these challenges. This paper utilizes panel data from Chinese listed companies to explore how ESG rating discrepancies affect the efficiency of green transformation in energy enterprises and further analyze their profound impact on sustainable development. The study employs benchmark regression and double machine-learning model regression to test hypotheses and strengthen the credibility of the research through robustness checks. This paper not only replaces regression methods and core explanatory variables but also adds control variables such as macroeconomic policy uncertainty and the number of employees in energy enterprises. Additionally, environmental violation cases are used as instrumental variables to address endogeneity issues. Finally, the study examines the impact of ESG rating discrepancies on energy enterprises, including their effect on corporate uncertainty (e.g., stock price volatility) and corporate performance (e.g., financial performance) [79]. The empirical results reveal several key conclusions. First, ESG rating discrepancies significantly enhance the efficiency of green transformation in energy enterprises. Second, this positive impact is more pronounced in energy enterprises with a more balanced board structure. Lastly, energy enterprises with higher market attention can also facilitate this positive effect. These hypotheses provide clear directions for empirical research. Finally, through robustness checks, mechanism analyses, and further investigations into corporate impacts, the study enhances the credibility and interpretability of the results. By replacing regression methods, adding control variables, substituting core explanatory variables, and employing instrumental variable methods to address endogeneity, the robustness of the findings is ensured.

6. Conclusions

In summary, our research is the first to incorporate the comprehensive indicator of green transformation efficiency in energy enterprises, innovatively exploring the moderating role of internal board structure on ESG rating discrepancies [80]. This reveals a new dimension of the impact of ESG rating discrepancies on environmental factors in energy enterprises. This theoretical contribution not only expands the boundaries of research on ESG rating discrepancies and understanding of their effects on environmental factors in energy enterprises but also provides a new assessment perspective for energy companies in their responses to climate change and dual carbon policies. Furthermore, previous studies have rarely considered the impact of internal governance structures on environmental performance in energy enterprises [81–83]. This finding offers a new research perspective for optimizing governance structures in energy enterprises and has significant practical implications for their sustainable development. Additionally, the use of double machine-learning model regression allows for more accurate identification of the complex causal relationship between green transformation efficiency and ESG rating discrepancies compared to traditional linear models. This methodological innovation enables a deeper understanding of how comprehensive factors influence the sustainable development of energy enterprises. Moreover, by exploring the mechanism of green innovation, this paper analyzes how ESG rating discrepancies affect the efficiency of green transformation in energy enterprises. Finally, by examining the effects of ESG rating discrepancies on corporate uncertainty and financial performance, it reveals the comprehensive impact on the overall performance of energy enterprises.

At the same time, this research provides new insights into understanding and improving the green transformation efficiency of energy enterprises in ESG ratings. On the one hand, energy enterprises should emphasize green transformation efficiency as a critical indicator of their sustainable development, which not only responds to global climate change and dual carbon policy requirements but also significantly enhances their ESG ratings [84]. By optimizing energy use, energy enterprises can reduce energy consumption, lower operational costs, and improve production efficiency, effectively boosting their ESG scores and enhancing market competitiveness [85]. On the other hand, investors and rating agencies should consider incorporating green transformation efficiency into ESG assessment frameworks to evaluate the sustainable development capabilities of energy enterprises more comprehensively [86]. This can help alleviate the information asymmetry issues arising from ESG rating discrepancies and promote the healthy development of capital markets.

The verification of hypotheses and experimental results in this study also reveals several important management insights. The validation of H1 indicates that ESG rating discrepancies significantly promote the efficiency of green transformation in energy enterprises, emphasizing that under different rating standards for various energy companies, ESG discrepancies can help enterprises monitor themselves more comprehensively. For large-scale energy enterprises with high emissions, discrepancies in ESG ratings can attract broader market attention, thereby motivating these companies to adopt more proactive emission reduction measures and green transformation strategies. Consequently, energy enterprises should pay greater attention to ESG ratings, increasing transparency and enhancing communication with investors to reduce rating discrepancies and gain trust and support from the capital market. The validation of H2 and H3 indicates that a balanced board structure and high capital market attention can enhance this positive effect. Therefore, energy enterprises should strive to optimize their board structures, increasing the proportion of independent directors, achieving gender balance, and reducing the proportion of the largest shareholder to enhance the board's oversight and decision-making capabilities, enabling more effective responses to ESG rating discrepancies. Finally, through further analysis of the green innovation mechanism, we find that green innovation is an essential mediator linking ESG rating discrepancies and the enhancement of green transformation efficiency in energy enterprises. ESG rating discrepancies can prompt companies to increase R&D investment in clean energy technologies, energy efficiency improvements, and emission reduction technologies, driving innovation in green technologies and products, thus improving green transformation efficiency [87]. Therefore, energy enterprises should regard green innovation as a key strategy for enhancing ESG performance and addressing climate change, accelerating the commercialization of green technologies.

7. Research Limitations

While this study provides valuable insights into the role of ESG rating discrepancies in promoting green transformation efficiency, there are several limitations that should be noted:

(1) The study focuses on Chinese listed energy companies, which may limit the generalizability of the findings to other regions or industries. Future research could consider cross-country comparisons or other sectors.
(2) The proxy variable for green transformation efficiency may not capture all dimensions of corporate sustainability performance. Incorporating alternative metrics could enhance robustness.
(3) Although the use of double machine learning mitigates endogeneity, potential biases from unobserved variables cannot be entirely ruled out.
(4) The data spans from 2007 to 2022. Future studies could explore more recent data to assess the evolving dynamics of ESG performance.

Author Contributions: Conceptualization, Y.W. (Yuan Wang); Software, Y.W. (Yuan Wang); Investigation, J.W.; Resources, Y.W. (Yuan Wang); Data curation, Y.W. (Yuejia Wang); Writing—original draft, Y.W. (Yuejia Wang); Writing—review and editing, Y.W. (Yuan Wang); Supervision, Y.W. (Yuan Wang); Project administration, Y.W. (Yuan Wang); Funding acquisition, J.W. All authors have read and agreed to the published version of the manuscript.

Funding: This research was funded by [University Innovation and Entrepreneurship Program of the Ministry of Education] grant number [202410559018] and The APC was funded by [University Innovation and Entrepreneurship Program of the Ministry of Education].

Data Availability Statement: The data presented in this study are available on request from the corresponding author due to database permissions.

Conflicts of Interest: The authors declare no conflict of interest.

Appendix A

Table A1. Random forest algorithm results.

	(1) Z1ML	(2) T1ML	(3) T2ML	(4) T3ML	(5) T4ML
ESGdif	0.013 *** (0.001)				
ESGdif6:Indrcrat2		0.001 *** (0)			
ESGdif6:Feldrcrat			0.002 *** (0)		
ESGdif6:LrgHldRt				0.001 *** (0)	
ESGdif6:gaze					0.001 *** (0)

Note. Unstandardized regression coefficients are displayed, with standard errors in parentheses. *** $p < 0.001$.

Table A2. Bagging model algorithm results.

	(1) Z1ML	(2) T1ML	(3) T2ML	(4) T3ML	(5) T4ML
ESGdif	0.116 *** (0.023)				
ESGdif6:Indrcrat2		0.003 *** (0.001)			
ESGdif6:Feldrcrat			0.002 * (0.001)		
ESGdif6:LrgHldRt				0.002 *** (0)	
ESGdif6:gaze					0.005 *** (0.001)

Note. Unstandardized regression coefficients are displayed, with standard errors in parentheses. * $p < 0.05$. *** $p < 0.001$.

Table A3. Adjust the variable quantization method to change the results.

	(1) Z1ML	(2) T1ML	(3) T2ML	(4) T3ML	(5) T4ML
ESGdif	0.005 *** (0.001)				
ESGdif6:Indrcrat2		0.001 *** (0.000)			
ESGdif6:Feldrcrat			0.001 *** (0)		
ESGdif6:LrgHldRt				0.001 *** (0)	
ESGdif6:gaze					0.003 *** (0.001)

Note. Unstandardized regression coefficients are displayed, with standard errors in parentheses. *** $p < 0.001$.

Table A4. The quantization method of dependent variable changes the results.

	(1) Z1ML	(2) T1ML	(3) T2ML	(4) T3ML	(5) T4ML
ESGdif	0.013 *** (0.001)				
ESGdif6:Indrcrat2		0.001 *** (0)			
ESGdif6:Feldrcrat			0 *** (0)		
ESGdif6:LrgHldRt				0.002 *** (0)	
ESGdif6:gaze					0.001 *** (0)

Note. Unstandardized regression coefficients are displayed, with standard errors in parentheses. *** $p < 0.001$.

Table A5. The quantization method of the argument changes the results.

	(1) Z1ML	(2) T1ML	(3) T2ML	(4) T3ML	(5) T4ML
ESGdif	0.019 ** (0.009)				
ESGdif6:Indrcrat2		0.001 *** (0)			
ESGdif6:Feldrcrat			0 *** (0)		
ESGdif6:LrgHldRt				0 *** (0)	
ESGdif6:gaze					0.001 *** (0)

Note. Unstandardized regression coefficients are displayed, with standard errors in parentheses. ** $p < 0.01$. *** $p < 0.001$.

References

1. Singh, R.L.; Singh, P.K. Global environmental problems. In *Principles and Applications of Environmental Biotechnology for a Sustainable Future*; Springer: Singapore, 2017; pp. 13–41.
2. Gazzola, P.; Del Campo, A.G.; Onyango, V. Going green vs going smart for sustainable development: Quo vadis? *J. Clean. Prod.* **2019**, *214*, 881–892. [CrossRef]
3. Zhou, Y.; Hao, F.; Meng, W.; Fu, J. Scenario analysis of energy-based low-carbon development in China. *J. Environ. Sci.* **2014**, *26*, 1631–1640. [CrossRef] [PubMed]
4. Jiang, Z.; Lyu, P.; Ye, L.; Zhou, Y.W. Green innovation transformation, economic sustainability and energy consumption during China's new normal stage. *J. Clean. Prod.* **2020**, *273*, 123044. [CrossRef]
5. Huang, M.T.; Zhai, P.M. Achieving Paris Agreement temperature goals requires carbon neutrality by middle century with far-reaching transitions in the whole society. *Adv. Clim. Chang. Res.* **2021**, *12*, 281–286. [CrossRef]
6. Plotkin, J.; Levchenko, N.; Shyshkanova, G.; Levchenko, S. Development of Energy Enterprises in the Context of Green Transformation. *J. Eng. Sci.* **2023**, *10*, G22–G33. [CrossRef]
7. Kozar, Ł.J.; Sulich, A. Energy Sector's Green Transformation towards Sustainable Development: A Review and Future Directions. *Sustainability* **2023**, *15*, 11628. [CrossRef]
8. Latapí Agudelo, M.A.; Johannsdottir, L.; Davidsdottir, B. Drivers that motivate energy companies to be responsible. A systematic literature review of Corporate Social Responsibility in the energy sector. *J. Clean. Prod.* **2020**, *247*, 119094. [CrossRef]
9. Baratta, A.; Cimino, A.; Longo, F.; Solina, V.; Verteramo, S. The Impact of ESG Practices in Industry with a Focus on Carbon Emissions: Insights and Future Perspectives. *Sustainability* **2023**, *15*, 6685. [CrossRef]
10. UN Global Compact. *Who Cares Wins: Connecting Financial Markets to a Changing World*; The Global Compact: New York, NY, USA, 2004.
11. Sibarani, S. ESG (Environmental, Social, and Governance) Implementation to Strengthen Business Sustainability Pt. Migas–North Field. *Eur. J. Bus. Manag. Res.* **2023**, *8*, 147–150. [CrossRef]
12. Zhu, Y.; Yang, H.; Zhong, M. Do ESG Ratings of Chinese Firms Converge or Diverge? A Comparative Analysis Based on Multiple Domestic and International Ratings. *Sustainability* **2023**, *15*, 12573. [CrossRef]
13. Yi, Z.; Li, H. Environmental, Social and Governance Investing Assessment under Incomplete Information by the Power Aggregation Model on Probabilistic Linguistic Terms. *IEEE Access* **2021**, *9*, 120442–120450. [CrossRef]
14. Billio, M.; Costola, M.; Hristova, I.; Latino, C.; Pelizzon, L. Inside the ESG ratings: (Dis)agreement and performance. *Corp. Soc. Responsib. Environ. Manag.* **2021**, *28*, 1426–1445. [CrossRef]
15. Zhang, L. ESG Rating Divergence: Beauty Is in the Eye of the Beholder. *J. Index Investig.* **2021**, *12*, 53–63. [CrossRef]
16. Jacobs, B.I.; Levy, K.N. The Challenge of Disparities in ESG Ratings. *J. Impact ESG Investig.* **2022**, *2*, 107–111. [CrossRef]
17. Kiesel, F.; Lücke, F. ESG in credit ratings and the impact on financial markets. *Financ. Mark. Inst. Instrum.* **2019**, *28*, 263–290. [CrossRef]
18. Zumente, I.; Lāce, N. Esg rating—Necessity for the investor or the company? *Sustainability* **2021**, *13*, 8940. [CrossRef]
19. Chininga, E.; Alhassan, A.L.; Zeka, B. ESG ratings and corporate financial performance in South Africa. *J. Account. Emerg. Econ.* **2024**, *14*, 692–713. [CrossRef]
20. Drempetic, S.; Klein, C.; Zwergel, B. The Influence of Firm Size on the ESG Score: Corporate Sustainability Ratings Under Review. *J. Bus. Ethics* **2020**, *167*, 333–360. [CrossRef]
21. Gao, S.; Meng, F.; Wang, W.; Chen, W. Does ESG always improve corporate performance? Evidence from firm life cycle perspective. *Front. Environ. Sci.* **2023**, *11*, 1105077. [CrossRef]
22. Kim, S.; Li, Z. Understanding the impact of esg practices in corporate finance. *Sustainability* **2021**, *13*, 3746. [CrossRef]
23. Yu, H.; Zhang, L. On the Mechanics of sustainability: ESG Rating and company performance. *SSRN Electron. J.* **2021**. [CrossRef]
24. Zhang, K.; Md Kassim, A.A.; Guo, Y. Research on Corporate Sustainability from an ESG Perspective. *Front. Bus. Econ. Manag.* **2023**, *8*, 185–191. [CrossRef]
25. Clément, A.; Robinot, É.; Trespeuch, L. Improving ESG Scores with Sustainability Concepts. *Sustainability* **2022**, *14*, 13154. [CrossRef]
26. Amir, A.Z.; Serafeim, G. Why and how investors use ESG information: Evidence from a global survey. *Financ. Anal. J.* **2018**, *74*, 87–103. [CrossRef]
27. Akgun, O.T.; Mudge, T.J.; Townsend, B. How Company Size Bias in ESG Scores Impacts the Small Cap Investor. *J. Impact ESG Investig.* **2021**, *1*, 31–44. [CrossRef]
28. Que, X. Optimization path of corporate ESG performance and compensation performance management under carbon neutral target. *Appl. Math. Nonlinear Sci.* **2023**, *9*, 1–14. [CrossRef]
29. Gregory, R.P. The influence of firm size on ESG score controlling for ratings agency and industrial sector. *J. Sustain. Financ. Investig.* **2024**, *14*, 86–99. [CrossRef]

30. De Souza, R.; Casisi, M.; Micheli, D.; Reini, M. A review of small–medium combined heat and power (CHP) technologies and their role within the 100% renewable energy systems scenario. *Energies* **2021**, *14*, 5338. [CrossRef]
31. Henriques, J.; Catarino, J. Motivating towards energy efficiency in small and medium enterprises. *J. Clean. Prod.* **2016**, *139*, 42–50. [CrossRef]
32. Sayed, E.T.; Wilberforce, T.; Elsaid, K.; Rabaia, M.K.H.; Abdelkareem, M.A.; Chae, K.J.; Olabi, A.G. A critical review on environmental impacts of renewable energy systems and mitigation strategies: Wind, hydro, biomass and geothermal. *Sci. Total Environ.* **2021**, *766*, 144505. [CrossRef]
33. Christensen, D.M.; Serafeim, G.; Sikochi, A. Why is Corporate Virtue in the Eye of The Beholder? The Case of ESG Ratings. *Account. Rev.* **2022**, *97*, 147–175. [CrossRef]
34. Widyawati, L. Measurement concerns and agreement of environmental social governance ratings. *Account. Financ.* **2021**, *61*, 1589–1623. [CrossRef]
35. Capizzi, V.; Gioia, E.; Giudici, G.; Tenca, F. The divergence of ESG ratings: An analysis of Italian listed companies. *J. Financ. Manag. Mark. Inst.* **2021**, *9*, 2150006. [CrossRef]
36. Berg, F.; Kölbel, J.F.; Rigobon, R. Aggregate Confusion: The Divergence of ESG Ratings*. *Rev. Financ.* **2022**, *26*, 1315–1344. [CrossRef]
37. Hughes, A.; Urban, M.A.; Wójcik, D. Alternative esg ratings: How technological innovation is reshaping sustainable investment. *Sustainability* **2021**, *13*, 3551. [CrossRef]
38. Avramov, D.; Cheng, S.; Lioui, A.; Tarelli, A. Sustainable investing with ESG rating uncertainty. *J. Financ. Econ.* **2022**, *145*, 642–664. [CrossRef]
39. Kim, R.; Koo, B. The impact of ESG rating disagreement on corporate value. *J. Deriv. Quant. Stud.* **2023**, *31*, 219–241. [CrossRef]
40. Schumacher, K. Environmental, Social, and Governance (ESG) Factors and Green Productivity: The Impacts of Greenwashing and Competence Greenwashing on Sustainable Finance and ESG Investing. *SSRN Electron. J.* **2023**. [CrossRef]
41. Cohen, B.; Winn, M.I. Market imperfections, opportunity and sustainable entrepreneurship. *J. Bus. Ventur.* **2007**, *22*, 29–49. [CrossRef]
42. Gilberto Gomes, S.; Fabio Henrique, P. ESG (Environmental, Social and Governance) in construction civil: Concept that can contribute to the sector. *Ann. Civ. Environ. Eng.* **2022**, *6*, 064–065. [CrossRef]
43. Epstein, M.J.; Buhovac, A.R.; Elkington, J.; Leonard, H.B.D. Making sustainability work: Best practices in managing and measuring corporate social, environmental and economic impacts. In *Making Sustainability Work: Best Practices in Managing and Measuring Corporate Social, Environmental and Economic Impacts*; Taylor and Francis: Abingdon, UK, 2017. [CrossRef]
44. Friede, G.; Busch, T.; Bassen, A. ESG and financial performance: Aggregated evidence from more than 2000 empirical studies. *J. Sustain. Financ. Investig.* **2015**, *5*, 210–233. [CrossRef]
45. Asni, N.; Agustia, D. Does corporate governance induce green innovation? An emerging market evidence. *Corp. Gov.* **2022**, *22*, 1375–1389. [CrossRef]
46. Utami, P.; Gandakusuma, I. Sustainability's influence on energy firm performance: An ESG perspective. *J. Apl. Akunt.* **2023**, *8*, 207–221. [CrossRef]
47. Clement, R.W. The lessons from stakeholder theory for U.S. business leaders. *Bus. Horiz.* **2005**, *48*, 255–264. [CrossRef]
48. Markovich, K.A.; Alekseevna, E.A.; Alekseevna, C.D. Ecological, Social and Governance Impact on the Company's Performance: Information Technology Sector Insight. *Procedia Comput. Sci.* **2022**, *214*, 1065–1072. [CrossRef]
49. Davis, S.J.; Lewis, N.S.; Shaner, M.; Aggarwal, S.; Arent, D.; Azevedo, I.L.; Benson, S.M.; Bradley, T.; Brouwer, J.; Chiang, Y.M.; et al. Net-zero emissions energy systems. *Science* **2018**, *360*, eaas9793. [CrossRef]
50. Alsayegh, M.F.; Rahman, R.A.; Homayoun, S. Corporate economic, environmental, and social sustainability performance transformation through ESG disclosure. *Sustainability* **2020**, *12*, 3910. [CrossRef]
51. Sharma, S.; Ruud, A. On the path to sustainability: Integrating social dimensions into the research and practice of environmental management. *Bus. Strategy Environ.* **2003**, *12*, 205–214. [CrossRef]
52. Zhou, D.; Zhou, R. Esg performance and stock price volatility in public health crisis: Evidence from COVID-19 pandemic. *Int. J. Environ. Res. Public Health* **2022**, *19*, 202. [CrossRef] [PubMed]
53. Wang, N.; Pan, H.; Feng, Y.; Du, S. How do ESG practices create value for businesses? Research review and prospects. *Sustain. Account. Manag. Policy J.* **2023**, *15*, 1155–1177. [CrossRef]
54. Kang, S. Experienced Independent Directors. *SSRN Electron. J.* **2013**. [CrossRef]
55. Yu, E.P.-y.; Van Luu, B. International variations in ESG disclosure—Do cross-listed companies care more? *Int. Rev. Financ. Anal.* **2021**, *75*, 101731. [CrossRef]
56. Tibiletti, V.; Marchini, P.L.; Furlotti, K.; Medioli, A. Does corporate governance matter in corporate social responsibility disclosure? Evidence from Italy in the "era of sustainability". *Corp. Soc. Responsib. Environ. Manag.* **2021**, *28*, 896–907. [CrossRef]
57. Giannarakis, G.; Andronikidis, A.; Sariannidis, N. Determinants of environmental disclosure: Investigating new and conventional corporate governance characteristics. *Ann. Oper. Res.* **2020**, *294*, 87–105. [CrossRef]

58. Disli, M.; Yilmaz, M.K.; Mohamed, F.F.M. Board characteristics and sustainability performance: Empirical evidence from emerging markets. *Sustain. Account. Manag. Policy J.* **2022**, *13*, 929–952. [CrossRef]
59. Iren, P. Gender diversity of boardrooms and firm financial performance. *Risk Gov. Control Financ. Mark. Inst.* **2016**, *6*, 28–33. [CrossRef]
60. Attah-Boakye, R.; Adams, K.; Kimani, D.; Ullah, S. The impact of board gender diversity and national culture on corporate innovation: A multi-country analysis of multinational corporations operating in emerging economies. *Technol. Forecast. Soc. Chang.* **2020**, *161*, 120247. [CrossRef]
61. Wicker, P.; Feiler, S.; Breuer, C. Board gender diversity, critical masses, and organizational problems of non-profit sport clubs. *Eur. Sport Manag. Q.* **2020**, *22*, 251–271. [CrossRef]
62. Tamimi, N.; Sebastianelli, R. Transparency among S&P 500 companies: An analysis of ESG disclosure scores. *Manag. Decis.* **2017**, *55*, 1660–1680. [CrossRef]
63. Ng, Y.K. Individual irrationality and social welfare. *Soc. Choice Welf.* **1989**, *6*, 87–101. [CrossRef]
64. Que, J.; Zhang, X. Money chasing hot industries? Investor attention and valuation of venture capital backed firms. *J. Corp. Financ.* **2021**, *68*, 101949. [CrossRef]
65. Cordeiro, J.J.; Tewari, M. Firm Characteristics, Industry Context, and Investor Reactions to Environmental CSR: A Stakeholder Theory Approach. *J. Bus. Ethics* **2015**, *130*, 833–849. [CrossRef]
66. Beatty, R.P.; Ritter, J.R. Investment banking, reputation, and the underpricing of initial public offerings. *J. Financ. Econ.* **1986**, *15*, 213–232. [CrossRef]
67. Stiglitz, J.E. Credit Markets and the Control of Capital. *J. Money Credit. Bank.* **1985**, *17*, 133–152. [CrossRef]
68. Feng, Y.; Gao, Y.; Hu, S.; Sun, M.; Zhang, C. How does digitalization affect the green transformation of enterprises registered in China's resource-based cities? Further analysis on the mechanism and heterogeneity. *J. Environ. Manag.* **2024**, *365*, 121560. [CrossRef]
69. Stoian, A.; Iorgulescu, F. Sustainable Capital Market. In *Financing Sustainable Development: Key Challenges and Prospects*; Springer: Berlin/Heidelberg, Germany, 2019. [CrossRef]
70. Chernozhukov, V.; Chetverikov, D.; Demirer, M.; Duflo, E.; Hansen, C.; Newey, W.; Robins, J. Double/debiased machine learning for treatment and structural parameters. *Econom. J.* **2018**, *21*, C1–C68. [CrossRef]
71. McSharry, P.E.; Smith, L.A.; Tarassenko, L. Comparison of Predictability of Epileptic Seizures by a Linear and a Nonlinear Method. *IEEE Trans. Biomed. Eng.* **2003**, *50*, 628–633. [CrossRef] [PubMed]
72. Risuleo, R.S.; Bottegal, G.; Hjalmarsson, H. Identification of Linear Models from Quantized Data: A Midpoint-Projection Approach. *IEEE Trans. Autom. Control* **2020**, *65*, 2801–2813. [CrossRef]
73. Salvioni, D.M.; Gennari, F. Corporate Governance, Sustainability and Capital Markets Orientation. *Int. J. Manag. Sustain.* **2014**, *3*, 469–483. [CrossRef]
74. Banks, E. External Governance Mechanisms: Systemic Accountability. In *Corporate Governance*; Palgrave Macmillan: London, UK, 2004.
75. Salvioni, D.M.; Astori, R. Sustainable Development and Global Responsibility in Corporate Governance. *Symphonya Emerg. Issues Manag.* **2013**, *1*, 28–52. [CrossRef]
76. Hegarty, W.H.; Hoffman, R.C. Who influences strategic decisions? *Long Range Plan.* **1987**, *20*, 76–85. [CrossRef]
77. Gristo, P. President's Page: Energy transition as an opportunity for Latin American geoscientists, companies, and the people. *Lead. Edge* **2023**, *42*, 234–235. [CrossRef]
78. Cherepovitsyn, A.; Rutenko, E. Strategic Planning of Oil and Gas Companies: The Decarbonization Transition. *Energies* **2022**, *15*, 6163. [CrossRef]
79. Watson, T.J. Ethical choice in managerial work: The scope for moral choices in an ethically irrational world. *Hum. Relat.* **2003**, *56*, 167–185. [CrossRef]
80. Aich, S.; Thakur, A.; Nanda, D.; Tripathy, S.; Kim, H.C. Factors affecting esg towards impact on investment: A structural approach. *Sustainability* **2021**, *13*, 10868. [CrossRef]
81. Acharya, V.V.; Myers, S.C.; Rajan, R.G. The Internal Governance of Firms. *J. Financ.* **2011**, *66*, 689–720. [CrossRef]
82. Cheng, Q.; Lee, J.; Shevlin, T. Internal governance and real earnings management. *Account. Rev.* **2016**, *91*, 1051–1085. [CrossRef]
83. Jaggi, B. Corporate Governance: Structure and Consequences. In *Encyclopedia of Finance*, 3rd ed.; Springer International Publishing: Cham, Switzerland, 2022. [CrossRef]
84. Miatishkin, G.; Mironenkova, A. Energy Efficiency Index of Industrial Energy Consumers for Achieving the Sustainable Development Goals. In Proceedings of the 2021 International Scientific and Technical Engine Conference, EC 2021, Samara, Russia, 23–25 June 2021. [CrossRef]
85. Hu, W.; Wang, P.; Gooi, H.B. Toward optimal energy management of microgrids via robust two-stage optimization. *IEEE Trans. Smart Grid* **2018**, *9*, 1161–1174. [CrossRef]

86. Dosanu, M.; Ciobanu, A.; Gheorghe, S. Exploring the Environmental, Social and Governance Measures in Real Estate—The Impact in Energy Management. In Proceedings of the 2023 11th International Conference on ENERGY and ENVIRONMENT, CIEM 2023, Bucharest, Romania, 26–27 October 2023. [CrossRef]
87. Perotti, S.; Colicchia, C. Greening warehouses through energy efficiency and environmental impact reduction: A conceptual framework based on a systematic literature review. *Int. J. Logist. Manag.* **2023**, *34*, 199–234. [CrossRef]

energies

MDPI

Article

Impacts of Natural Gas Pipeline Congestion on the Integrated Gas–Electricity Market in Peru

Richard Navarro [1], Hugo Rojas [1], Jaime E. Luyo [1], Jose L. Silva [2] and Yuri P. Molina [2,*]

[1] Faculty of Mechanical Engineering, National University of Engineering, Lima 15333, Peru; richard.navarro.r@uni.pe (R.N.); hrojase@uni.pe (H.R.); jeluyo@yahoo.es (J.E.L.)
[2] Department of Electrical Engineering, Federal University of Paraíba, João Pessoa 58051-900, Paraíba, Brazil; joseleandro.silva@estudante.cear.ufpb.br
* Correspondence: molina.rodriguez@cear.ufpb.br

Abstract: This paper investigates the impact of natural gas pipeline congestion on the integrated gas–electricity market in Peru, focusing on short-term market dynamics. By simulating congestion by reducing the primary natural gas pipeline's capacity, the study reveals significant patterns in production costs and load flows within the electrical network. The research highlights the critical interdependencies between natural gas and electricity systems, emphasizing how constraints in one network can directly affect the other. The findings underscore the importance of coordinated management of these interconnected systems to optimize economic dispatch and ensure the reliability of both gas and electricity grids. The study also proposes strategic public policy interventions to mitigate the financial and physical impacts of pipeline congestion, contributing to more efficient and resilient energy market operations.

Keywords: economic dispatch; electricity grid; gas grid; natural gas; thermal power stations; optimization

1. Introduction

Despite the traditionally well-defined interconnection between electricity and gas networks, these infrastructures have typically been managed as separate systems. However, it is crucial to acknowledge the vital role of gas-fired generators in electricity production, as highlighted in recent research [1]. Moreover, with the increasing need to balance the variability of renewable energy sources and the potential of green hydrogen for decarbonizing the energy sector, a significant shift is occurring, as detailed in [2]. This shift marks a major transformation towards a more integrated management of gas and electricity networks.

The integration of natural gas and electricity systems has garnered significant interest among researchers [3–5]. These studies offer comprehensive analyses of the operational coordination between gas and electrical systems, emphasizing the critical importance of flexibility. Rubio-Barros [6] explored the planning of joint operations, underscoring the intricate interdependence between natural gas and electricity networks. Additionally, to address supply security in integrated systems, researchers such as [7] have developed simulation models. This research is particularly relevant given the substantial consumption by natural gas thermal generation plants (CGTGNs), which significantly affects overall system gas consumption.

Hibbard et al. [8] provided a comprehensive analysis of the evolving interdependence between natural gas and electricity systems, exploring key influencing factors. Ventosa et al. [9] presented a modeling approach designed to evaluate the technical constraints affecting the operations and availability of natural gas networks. These studies underscore the direct impact on natural gas demand, energy dispatch, and, fundamentally, the reliability and safety of energy systems. Traditionally, research into natural gas and electricity systems has been compartmentalized, reflecting their unique operational characteristics. Moreover, coordination between natural gas and electricity dispatch has generally been conducted independently, with minimal cross-communication. Despite some studies addressing operational constraints and strategic planning, the integration of gas and electricity markets, along with their models, presents a significant opportunity for further research and development in this critical area.

In Peru, the natural gas (NG) and electricity markets are pursuing initiatives to integrate various energy systems into a unified structure, which is crucial for the overall transformation of the energy sector [10]. This integration is imperative because these systems are interdependent; the risks and uncertainties affecting one system inevitably impact the others. Consequently, it is essential to identify and quantify these risks and develop strategies to mitigate their effects.

Although Peru has taken steps in this direction with the approval of D.S No. 012-2021-EM, which established regulations to optimize the use of natural gas and created the "Natural Gas Manager"—an entity responsible for gathering data on the availability and capacity of NG volumes for short-, medium-, and long-term forecasting—this entity does not have responsibilities for real-time dispatch operations. Therefore, despite these efforts, Peru remains far from achieving the comprehensive integration detailed in [1].

Our objective is to analyze the impact of congestion within the natural gas network on the short-term electricity market within an integrated gas–electric system. We aim to explore the interactions among network parameters, anticipate potential contingency issues in an integrated market, and ensure security and reliability in a competitive market environment.

The main contributions of this research are:

- A comprehensive quantitative and qualitative analysis of the interdependencies between electricity systems and natural gas (NG) networks within an integrated market context. This involves examining how these two energy systems interact and influence each other, identifying key factors that affect their performance and reliability.
- The development and presentation of a reduced equivalent electrical network model for the SEIN (Interconnected Electrical System of Peru) at the 500 kV level. This model simplifies the complex electrical network, making it easier to study and understand. Additionally, an equivalent model for the Peruvian natural gas pipeline system is provided, facilitating a better understanding of the gas network's behavior and its interaction with the electrical system.
- Insights into the economic and physical impacts of congestion in natural gas pipelines on both the short-term electricity market and the transmission network. This includes analyzing how bottlenecks in the gas pipeline system can affect electricity prices, supply reliability, and the overall efficiency of the energy market.
- Proposals for mitigation mechanisms through strategic public policies aimed at alleviating the economic and physical effects of congestion. These mechanisms are designed to ensure the reliable and efficient operation of the integrated gas–electricity system by addressing the root causes of congestion and improving coordination between the gas and electricity sectors.

The structure of this work is as follows: Section 2 presents a model of the integrated gas–electricity market and includes an integrated analysis of the parameters and equations governing both networks. Section 3 provides a detailed examination of the economic and physical impacts resulting from gas network congestion on the short-term market and electrical transmission system. Finally, Section 4 concludes the study by summarizing the key insights and implications derived from the research.

2. The Model Proposed for the Integrated Analysis of the Electricity and Gas Systems

The objective of this research is to evaluate the impact of congestion in the natural gas network on the short-term electricity market dynamics within an integrated gas–electric system. We investigate potential contingencies within this integrated market framework, with a focus on ensuring safety and reliability in a competitive market environment. To achieve this objective, we have refined and extended the models previously developed in [11,12], placing particular emphasis on the interdependencies between the natural gas and electricity markets. Our findings highlight how congestion in natural gas pipelines influences electric power transmission, deepening our understanding of the interconnected operations of these energy systems. This research emphasizes the need for coordinated strategies to maintain system reliability and market competitiveness, making a significant contribution to the body of knowledge on effectively managing interconnected energy systems.

Our research demonstrates the effects of congestion in the natural gas pipeline on electric power transmission networks in short-term operations.

2.1. Gas–Electricity Optimization Model

The planning of wholesale energy markets, often referred to as short-term markets, revolves around economically determining the optimal cost for energy production. This involves a comprehensive consideration of various factors, including minimizing operating costs, dispatching loads from different energy sources, accounting for associated marginal costs, and addressing network constraints [1,13].

A complex interplay of variables and constraints shapes the selection of energy sources and the technologies used for electricity production. These systems are exposed to risks and uncertainties, including the variability of energy sources and the inherent limitations of the technologies employed. Renewable energy plants, in particular, add significant complexity to the planning and operation of energy systems due to their heavy reliance on variable weather conditions. These stochastic elements require sophisticated models and strategies to ensure efficient and reliable energy production under intermittent supply conditions.

In our study, the complex interaction between natural gas sources, the natural gas network, combined-cycle gas turbine generators (CCGTs), and the constraints of transmission network parameters introduces a high degree of interdependence [14]. This interdependence presents significant challenges in calculating optimal dispatch values [15]. Our research focuses on optimizing load dispatch within an integrated gas–electricity system to effectively address these challenges. We utilize a reduced equivalent model at the 500 kV voltage level of the SEIN (National Interconnected Electric System) in Peru, along with an equivalent model for the Peruvian natural gas pipeline. These models, thoroughly detailed by Navarro [11] and Rojas [12], serve as the foundation for our analysis and optimization efforts, aiming to enhance efficiency and reliability in energy dispatch within the integrated system.

2.1.1. Equivalent Reduced Model of the SEIN Peru

The reduced electric model of the SEIN, configured with 12 buses, is characterized by elements represented through their admittances in per unit (p.u.) values. This simplified electrical model of the 12-bus SEIN is designed to make the complex network more manageable for analytical purposes while preserving the crucial characteristics and interactions of the system.

This model simplifies the intricate electrical network into a more tractable form, allowing for detailed study while retaining the essential electrical properties and relationships between the buses. It enables thorough investigations into transmission behaviors, load distributions, and potential bottlenecks. The admittance matrix, as detailed in Table 1, encapsulates the electrical connectivity and characteristics of the buses, providing a quantitative foundation for further analysis and optimization within the integrated gas–electricity framework. This approach facilitates an improved understanding and management of the interconnected energy system.

The outcomes of the power flow analysis conducted on the 12-bus reduced equivalent model of the SEIN Peru at 500 kV are detailed in Table 2.

Table 1. Admittance parameters of the 12-bus reduced equivalent model of the SEIN at 500 kV (in p.u.).

Name Line (500 kV)	From (Bus)	To (Bus)	Admittance Y	PHI (°)	R (p.u.)	X (p.u.)	LIMIT (MVA)
LT La Niña—Trujillo 500 kV	1	2	28.33	−1.49	0.0027	0.0352	420.3
LT Trujillo—Chimbote 500 kV	2	3	62.38	−1.51	0.001	0.016	438.6
LT Chimbote—Carabayllo 500 kV	3	4	23.15	−1.5	0.003	0.0431	376.6
LT Carabayllo—Carapongo 500 kV	4	12	292.1	−1.45	0.0004	0.0034	1379.7
LT Carapongo—ChilcaCTM 500 kV	12	6	129.34	−1.48	0.0007	0.0077	780
LT Chilca—Poroma 500 kV	6	5	20.99	−1.49	0.0037	0.0475	646.3
LT Poroma—Ocoña 500 kV	5	9	71.51	−1.37	0.0028	0.0137	491
LT Ocoña—San José 500 kV	9	8	140.08	−1.37	0.0014	0.007	267.5
LT San José—Montalvo 500 kV	8	7	166.09	−1.37	0.0012	0.0059	0
LT Montalvo—Yarabamba 500 kV	7	11	76.16	−1.5	0.0009	0.0131	584.2
LT Yarabamba—Poroma 500 kV	11	5	35.48	−1.43	0.004	0.0279	0
LT Poroma—Colcabamba 500 kV	5	10	47.42	−1.44	0.0028	0.0209	889.2

Table 2. Load flow analysis results for the 12-bus reduced equivalent model of the SEIN Peru at 500 kV.

Number Bus	Name Bus	Voltage p.u.	Angle (°)	Load MW	Load MVAR	Gen MW	Gen MVAR	Base kV
B1	La Nina 500 kV	1.000	−40.16	420.3	230	96.3	32.7	500
B2	Trujillo 500 kV	1.000	−33.54	438.6	224.7	30.3	39.5	500
B3	Chimbote 500 kV	1.000	−26.73	376.6	87.1	333.5	167.8	500
B4	Carabayllo 500 kV	1.000	−6.63	1379.7	900.5	908.6	643.6	500
B5	Poroma 500 kV	0.969	−7.39	780	101.7	0	0	500
B6	Chilca 500 kV	1.000	0	646.3	517.3	1841	735.5	500
B7	Montalvo 500 kV	1.000	−10.49	491	268.6	167	150.7	500
B8	San Jose 500 kV	0.999	−10.67	267.5	70.5	0	316	500
B9	Ocoña 500 kV	0.986	−9.60	0	−96.8	0	0	500
B10	Colcabamba 500 kV	1.000	−1.74	584.2	452.5	1332.6	307.6	500
B11	Yarabamba 500 kV	1.000	−7.64	0	−150	383	167	500
B12	Carapongo 500 kV	1.000	−4.11	889.2	646.8	1251.3	−1.3	500

Figure 1 depicts the 12-bus reduced equivalent model of the SEIN at the 500 kV voltage level.

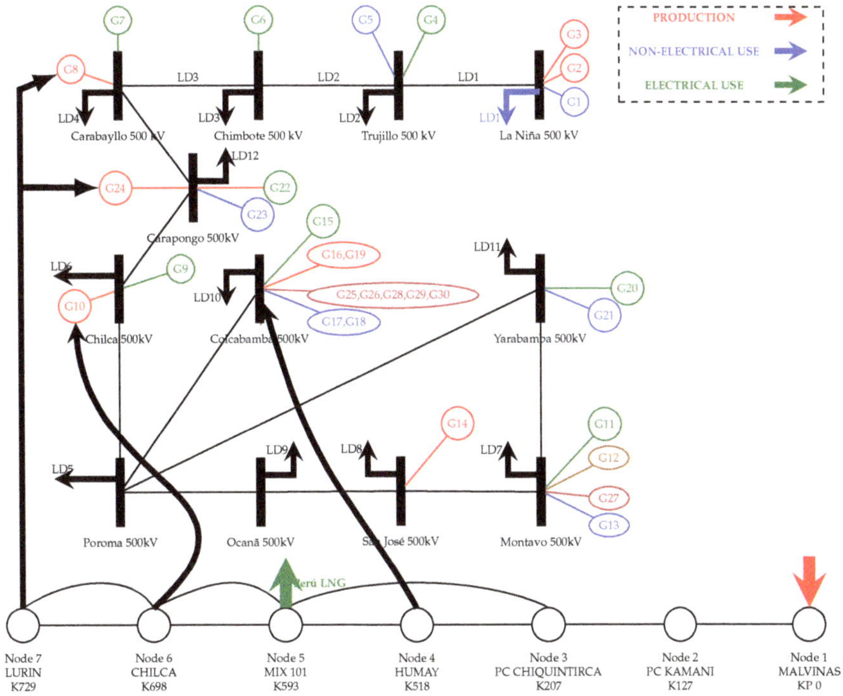

Figure 1. Interaction diagram of the Peruvian integrated gas–electricity system.

2.1.2. Equivalent Model of the Peruvian Natural Gas Pipeline Network

The equivalent model of the natural gas (NG) pipeline network is designed to conform to the parameters and constraints governing the relationships between gas flows and pressures. Addressing the challenge presented by the non-linear nature of these flow-pressure relationships, we have developed a linear approximation [16]. This approximation is based on an equivalent model of the NG network that incorporates all relevant network parameters, facilitating an accurate representation of these critical interactions.

Table 3 displays the lengths of the main natural gas pipeline sections, organized according to their pipe diameters. This table includes segments owned by various operators, such as Peru LNG and the Loop Coast, offering a comprehensive overview of the network's physical structure.

Table 3. Main parameters of the equivalent model of the Peruvian gas pipeline network (adapted from Osinergmin).

Line	Node Start	Node End	Has Compresor	Diameter (Inch)	Length km	Accountability
1	Node 1 (Malvinas)	Node 2 (Kamani)	YES	32	126.7	main network
2	Node 2 (Kamani)	Node 3 (Chiquintirca)	YES	32	81.27	main network
3	Node 3 (Chiquintirca)	Node 4 (Humay)	NO	24	309.93	main network
4	Node 4 (Humay)	Node 5 (Mix 101)	NO	18	75.24	main network
5	Node 3 (Chiquintirca)	Node 5 (Mix 101)	NO	34	406	Peru LNG
6	Node 5 (Mix 101)	Node 6 (Chilca)	NO	18	105	main network
7	Node 5 (Mix 101)	Node 6 (Chilca)	NO	24	105	Loop coast
8	Node 6 (Chilca)	Node 7 (Lurín)	NO	18	31.16	main network
9	Node 6 (Chilca)	Node 7 (Lurín)	NO	24	31.16	Loop coast

Table 4 provides the URL coordinates for the nodes, while Table 5 presents the maximum and minimum natural gas flows, taking into account both electrical and non-electrical demands.

Table 4. URL coordinates of the nodes of the Peruvian gas pipeline network (Adapted from https://observatorio.osinergmin.gob.pe/ accessed on 15 March 2023).

Node	Name	KP	Zone	North	East
1.	Malvinas	KP 0	18	8,689,907.55	724,013.26
2.	Kamani	KP 127	18	8,599,403.00	691,070.00
3.	Chiquintirca	KP 207	18	8,599,333.00	691,017.00
4.	Humay	KP 517.9	18	8,480,890.00	404,048.00
5.	Mix 101 (CUA)	KP 594.9	18	8,536,618.00	361,063.00
6.	Chilca	KP 699	18	8,614,339.00	314,286.00
7.	Lurin	KP 729	18	8,640,336.00	300,970.00

Table 5. Maximum and minimum volumes, considering electrical and non-electrical demands (adapted from Osinergmin).

Node	Name	Level Productiốn		Use Max (Elect/No Elect)		Level of Pressure	
		Min	Max			Min (bar)	Max (bar)
		MMPCD	MMPCD	MMPCD	MMPCD		
1.	MALVINAS	460	1605	0	0	147	147
2.	KAMANI	460	1605	0	0	136	147
3.	CHIQUINTIRCA	225	435	0	0	109	147
4.	HUMAY	196	386	49	0	120	135
5.	MIX 101	416	936	0	620	113	102
6.	CHILCA	203	516	420	0	104	54
7.	LURIN	203	516	0	516	104	46

Figure 1 showcases the integration of the 12-bus reduced equivalent model of the SEIN with the equivalent model of the Peruvian natural gas pipeline network.

2.2. *Objetive Function*

Building on the foundational research presented in [11,12], this paper extends its models to delve deeper into the complex interdependencies between the natural gas and electricity markets. We have developed an optimization function specifically tailored for the competitive dynamics of an integrated gas–electricity market. The primary objective of this optimization is to maximize social profit, which effectively translates into minimizing the overall social costs associated with energy provision. This includes reducing the production costs of electricity, decreasing expenditures on natural gas supplies, and lowering the operational costs of the natural gas network. Through this framework, our goal is to enhance the cost-effectiveness and efficiency of the energy supply system, thereby delivering economic and societal benefits.

The optimization model introduced in this study features the role of a gas–electricity system operator, who is tasked with the real-time management of both the electrical grid and the natural gas pipeline network. The fundamental aim of this model is to guide the operator in minimizing costs by carefully balancing operating expenses, pipeline transportation costs, and fuel costs for natural gas-fired power plants. This comprehensive approach to cost management emphasizes the operator's critical role in optimizing the efficiency and economic performance of the integrated system.

To support the effective implementation and analysis of this model, we establish the following assumptions:

- Power Balance: There must be a continuous balance between energy supply and demand, ensuring that load flows are within network restrictions.

- Individual Generator Constraints: Each generator operates under specific constraints, including dispatch price, generation limits, and other operational limitations.
- Power Transmission Constraints: The transmission system is governed by constraints to ensure its reliable and secure operation. These include line capacities, voltage limits, and other transmission-related factors.
- Natural Gas Source and Pipeline Limits: The model considers limitations on the availability of natural gas and the capacity of pipelines, ensuring that the supply remains within specified limits.
- Natural Gas Network Constraints: The natural gas network faces operational constraints, such as pressure limits, flow capacity limits, and other network-related constraints.
- Coupling Constraints: There are specific constraints on the coupling between the electrical and gas systems, ensuring coordinated operation while respecting the operational limits and capabilities of each system.
- Hydroelectric Power Plants: The model assumes a linear cost function for hydroelectric power plants, which simplifies representation and excludes a hydro-thermal coupling.
- Time Resolution: The model uses a one-hour time resolution to analyze a 24-h day, allowing it to capture temporal dynamics and variations in demand and generation profiles.
- Transmission Costs: The objective function of the model excludes transmission costs, focusing primarily on minimizing generation and natural-gas-related costs.

These assumptions and constraints shape the formulation and analysis of the optimization model within the integrated gas–electricity market framework.

$$FO = EC + GC + TC \tag{1}$$

$$EC = \sum_{g=1}^{Gen}\sum_{t=1}^{24}\left(b_g * P_g^2(t) + a_g * P_g(t) + c_g + C_{Diesel} + C_{rer} + C_{res}\right) \tag{2}$$

$$GC = \sum_{n=1}^{NGs}\sum_{t=1}^{24}\left(S_n(t) * \ p_{ng}\right) \tag{3}$$

$$TC = \sum_{l=1}^{PG}\sum_{t=1}^{24}\left(f_{nm}(t) * p_{tg}\right) \tag{4}$$

where:

$P_g(t)$: Active Power delivered by generator g at period t, measured in [MW].
a_g, b_g, *and* c_g: Characteristic constants for each gas and diesel thermal generator, used in the quadratic equation for computing the cost of energy production from each bus generator. These constants depend on the power output of the generator and are measured in [USD/MW^2], [USD/MW], and [USD], respectively.
C_{Diesel}: Cost of Generation with Diesel in [USD].
C_{rer}: Cost of generation with conventional and non-conventional renewable sources in [USD].
C_{res}: Cost of energy storage system in [USD] that remains fixed throughout the period.
S_n: NG supply at node n in period t, measured in [10^6 m^3/h].
P_{ng}: Unit price of NG, measured in [USD/m^3].
$f_{mn}(t)$: NG flow from node m to node n in period t, measured in [10^6 m^3/h].
p_{tg}: Unit cost of NG transportation in [USD/m^3].
PG: number of pipes in the gas network.
NGs: Number of supply points in the NG network.
Gen: Number of power plants in the network.

2.2.1. Electrical System

For a typical bus "i", the power flow equations can be articulated by considering the network's impedance and the potential difference between buses "i" and "j", expressed in polar coordinates. These fundamental equations encompass both components of power: the active power, denoted as P_i, and the reactive power, represented by Q_i. Specifically, the equations can be expressed as follows:

(a) Power flow:

$$P_i = \sum_{j=1}^{N} |Y_{ij} V_i V_j| \cos(\theta_{ij} + \delta_j - \delta_i) \qquad \forall\, t \tag{5}$$

$$Q_i = -\sum_{j=1}^{N} |Y_{ij} V_i V_j| \sin(\theta_{ij} + \delta_j - \delta_i) \qquad \forall\, t \tag{6}$$

Notation:

$$Y_{ij} = |Y_{ij}|\langle\theta_{ij} = |Y_{ij}|\cos\theta_{ij} + j|Y_{ij}|\sin\theta_{ij} = G_{ij} + jB_{ij} \tag{7}$$

$$V_i = |V_i|\langle\delta_i = |V_i|(\cos\delta_i + j\sin\delta_i) \tag{8}$$

$$-P_{km}^{max} \leq P_{km} \leq P_{km}^{max} \tag{9}$$

where:
P_i: Active power at bus i during period t, measured in megawatts (MW);
Q_i: Reactive power generated at bus i during period t, measured in megavolt-amperes reactive (MVAR);
Y_{ij}: Impedance of the line between bus i and j, incorporating both resistance and reactance components;
θ_{ij}: Angle of total line admittance between bus i and j, representing the phase difference that affects power flow;
δ_i: Voltage phase shift angle at bus i, indicating the difference in phase angle of the voltage relative to a reference point;
P_{km}^{max}: Maximum limit of active power transferable in the section from node k to node m, specified in megawatts (MW);
P_{km}: Active power in the section from bus k to bus m, measured in megawatts (MW);

(b) Losses in the electrical network:

$$P_L = \sum_{i=1}^{N} P_{gi,t} - \sum_{i=1}^{N} P_{di,t} \qquad \forall\, t \tag{10}$$

$$Q_L = \sum_{i=1}^{N} Q_{gi,t} - \sum_{i=1}^{N} Q_{di,t} \qquad \forall\, t \tag{11}$$

where:
P_L: Active power losses within the network during period t, measured in megawatts (MW);
Q_L: Reactive power losses within the network during period t, measured in megavolt-amperes reactive (MVAR);
$\sum_{i=1}^{N} P_{gi,t}$: Total active power delivered by the generators in the network during period t, aggregated across all N buses, in MW;
$\sum_{i=1}^{N} P_{di,t}$: Total active power demand at all N buses in the network during period t, in MW;
$\sum_{i=1}^{N} Q_{gi,t}$: Total reactive power supplied by the generators in the network during period t, aggregated across all N buses, in MVAR;
$\sum_{i=1}^{N} Q_{di,t}$: Total reactive power demand at all N buses in the network during period t, in MVAR;
N: The total number of buses in the electrical network.

(c) Power balance:

$$\sum_{g=1}^{G} P_{gi,t} - P_{di,t} = \sum_{j=1}^{N} |Y_{ij}V_{i,t}V_{j,t}| \cos\left(\theta_{ij} + \delta_{j,t} - \delta_{i,t}\right) \quad \forall\, i,\, t \tag{12}$$

$$\sum_{g=1}^{G} Q_{gi,t} - Q_{di,t} = -\sum_{j=1}^{N} |Y_{ij}V_{i,t}V_{j,t}| \sin\left(\theta_{ij} + \delta_{j,t} - \delta_{i,t}\right) \quad \forall\, i,\, t \tag{13}$$

$$V_{min} \leq V_{i,t} \leq V_{max} \ldots \forall\, i,\, t \tag{14}$$

$$\delta_{min} \leq \delta_{i,t} \leq \delta_{max} \ldots \forall i,\, t \tag{15}$$

where:
$\sum_{g=1}^{G} P_{gi,t}$: Total variable active power delivered by generator g to bus i at time t, measured in megawatts (MW);
$\sum_{g=1}^{G} Q_{gi,t}$: Total variable reactive power delivered by generator g to bus i at time t, measured in megavolt-amperes reactive (MVar);
$P_{di,t}$: Total active power load on bus i during period t, measured in MW;
$Q_{di,t}$: Total reactive power load on bus i during period t, measured in MVar;
The units of measurement for these variables are specified as follows: active power is expressed in megawatts (MW) and reactive power in megavolt-amperes reactive (MVar), aligning with the standard units of power in electrical engineering.
The units are in values per unit (p.u.), the units of active power are in MW, and the units of reactive power are in MVar.

2.2.2. Natural Gas Network

(a) Energy Dispatch Equations for Thermal Power Plants Fueled with Natural Gas (NGFTPUs):
Taking into account a pipeline network composed of various nodes, each with a defined requirement for natural gas (NG) aimed at generating electricity, the formulas documented in [17] serve to characterize the operation of each Natural Gas Fueled Thermal Power Unit ($NGFTPUs$).

$$P_{gi,min} \leq P_{gi,t} \leq P_{gi,max} \ldots \forall i,\, t \tag{16}$$

$$\sum_{n=1}^{NGs} P_{gi}(e_n) = P_{di} \ldots \ \forall\, t \tag{17}$$

$$P_{gi} = \eta(e_n) * LHV * e_n \tag{18}$$

where:
P_{gi}: Power generated in MW by NG thermal plant;
e_n: Flow of NG from node n to the generator, measured in cubic meters per second (m^3/s);
η: Efficiency of the thermal power plant, a dimensionless coefficient;
LHV: Low Heating Value Constant equivalent to 35.07 $MW/(m^3/s)$.
Equation (16) delineates the capacity constraints for the Natural Gas Fueled Thermal Power Units (NGFTPUs) situated at node n within the pipeline network. It is essential to highlight that NGFTPUs operate with a minimum non-zero capacity to maintain operational stability. This requirement is crucial for ensuring the reliable and secure operation of the NGFTPUs.
Equation (17) specifies the electrical power demand P_{gi}, highlighting the system's reliance on thermal generation using natural gas. It demonstrates how Natural Gas Fueled Thermal Power Units (NGFTPUs) dynamically manage energy dispatch in response to system demand, contingent upon the natural gas network's capacity to meet gas requirements. The aim is to optimize operation efficiency, thereby reducing both energy production and gas transportation costs within the pipeline network.

Equation (18) quantifies the relationship between the input of NG fuel and the resulting production of electrical energy, reflecting the efficiency and operational characteristics of the combined-cycle NGFTPUs.

Equation (16) outlines the generation capacity constraints of the Natural Gas Fueled Thermal Power Units (NGFTPUs) at node n. It is critical to recognize that the correlation between energy generation and the flow of natural gas is complex and non-linear, influenced by a multitude of factors. These encompass the specific attributes of the thermal power plant, the inherent properties of the natural gas, and environmental variables pertinent to the location, such as altitude (measured in meters above sea level) and climatic conditions (including temperature and relative humidity). As documented in [17], this intricate interplay means that energy output in a combined cycle plant is best represented through a cubic function of natural gas flow. Thus, this nuanced relationship can be encapsulated in the subsequent mathematical expression.

$$P_{gi}(e_n) = k_3 e_n^3 + k_2 e_n^2 + k_1 e_n \tag{19}$$

In Equation (19), the coefficients k_3, k_2, and k_1 depend on the specific characteristics of the natural gas thermal generation plants. However, determining these coefficients can be a complex task. As a result, for the purposes of the research work conducted in [17,18], a simplified approximation was employed. The simplified Equation (19) was applied to the test networks in the research work, assuming that for each unit of power produced in megawatts (MW), approximately 0.05 m^3/s or 4320 m^3/day of natural gas is required.

$$P_{gi}(e_n) = k_1 * e_n \tag{20}$$

where:

k_1 = 0.0023148 MWatts·day/106 m^3

(b) NG flow equations:

As discussed in Section 2.1.2, we established that the pipeline network's equivalent model consists of nodes for importing or exporting natural gas (*NG*), adhering to the principle of mass balance at each node. Consequently, the *NG* injection flows into a node are balanced by the *NG* flows allocated for both electrical generation and non-electrical uses, as supported by the findings in [17].

$$S_n + \sum_m f_{mn} = \sum_j f_{no} + d_n + e_n \qquad \forall\, t \tag{21}$$

where:

S_n: NG supply at node n, measured in cubic meters per hour (m^3/h) or cubic meters per day (m^3/day);

f_{mn}: NG flow from node m to node n, expressed in cubic meters per hour (m^3/h) or cubic meters per day (m^3/day);

e_n: NG flow from node n to the generator, measured in cubic meters per second (m^3/s);

d_n: NG demand for non-electric use at node n, specified in cubic meters per hour (m^3/h) or cubic meters per day (m^3/day).

Equation (21) shows the natural gas flow balance at node "n". Likewise, it is important to consider that these flows are influenced by various characteristics associated with natural gas, including the natural gas pressure at the inlet and outlet of the node, as well as the pipe section. To describe the relationship between pressure and natural gas flow, the Weymouth Equation (22) is employed. This equation defines the relationship between pressures and the flow of NG, providing a valuable tool for analyzing and understanding the behavior of natural gas flow within the system.

$$Sign(f_{nm}){f_{nm}}^2 = {C_{nm}}^2(p_n^2 - p_m^2) \qquad \forall t \tag{22}$$

$${f_{nm}}^2 \leq -{C_{nm}}^2(p_n^2 - p_m^2) \qquad \forall t \tag{23}$$

$$p_{n,min} \leq p_n \leq p_{n,max} \qquad \forall t \tag{24}$$

$$f_{nm,min} \leq f_n \leq f_{nm,max} \qquad \forall t \tag{25}$$

$$S_{n,min} \leq S_n \leq S_{n,max} \qquad \forall t \tag{26}$$

where:
p_n: Pressure at node n, measured in bars;
p_m: Pressure at node m, in bars;
C_{nm}^2: Constant reflecting the chemical composition of the NG and the characteristics of the nm section of the NG pipeline, given in cubic meters to the sixth power per square bars (m^6/bars^2).
Subject to the following constraints:
$f_{nm,max}$: Maximum NG flow from node n to node m, specified in cubic meters per hour (m^3/h);
$f_{nm,min}$: Minimum NG flow from node n to node m, in cubic meters per hour (m^3/h);
$p_{n,max}$: Maximum pressure allowed at node n, measured in bars;
$p_{n,min}$: Minimum pressure allowed at node n, in bars;
$S_{n,max}$: Maximum limit of the NG supply delivered by the producer or importer node n (m^3/h);
$S_{n,min}$: Minimum limit of the NG supply delivered by the producer or importer node n (m^3/h)
The constant C_{nm} depends on the characteristics of the NG pipeline, such as the diameter, length, and roughness of its walls, among other aspects, and the chemical composition of the NG.
The constant C_{nm} is influenced by several factors related to the natural gas (NG) pipeline, including its diameter, length, the roughness of its internal walls, among other physical characteristics, as well as the chemical composition of the NG itself.
If $f_{nm} > 0$, this indicates that the natural gas (NG) flow is directed from node n to node m. Conversely, if $f_{nm} < 0$, it signifies that the NG flow reverses, moving from node m to node n. Equation (23) models the NG flow within the pipeline network as a quadratic function of the pressure at the respective end nodes. Specifically, when the inlet pressure at node n exceeds the outlet pressure at node m ($p_n > p_m$), it results in natural gas flowing from node n to node m ($f_{nm} > 0$).
Equation (23) imposes a unique constraint at each node, often necessitating the use of compressors. These devices are crucial for boosting the pressure at specific nodes where an increase in natural gas (NG) pressure is required. By enhancing the pressure, compressors enable a greater flow of NG than what would be possible under standard conditions, thereby aiding in the efficient redirection of NG throughout the network. For instance, in scenarios like in Peru, elevating the NG pressure is essential for transporting gas across challenging terrains such as the Andes mountains. Consequently, compressors allow for bypassing the maximum flow constraints, facilitating the injection of larger volumes of NG into the gas pipeline transmission networks. Nonetheless, it remains critical to ensure that pressure levels are kept within their prescribed limits to preserve the operational integrity and safety of the system.
Similarly, constraint (26) regarding the supply of natural gas S_n adapts dynamically. If $S_n > 0$ at node n, it signifies that the node acts as a producer or importer of natural gas. On the other hand, if $S_n < 0$ at node n, it denotes that the node serves as a consumer of natural gas, encompassing both electrical and non-electrical

consumption. This constraint effectively captures the varied roles and functions of nodes within the natural gas network, accounting for the intricate balance of supply and demand.
The function $Sign(f)$, introduced in Equation (22), is utilized to determine the direction of natural gas (NG) flow.

$$Sign(x) = \begin{cases} 0, & if\ x = 0 \\ 1, & if\ x > 0 \\ -1, & if\ x < 0 \end{cases} \tag{27}$$

For instance, if the flow from node m to node n is designated as positive, and the outcome of the simulation yields $f_{mn} > 0$, this indicates that the natural gas (NG) flow direction is indeed from node m to node n. Conversely, should the simulation result in $f_{mn} < 0$, it implies that the NG flows in the pipeline are directed from node n to node m.
The inequality presented in Equation (25) highlights the physical constraints inherent to gas pipelines, specifically addressing the maximum and minimum flow limits of natural gas (NG) that delineate flow directionality within the system. Under typical circumstances, the lower limit is set to zero, indicating no reverse flow is allowed. However, in the presence of compressors within the pipeline, a minimum flow rate greater than zero may be established to ensure effective gas transportation. Conversely, the maximum flow rate is determined by the pipeline's capacity and can be quantified through the subsequent equation:

$$f_{nm} = \sqrt{C_{nm}^2 (p_n^2 - p_m^2)} \tag{28}$$

If $f_{nm} > 0$, it indicates that the natural gas (NG) flow is directed from node i to node j. Conversely, if $f_{nm} < 0$, the NG flow reverses, moving from node j to node i. To accurately model this behavior, a binary variable is incorporated into the equation, transforming the formulation into a combinatorial problem.
The variable C_{nm}, featured in Equations (29) and (30), is defined as follows:

$$C_{nm} = 96.074830 * 10^{-15} \frac{D_{nm}^5}{\lambda_{nm} * z * T * L_{nm}\delta} \tag{29}$$

$$\frac{1}{\lambda_{nm}} = 9\left[2\,log\,\left(\frac{3.7\,D_{nm}}{\varepsilon}\right)\right]^2 \tag{30}$$

where:
D^{nm}: Internal diameter of the natural gas (NG) duct, measured in millimeters (mm);
z: NG compressibility factor, a dimensionless unit valued at 0.8;
T: Constant temperature for NG, set at 281.15 Kelvin (K);
Lnm: Length of the NG pipeline section from node n to node m, in kilometers (km);
δ: Density of NG relative to air, a dimensionless value of 0.6106 and;
ε: Absolute roughness of the NG duct, quantified as 0.05 mm.

3. Case Study and Results Analysis

The set of equations described above defines a mixed non-linear optimization problem. This is primarily due to the presence of binary terms, non-linear relationships, and the constraints associated with both the gas pipeline system and the thermoelectric generation system included in the proposed model, as illustrated in Figure 1. The combination of these elements results in a complex optimization problem that requires specialized techniques and algorithms to obtain optimal solutions. The integrated model encompasses the interdependencies between the gas pipeline system and the thermoelectric generation system, enabling a comprehensive analysis and optimization of the integrated energy system.

To simplify practical considerations, it was determined necessary to group the 182 generation plants in the system into 24 plants with distinct technologies but similar operating costs. This grouping helps streamline the analysis and management of the system, as it reduces the complexity associated with individual plant-level considerations. By categorizing the plants based on their technology and comparable operating costs, it becomes more feasible to implement effective optimization strategies and make informed decisions regarding the operation and planning of the integrated energy system.

To assess the impact of congestion on the natural gas network, it was necessary to deliberately impose constraints on the pipeline's capacity. By simulating congestion and analyzing the network's behavior, the effects on bus prices and the resulting low energy production costs can be studied. This analysis considers the demand of a typical day, as indicated in Table 6. By examining this scenario, a comprehensive understanding of the system's response to congestion and its implications on energy production costs can be obtained.

Table 6. Electrical demand SEIN (GW) adapted from Coes.

Hour	16 August 2022	Hour	16 August 2022	Hour	16 August 2022
01:00	5.99	09:00	6.61	17:00	6.97
02:00	5.87	10:00	6.87	18:00	6.80
03:00	5.75	11:00	6.99	19:00	7.06
04:00	5.76	12:00	7.11	20:00	7.09
05:00	5.83	13:00	6.94	21:00	7.05
06:00	5.96	14:00	6.90	22:00	6.92
07:00	6.19	15:00	7.01	23:00	6.66
08:00	6.35	16:00	7.06	00:00	6.29

3.1. Congestion Effects of a Natural Gas Pipeline

To analyze the behavior of the integrated gas–electricity system under congestion scenarios, it is important to recognize that operational congestion events are often not prevalent under typical real-world conditions, as noted in [11]. To simulate and study these hypothetical conditions, we deliberately reduce the maximum capacity of the primary natural gas pipeline within the system. By imposing this constraint, we can effectively induce congestion scenarios for a typical demand profile corresponding to the year 2022. These capacity reductions could result from failures or scheduled maintenance.

To explore a range of congestion levels, we progressively reduce the maximum capacity of the natural gas transmission network by varying percentages, starting from a 10% reduction. The reduction continues until reaching a percentage where the results no longer converge or remain stable. This approach enables us to observe the system's response under different levels of congestion and evaluate its impact on the integrated gas–electricity system.

The simulation outcomes, depicted in Figure 2, elucidate the impact of reducing the main pipeline's maximum capacity on the electricity production costs within the integrated system. Initially, when the pipeline capacity is diminished to 50% of its original size, the production costs remain unaffected. However, further reductions to 60% and 70% lead to a noticeable increase in production costs. This pattern reveals that the system possesses a certain resilience to reductions in pipeline capacity without incurring additional costs. Yet, surpassing a critical threshold, estimated between a 60% to 70% reduction, congestion-induced constraints begin to adversely affect the system's efficiency and the cost-effectiveness of electricity production. These results highlight the critical need for maintaining an optimal balance between pipeline capacity and electricity production costs to ensure the efficient functioning of the integrated gas–electricity network.

Figure 2 illustrates the cost implications for electricity production as a function of reducing the maximum capacity of the main natural gas pipeline.

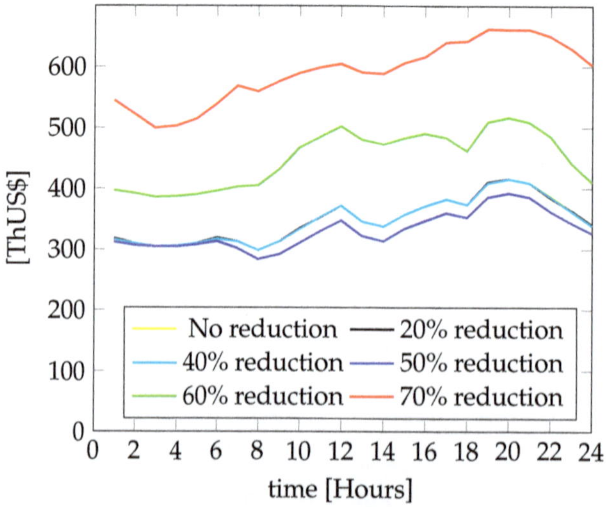

Figure 2. Cost of electricity production considering reduction of the maximum capacity of the main NG pipeline.

Figure 3 depicts the relationship between the costs of natural gas consumption and the demand for electricity generation under congestion scenarios. It reveals an inverse correlation between the cost of electricity production and the costs associated with natural gas consumption. Notably, as the maximum capacity of the main natural gas pipeline is curtailed by over 50%, we observe a decrease in the costs of natural gas for electricity generation. Moreover, a reduction surpassing 70% leads to a scenario where the costs associated with natural gas consumption effectively drop to zero. This illustrates the complex dynamics between pipeline capacity limitations and the economic aspects of energy production within the system.

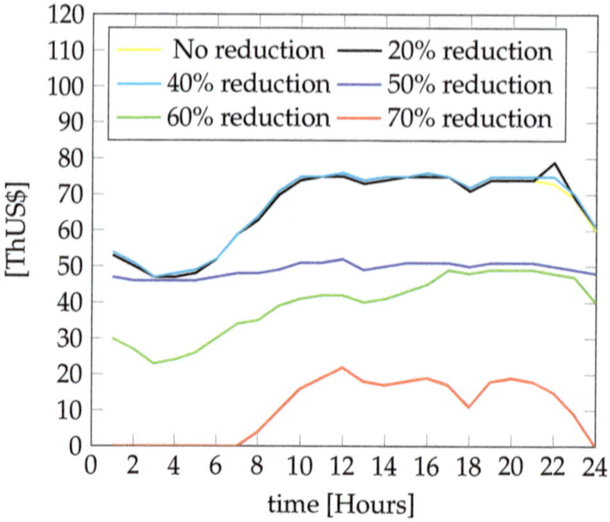

Figure 3. Natural gas consumption costs to cover electricity demand with reduction of the maximum capacity of the main natural gas pipeline.

Figures 2 and 3 provide a comprehensive view of the energy production costs within the integrated gas–electricity system, under scenarios of varying reductions in the maximum capacity of the main natural gas pipeline. The analyses reveal that production costs within the system begin to escalate once the pipeline capacity reduction surpasses 50%. This increase in costs can be attributed to a multitude of factors, such as the necessity to resort to generation plants with higher marginal costs, the lack of contribution from hydroelectric generation plants, and the constrained availability of natural gas, which limits the operational capacity of NG-fueled plants to meet the existing demand. These insights underscore the critical interdependencies between pipeline capacity and the operational economics of the integrated energy system.

Based on the comprehensive analysis of congestion effects within the natural gas pipeline network, and taking into account the implications of reducing the main natural gas pipeline's maximum capacity, significant disparities have been identified under specific conditions across two predefined scenarios (Scenario 1 and Scenario 2). These conditions highlight the differential impact of pipeline capacity constraints on the system's efficiency and cost-effectiveness, underscoring the nuanced dynamics that characterize each scenario's unique challenges and opportunities.

Conditions
No reduction in the capacity of the NG pipeline
Reduction of the capacity of the NG pipeline to 60%
Reduction of the capacity of the NG pipeline to 70%

Hence, we will now discuss the effects on the behavior of the plants, the effects on the bus prices, and the effects on the load flows in these three conditions for each scenario.

3.2. Impact of Reduced NG Pipeline Capacity on Generation Plant Operations

The operational dynamics of generation plants, particularly hydroelectric plants, undergo significant changes as the capacity of the main natural gas (NG) pipeline is reduced beyond a 50% threshold. Under such conditions, hydroelectric plants are called upon to increase their dispatch to meet the persistent demand for electricity. Nonetheless, their capacity to compensate for the reduced NG supply is bounded by their maximum operational volumes and the limitations posed by the capacities of the transmission lines to which they are connected. Figure 4 delineates the operational behavior of hydroelectric plants within these constraints, providing insight into how transmission line capacities further influence their ability to respond to demand.

The analysis of both scenarios reveals a consistent pattern: there is a notable dispatch overlap for hydroelectric plants between situations where there is no reduction in NG pipeline capacity and scenarios with a 60% reduction. This observation underscores the resilience of hydroelectric plants in contributing towards electricity demand satisfaction, even amid substantial reductions in the availability of natural gas. Such findings highlight the critical role of hydroelectric plants in maintaining system stability and mitigating the impacts of NG supply constraints.

It is reasonable to anticipate that with a lower supply of natural gas, resulting from the reduction in the maximum capacity of the main natural gas pipeline, the demand must be met by plants with higher operating costs, such as Diesel plants. These plants quickly increase their dispatch, leading to higher energy production costs, as illustrated in Figure 4.

Conversely, natural gas thermal generation decreases due to the effect of reducing the maximum capacity of the natural gas main pipeline. Notably, as congestion intensifies and reaches the 70% threshold, dispatch tends to reach zero.

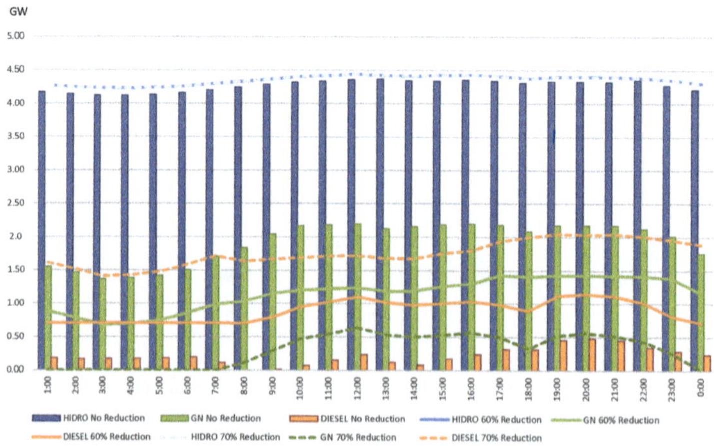

Figure 4. Behavior by kind of electricity generation considering the reduction of the maximum capacity of the main natural gas pipeline.

3.3. *Variation of Bus Prices in Scenarios Considering the Reduction of the Maximum Capacity of the Main Natural Gas Pipeline*

As the reduction in the maximum capacity of the natural gas pipeline increases, generation through natural gas thermal plants decreases, resulting in an increase in bus prices. While some bus prices, like those of bus 4, exhibit minimal variations, there are cases where significant bus price fluctuations are observed.

For instance, in Figure 5, bus 6 displays a trend where, as the reduction in the maximum capacity of the main natural gas pipeline increases, the bus price approaches its limit of 50 USD/MW.

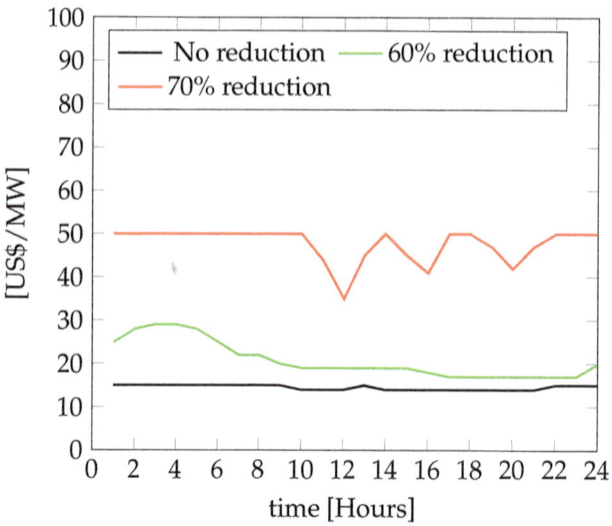

Figure 5. Variation of the price in bus 6 considering the reduction of the maximum capacity of the main natural gas pipeline.

In the case of bus 12, as depicted in Figure 6, it is evident that as congestion intensifies, the price on the bus approaches its upper limit of 80 USD/MW.

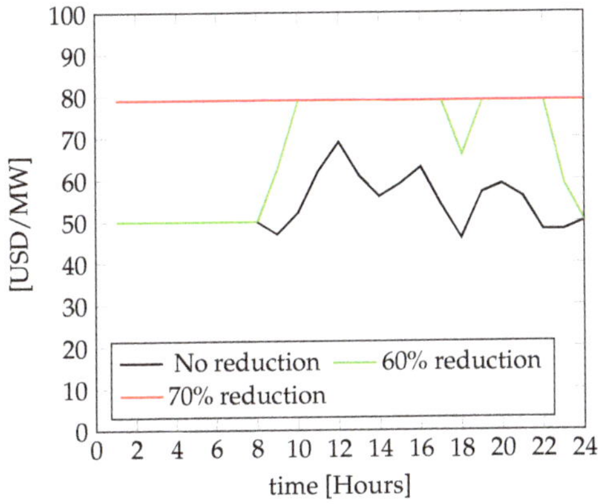

Figure 6. Variation of the price in bus 12 considering the reduction of the maximum capacity of the main natural gas pipeline.

3.4. Effects on Load Flows in Congestion Scenarios of a Natural Gas Pipeline

The energy flows between various buses exhibit distinct behavior when the reduction of the maximum capacity of the main natural gas pipeline approaches approximately 70% of its total capacity. At this juncture, the flows tend to reverse their direction or, in the absence of reversal, saturate the transmission network.

This behavior aligns with expectations, as the model's objective is to optimize production costs while adhering to network constraints and achieving an optimal dispatch strategy.

Load flows display notable variability in response to even minor demand signals, leading to significant fluctuations in power flows between bus 12 and 4. This dynamic response is visible in Figure 7, illustrating load flows between bus 12 and 4 under varying demand scenarios. These fluctuations underscore the system's sensitivity to even slight changes in load requirements.

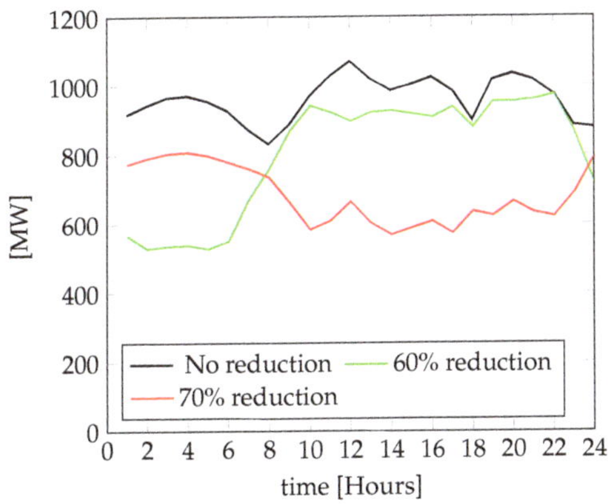

Figure 7. Load flow changes from bus 12 to 4 with reductions in the main natural gas pipeline's capacity

The dynamics of power flows from bus 6 to bus 12, as depicted in Figure 8, highlight the impact of natural gas pipeline capacity on the electrical grid. When the main natural gas pipeline operates at full capacity, the line between bus 6 and 12 typically reaches its maximum load limit. However, with increasing reductions in the pipeline's capacity, there is a noticeable decline in the flow through this line. This trend is directly linked to the diminishing natural gas supply, which, in turn, affects the power generation and subsequent flows between these buses. Thus, while unrestricted pipeline capacity allows for optimal line utilization, any constraints on the natural gas supply lead to a proportional decrease in power flows, illustrating the tight interconnection between natural gas availability and electrical grid performance.

Figure 8 presents the variation of the load flow between bus 6 and 12 considering the reduction in the maximum capacity of the main natural gas pipeline.

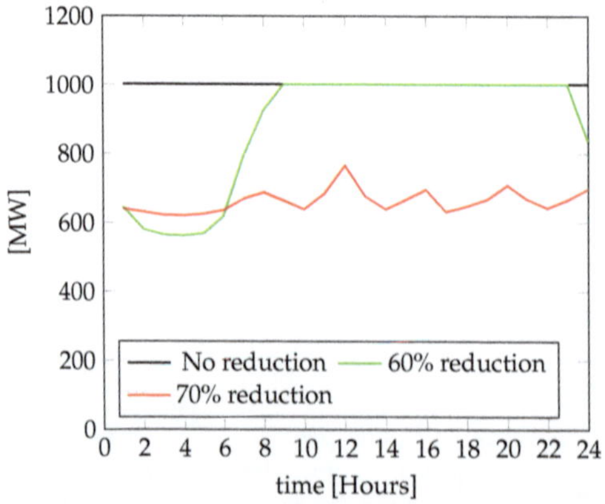

Figure 8. Variation of the load flow between bus 6 to 12 considering reduction of the maximum capacity of the main natural gas pipeline

3.5. Discussion

The results shed light on the implications of congestion in the natural gas network, particularly regarding system operating costs, bus prices, and power flow.

In scenarios where the maximum capacity of the primary natural gas pipeline is reduced by up to 50%, no significant variations are observed. Nevertheless, when reductions exceed 50% of the pipeline's maximum capacity, there is a noteworthy increase in energy production costs. These costs can escalate by as much as 69% in comparison to scenarios without congestion. Similarly, there is a significant increase in bus prices, often reaching their maximum values, accompanied by substantial fluctuations in load flows. In many instances, load flows approach or even exceed the maximum levels of transmission capacity.

Furthermore, the analysis uncovers that when congestion levels exceed roughly 70% of the pipeline's capacity, the results fail to converge. This phenomenon arises from a combination of factors, encompassing the installed capacity of the SEIN without gas generation, the constraints of transmission capacities, and the inherent limitations of network parameters. Consequently, it becomes imperative to implement load rejection systems to uphold system stability, as the system's demand cannot be met.

While the primary focus of this analysis is on natural gas pipeline congestion, it is important to note that similar hypotheses and results can be extrapolated to various other scenarios. These may include reduced natural gas supply, a gradual decline in natural gas production at the wellhead, diminishing proven reserves in the natural gas wells that supply the pipeline, and a range of other conditions.

4. Conclusions

This paper emphasizes the critical role of coordinating the electricity and natural gas markets to enhance operational efficiency and reduce associated costs, particularly in scenarios where pipeline co-management is involved.

The presented results provide valuable insights for shaping policies aimed at mitigating congestion effects in natural gas pipelines and addressing constraints in electrical transmission. These findings emphasize the significance of improved coordination between both systems under a unified market operator.

The research demonstrates the effects of congestion scenarios on production costs in the gas–electricity system. Notably, scenarios with reductions in the maximum capacity of the main pipeline below 50% show no observable impact on energy production costs, bus prices, or load flows. However, when the reduction exceeds 50%, substantial increases in energy production costs, reaching up to 69% higher costs compared to congestion-free scenarios, are observed. Furthermore, reductions exceeding 70% in the maximum capacity of the main pipeline result in non-convergent results due to capacity limitations and network constraints, rendering the system unable to meet demand.

Finally, this paper identifies several areas that warrant further investigation in future research. These include the integration of additional natural gas sources, expansion of natural gas networks, incorporation of low-emission gases into the natural gas system, integration of renewable sources, deployment of flexible assets and operations, and the establishment of energy storage facilities. Addressing these aspects would require the inclusion of additional equations into the model.

Author Contributions: Conceptualization, R.N., H.R., J.E.L. and Y.P.M.; methodology, R.N. and H.R.; validation, J.E.L. and Y.P.M.; investigation, R.N. and H.R.; writing—review and editing, R.N., J.E.L. and Y.P.M.; visualization, J.L.S.; supervision, Y.P.M. All authors have read and agreed to the published version of the manuscript.

Funding: This research received no external funding.

Data Availability Statement: The original contributions presented in the study are included in the article, further inquiries can be directed to the corresponding author.

Acknowledgments: This research was supported by the Coordination for the Improvement of Higher Education Personnel (CAPES) through grant No. 1009/2024 and Process No. 88881.909446/2023-01, under the PRAPG Program. We thank the Faculty of Mechanical Engineering at the National University of Engineering, Lima, Peru, and the Department of Electrical Engineering at the Federal University of Paraíba, Brazil, for their indispensable support.

Conflicts of Interest: The authors declare no conflicts of interest.

Nomenclature

A	The electrical current flow measured in amperes (A)
AC	Alternating current
a_g, b_g, *and* c_g	These constants are integral to the quadratic equation used to compute the energy production cost from each bus generator. They vary based on the generator's power output and are denoted in [USD/MW2], [USD/MW], and [USD], respectively.
C_{Diesel}	The cost associated with generating electricity using diesel, expressed in [USD].
C_{rer}	The cost of generating electricity using both conventional and non-conventional renewable energy sources, in [USD].
C_{res}	Cost of energy storage system in [USD] that remains fixed throughout the period.
CDF	Conceptual data abstraction called Common Data Format for storing, manipulating, and accessing multidimensional datasets
C_{nm}^2	Constant dependent on the chemical structure of the NG, and the characteristics of the *nm* section of the NG duct ($m^6/bars^2$)
D_{nm}	Internal diameter of NG duct (mm)
δ_i	Voltage phase shift angle on bus *i*.
d_n	The demand for natural gas for non-electrical uses at node *n*, expressed in m^3/h or m^3/day.
e_n	The volume of natural gas flowing from node *n* to the generator, in m^3/s.

f_{mn}	The volume of natural gas transported from node m to node n, in m^3/h or m^3/day.
$f_{nm,max}$	The upper limit on the volume of natural gas flow from node n to node m, in m^3/h.
$f_{nm,min}$	The lower limit on the volume of natural gas flow from node n to node m, in m^3/h.
$f_{nm}(t)$	The flow of natural gas between nodes n and m during period t, measured in 10^6 m^3/h.
f_{no}	The volume of natural gas transported from Node n to Node o, in m^3/h or m^3/day.
Gen	The total number of power plants within the network.
$k1$	0.0023148 MWatts·day/106 m^3
LHV	Low Heating Value Constant equivalent to 35.07 $MW/(m^3/s)$
L_{nm}	NG pipeline length nm (km)
η	Thermal power plant efficiency
N	Bus number in the electrical network
NGs	Number of supply points in the NG network
P_{di}	The active power load on bus i during period t, measured in MW.
P_{GenBus}	The power generated at a bus, in MW.
PG	number of pipes in the gas network
p_g	Unit price of NG in $[USD/m^3USD]$
$P_g(t)$	Active Power delivered by the generator g in the period t in [MW].
P_{gi}	Power generated in MW by NG thermal plant
$P_{g,i}^{(t)}$	Variable Active Power delivered by the generator g on the bus i at time t in [MW]
PHI	Offset angle
P_{km}^{max}	Maximum limit of active power in the section k-m (MW))
P_{km}	Active Power in the section k-m (MW)
P_i	Active power on the bus i in the period t
P_L	Active power losses in period t
p_m	Pressure reached at node m (bars)
p_n	Pressure reached at node n (bars)
$p_{n,max}$	Maximum pressure limit at node n, (bars)
$p_{n,min}$	Minimum pressure limit at node n, (bars)
p_{tg}	Unit cost of NG transportation in $[USD/m^3]$
Q_{di}	Reactive Load on the bus i in the period t
Q_i	Reactive Power generated in bus i in the period t
Q_L	Reactive Power Losses in the period t
S_n	NG supply at node n in (m^3/h or m^3/day)
$S_{n,max}$	Maximum limit of the NG supply delivered by the producer or importer node n (m^3/h)
$S_{n,min}$	Minimum limit of the NG supply delivered by the producer or importer node n (m^3/h)
$S_n(t)$	Supply of NG supplied at node n in period t in [106 m^3/h]
T	Temperature constant of NG, 281.15 K
USD	US American Dollars
V	Volt unit of potential difference
$VFinBus$	Final bus voltage, p.u.
V_i	Bus voltage value i;
W	Power Unit in Watts
X	Electrical impedance
Y	Admittance
Y_{ij}	Impedance of the line between busbars i and j
z	compressibility factor of NG, 0.8 (dimensionalless)
δ	Density of NG with respect to air, 0.6106 (dimensionalless)
δ_i	Angle of the tension in bar i
ε	Absolute roughness of the NG duct, 0.05 mm
η	Efficiency of the thermoelectric plant
θ_{ij}	Angle of total line admittance between busbars i and bus j
$\sum_{i=1}^{N} P_{gi,t}$	Total energy generated on the bus i in the period t
$\sum_{i=1}^{N} P_{di,t}$	Total Active Energy in bus i in period t
$\sum_{i=1}^{N} Q_{gi,t}$	Total Reactive Energy in bus i in the period t

$\sum_{i=1}^{N} Q_{di,t}$	Reactive load on bus i in period t
$\sum_{g=1}^{G} P_{gi}$	Active Power of the thermogenerators connected on the bus i
$\sum_{g=1}^{G} Q_{gi}$	Total Reactive Power of the thermogenerators
z	NG compressibility factor, 0.8 (dimensionalless units)

Abbreviations

COES	Council for Electrical System Economy Operating
DNLP	Discrete Non-Linear Programming
GAMS	General Algebraic Modeling System
GW	Gigawatts
IEEE	International electrical and electronics engineers
Km	kilometers
KP	Key point
KV	Kilo volts
LNG	Liquefied natural gas
NG	Natural Gas
NGFTPUs	Natural-Gas-Fueled Thermoelectric Power Units
MMScm	Million Metric Standard Cubic Meters
MVA	Mega Volt-Amper
MW	Megawatts
SEIN	Peruvian National Interconnected System
SDDP	Stochastic Dual Dynamic Programming
NGFTPUs	Natural-Gas-Fueled Thermoelectric Power Units

References

1. Xing, J.; Wu, P. Optimal Planning of Electricity-Natural Gas Coupling System Considering Power to Gas Facilities. *Energies* **2021**, *14*, 3400. [CrossRef]
2. Saedi, I.; Mhanna, S.; Mancarella, P. Integrated electricity and gas system modelling with hydrogen injections and gas composition tracking. *Appl. Energy* **2021**, *303*, 117598. [CrossRef]
3. Ameli, H.; Qadrdan, M.; Strbac, G. Coordinated Operation of Gas and Electricity Systems for Flexibility Study. *Front. Energy Res.* **2021**, *8*, 17. [CrossRef]
4. Qin, G.; Yan, Q.; Kammen, D.M.; Shi, C.; Xu, C. Robust optimal dispatching of integrated electricity and gas system considering refined power-to-gas model under the dual carbon target. *J. Clean. Prod.* **2022**, *371*, 133451. [CrossRef]
5. AlHajri, I.; Ahmadian, A.; Elkamel, A. Techno-economic-environmental assessment of an integrated electricity and gas network in the presence of electric and hydrogen vehicles: A mixed-integer linear programming approach. *J. Clean. Prod.* **2021**, *319*, 128578. [CrossRef]
6. Rubio-Barros, R.; Ojeda-Esteybar, D.; Ano, O.; Vargas, A. Combined Operational Planning of Natural Gas and Electric Power Systems: State of the Art. In *Natural Gas*; Potocnik, P., Ed.; IntechOpen: Rijeka, Croatia, 2010; Chapter 12. [CrossRef]
7. Pambour, K.A.; Cakir Erdener, B.; Bolado-Lavin, R.; Dijkema, G.P.J. Development of a Simulation Framework for Analyzing Security of Supply in Integrated Gas and Electric Power Systems. *Appl. Sci.* **2017**, *7*, 47. [CrossRef]
8. Hibbard, P.J.; Schatzki, T. The Interdependence of Electricity and Natural Gas: Current Factors and Future Prospects. *Electr. J.* **2012**, *25*, 6–17. [CrossRef]
9. Ventosa, M.; Baillo, A.; Ramos, A.; Rivier, M. Electricity market modeling trends. *Energy Policy* **2005**, *33*, 897–913. [CrossRef]
10. Yang, L.; Zhao, X.; Li, X.; Feng, X.; Yan, W. An MILP-Based Optimal Power and Gas Flow in Electricity-gas Coupled Networks. *Energy Procedia* **2019**, *158*, 6399–6404. [CrossRef]
11. Navarro, R.; Rojas, H.; De Oliveira, I.S.; Luyo, J.E.; Molina, Y.P. Optimization Model for the Integration of the Electric System and Gas Network: Peruvian Case. *Energies* **2022**, *15*, 3847. [CrossRef]
12. Rojas, H. Despacho Integrado de Sistemas Eléctricos de Potencia y de Gas Natural considerando Energías Renovables y Sistemas de Almacenamiento. *Fondo Editor. UNI* **2022**, *1*, 147.
13. Jenkins, S.E. Interdependency of electricity and natural gas markets in the united states: A dynamic computational model. *Mass. Inst. Technol. Technol. Policy Program* **2014**, *1*, 84.
14. Erdener, B.C.; Pambour, K.A.; Lavin, R.B.; Dengiz, B. An integrated simulation model for analysing electricity and gas systems. *Int. J. Electr. Power Energy Syst.* **2014**, *61*, 410–420. [CrossRef]
15. Gil, M.; Dueñas, P.; Reneses, J. Electricity and Natural Gas Interdependency: Comparison of Two Methodologies for Coupling Large Market Models Within the European Regulatory Framework. *IEEE Trans. Power Syst.* **2016**, *31*, 361–369. [CrossRef]
16. De Wolf, D.; Smeers, Y. The Gas Transmission Problem Solved by an Extension of the Simplex Algorithm. *Manag. Sci.* **2000**, *46*, 1454–1465. [CrossRef]

17. An, S.; Li, Q.; Gedra, T. Natural gas and electricity optimal power flow. In Proceedings of the 2003 IEEE PES Transmission and Distribution Conference and Exposition (IEEE Cat. No.03CH37495), Dallas, TX, USA, 7–12 September 2003; Volume 1, pp. 138–143. [CrossRef]
18. Liu, C.; Shahidehpour, M.; Fu, Y.; Li, Z. Security-Constrained Unit Commitment with Natural Gas Transmission Constraints. *IEEE Trans. Power Syst.* **2009**, *24*, 1523–1536. [CrossRef]

energies

MDPI

Article

Structural Effects of Economic Shocks on the Macroeconomic Economy–Electricity–Emissions Nexus in India via Long-Term Cointegration Approach

Soumya Basu, Keiichi Ishihara, Takaya Ogawa and Hideyuki Okumura *

Graduate School of Energy Science, Kyoto University, Kyoto 606-8501, Japan;
soumya.basu.28e@st.kyoto-u.ac.jp (S.B.); ishihara.keiichi.6w@kyoto-u.ac.jp (K.I.);
ogawa.takaya.8s@kyoto-u.ac.jp (T.O.)
* Correspondence: okumura@energy.kyoto-u.ac.jp

Abstract: For developing nations to achieve net-zero targets, macroeconomic linkages impacting the decoupling of emissions from economic growth must account for non-linear business cycles and economic shocks. This study aims to delineate decarbonization policy pathways for the Indian electricity sector in the aftermath of COVID-19 by analysing the long-term evolution of the economy–electricity–emissions (3E) nexus during the 2008 financial crisis and during COVID-19, covering the period of 1996Q2 to 2020Q3. Upon testing multiple theoretical 3E systems, it was found that a model internalizing trade, inflation, and stochasticity was able to minimize the reproduction errors from growth to recession phases, as well as predict the rebound effect from an economic crisis. This was revealed to be due to more information within the coefficients in a trade stochastic model. Our results confirm the existence of electricity-associated emission decoupling with capital formation in the long-run, post-crisis, while economic growth and inflation increase CO_2 emissions. The main finding highlights the negative feedback loop of inflation->trade->emissions, which shows that GDP and emissions are not directly causal. This long-run macroeconomic dynamic death spiral causes decoupling to be inhibited, where fossil fuel imports should not be subsidized for economic shock rebound, and the risk hedging of energy transition investments should occur in the post-COVID-19 era.

Keywords: decoupling; business cycles; inflation; COVID-19; stochastic modelling; approximate entropy

1. Introduction

It is a challenge for modern economic policies to decouple energy production from carbon emissions in order to combat Human-Caused Climate Change (HCCC). The case for an economy in its developing phase is that energy use, specifically electricity generation, occurs primarily from fossil fuels (FFs), like coal and oil [1]. This is especially true due to the structural integration of FFs with economic growth ever since the industrial revolution, with all developed nations of the world banking on FF use to power their economies over the past century [2,3]. With HCCC having palpable and long-term detrimental effects on the environment, it is economically difficult to structurally decouple FFs and introduce alternative clean energy, such as renewable energy (RE), for electricity generation in developing economies, like India and China [1,4–6]. Both India and China have large populations and a fast-growing economy, which is dependent on tremendous amounts of energy generation to keep the economic machinery functioning [7]. Rapid structural change can lead to drastic short-term socioeconomic shocks, making energy transition a non-straightforward solution [3].

Decoupling is achievable not only by clean RE, which involves intermittent sources like wind and solar energy [3,8], but also by advancing cleaner generation from oil and gas and increasing production efficiency [9]. However, continual technological advancement needs innovation, which must be powered by economic growth, implying that there needs to be an incentive for the same. From a macroeconomic perspective, levels of CO_2 emissions can be an indicator of the propensity for energy transition [8]. As a result, most macroeconomic models measure decoupling as either energy intensity reduction or emission intensity reduction [3]. However, it has been argued in the literature that the Gross Domestic Product (GDP) is not necessarily an indicator for growth [10], with studies often pointing that the total capital is what determines the capability to innovate [11]. Thus, the first issue within an economy–electricity–emissions (3E) nexus is to establish the behaviour of how each macroeconomic determinant affects CO_2 emission levels with the complexity of the GDP, capital, and even international trade being part of dynamic feedback systems [12–14]. This paper attempts to create several control systems to approximate the real-world behaviour of the Indian 3E system to gauge the exact level of decoupling in electricity generation.

The second aspect of this paper tackles the main deficiency in the current literature. While GDP is treated by existing econometric models as linear or exponential while estimating the causality with emissions [3,14,15], higher-order phenomena are seen in economic growth. Specifically, bi-stage cyclic patterns of growth and recession in business cycles are prevalent [16]. At the peak of a business cycle, interest rates are the lowest, leading to expanding GDP growth and simultaneous energy use, which also causes inflation [17]. It may be pointed out that innovation also hits a peak at this point. However, to control inflation, interest rates are risen by federal banks, which contracts the GDP and production (ultimately energy use), termed as the recession phase [16]. To the best of our knowledge, the existing research fails to address how the decoupling of emissions in power generation changes within the business cycle. Thus, the second target of this paper is to empirically analyse various 3E control systems to identify the dynamics of CO_2 emissions in business cycle phases and their relationship to macroeconomic variables. For achieving decoupling in a developing economy, it is imperative to identify how the status of decarbonization changes in the higher-order behaviour of economic systems [6,9].

The third issue of measuring the decarbonization of electricity in developing countries is the impact of exogenous shocks on the incentive of decoupling, which has not been addressed in the macroeconomic literature. For example, the sub-prime mortgage crisis of 2008 was a global phenomenon that changed the course of RE development sharply, with Chinese solar panels overtaking Japanese and European production sharply [18]. At the same time, developing nations' carbon emissions and the emission intensity of electricity use sharply increased post-2008 crisis [18,19], implying that there was lesser incentive to generate clean power over FF power. While many previous studies have revealed how decarbonization patterns changed in the aftermath of the 2008 crisis [18,20–22], the macroeconomic linkages that accelerate CO_2 emissions are not known. In the rebound of COVID-19, we have also seen a rebound of emissions. While it may be argued that COVID-19 is vastly different from the sub-prime mortgage crisis, a macroeconomic lens will visualize both crises as a sudden shock to the GDP and consumer demand [18,23,24].

This paper focuses on the causalities of the decarbonization of power generation in India in relation to macroeconomic variables and how such causalities are impacted by higher-order economic behaviour and economic shocks. India was chosen as the case study because it is the most populous country, with a significant portion of the country being below the poverty line [25], which affects the incentive for decoupling. With FF penetration increasing from 11.7 TJ (2000) to 29.6 TJ (2020) [26], more than a third of the total power generation comes from coal [27], which suggests that a fast-growing economy is not necessarily boosting RE [28]. Simultaneously, economic growth has been noticeable, with the GDP increasing by 550% in this period [29,30], notwithstanding that inflation has increased by 5.8% [31] and cumulative emissions have increased by 250% [32] from 2000 to 2020. Secondly, unlike other fast-growing developing economies like China, the 2008 crisis did not significantly contract the GDP or cumulative emissions in India [33], but CO_2 emissions decreased by 6% in the year following COVID-19 [34]. This has drastic impacts on the 3E nexus' decarbonization efforts, which previous research has not addressed with higher-order GDP growth [3,9,20,21]. Thirdly, India is a much larger net importer of oil products than China [28], which raises the question as to how electricity-related emissions have changed in relation to Trade Openness (TROP) during higher-order movements and before and after economic shocks.

The importance of this paper is to delineate the interplay of the incentive for innovation in decoupling in the electricity sector of a fast-growing, developing economy during business cycle movements and the economic shocks of 2008 (financial crisis) and 2020 (COVID-19), such that economic policies for net-zero targets can be identified in a post-pandemic world. This research explores the key macroeconomic pathways that can eventually lead to a stable achievement of net-zero targets and the continued decarbonization of electricity generation [35]. The rest of the paper is organized as follows: Section 2 shows a literature review to introduce the theoretical basis of macroeconomic frameworks in order to build the hypothesis across economic events. Section 3 shows the methodology that is used to analyse the linkages within the 3E nexus, and Section 4 introduces the data of the analysis. Section 5 presents the results and discussions of policy implications, and Section 6 shows a robustness analysis of the results. The final section concludes the main findings of this paper.

2. Literature Review, Hypothesis Development, and Macroeconomic Frameworks

The theoretical basis for this paper lies on the inverted U-shaped Environmental Kuznets Curve (EKC) hypothesis. The 'inverted U' postulate of the EKC hypothesis implies that, in the initial stages of economic growth, CO_2 emissions increase, while upon economic maturity, emissions decrease [8,36]. The primary concern with this hypothesis is that there are several studies which do claim the existence of the EKC, specifically in high-income and high-RE-penetrated economies [6,8,21,24]. However, about 20% of the studies refute the existence of the EKC, mainly in developing countries [12,37–39]. A major gap in the literature is that most studies focus on total energy use across all sectors and do not focus specifically on electricity generation, which is responsible for the maximum energy use and emissions among the economic sectors in India [37].

The literature on the EKC can be divided into four strands. The first strand deals with reduction in the energy intensity of the economy, mainly comprising bivariate models employing energy use and GDP as the functional variables. The basis of this hypothesis lies in the fact that higher economic growth results in higher energy use, which leads to efficient methods of energy use, further increasing economic growth [40–44]. Granger causality has been extensively used in these studies, but mixed evidence has been found in such cases [43,44]. The second strand combines the first type with CO_2 emissions, focusing on the reduction in emission intensity. Normally, such studies are bivariate or trivariate with an energy–economy–emissions nexus approach [5,8,9,45,46], with very few studies focusing on the electricity aspect in developing countries [37,47]. One particular recent study confirmed that both economic growth and energy use significantly degraded the environment in eight developing Asian nations, including India [5]. However, the question of the extent of economic shock impact was not analysed, even though the time interval was studied extensively [5,9,45,48,49].

The third strand of the literature is a generational advancement of the EKC hypothesis. The dynamic relationships underscoring the pathways for decoupling are examined through the lens of the Cobb-Douglas production theory (which suggests that the GDP is a result of capital formation, labour, and productivity [50]). The main hypothesis of the studies is not only limited to emission intensity reduction in the GDP but is also extended to emission intensity reduction in capital formation [50]. Since productivity is ultimately a result of the efficiency of energy conversion, energy use acts as a proxy for factor productivity [45,51–54]. Labor is often proxied by employment [22,48,51,55], with one study attempting to evaluate the dynamic relationship between electricity-use and employment in India [56]. A major drawback of most studies in this strand is that the stochastic nature of labour and productivity variation with business cycles is not incorporated into these studies. More so, very few studies have tested the dynamic links with regard to electricity generation [55,56].

The final branch deals with the highly stochastic nature of trade (or TROP, as defined in most studies). TROP is non-deterministic and may have completely different trends from business cycle fluctuations, yet very few studies have addressed how the decarbonization of energy use varies with TROP during growth and recession phases. While considering a control system, multiple studies found that increased trade has a beneficial effect on the 3E nexus in the long-run, reducing emissions [57–60]. However, in terms of energy transition, increased imports can be a sign of a lack of energy security.

Within the 3E nexus, it has been stated that 'short-term' economic growth is aligned with 'long-term' decarbonization goals [18,61,62], but the macroeconomic pathway to achieve the same has not been provided. The authors of [62] argue that economic shocks change RE policy direction, but they fail to empirically show how exactly such a change is sustained in the long-run. While structural breaks do exemplify the change in policy directions [59,63], they do not explain which factor aids in the rebound of energy use and how the status of decoupling adapts. Moreover, GDP, TROP, capital, etc. needs to be pivoted to an economic shock to empirically determine the exact pathway of decarbonization in the post-crisis macroeconomy. Decarbonization has second priority compared to economic recovery after COVID-19 [23], and past studies have only shown the symptoms of COVID-19-related impact on electricity-related emissions [64,65]. A clear post-pandemic macroeconomic theory for ensuring decoupling is yet to be empirically determined.

The second gap in the literature is the focus on the specific macroeconomics of India, as the third highest global CO_2 emitter [32]. While there have been a few studies on the dynamics of electricity production and the GDP in India [4,37], no previous research has addressed how inflation affects the dynamics of electricity-related CO_2 emissions. As business cycles change inflation along with the GDP, it is imperative to also answer what the causality of CO_2 emissions and inflation is and how it changes with growth and recession phases. The existing models have mostly used a GDP deflator, ignoring how inflation specifically plays a part in decarbonization [8,9,44,47,59]. Specifically, inflation increases after the rebound from a shock [61,65], but previous studies have not delineated how the feedback from inflation to the GDP to CO_2 emissions plays out. This is a key contribution of this paper, showing the dynamic links between inflation and emissions, within production and trade models, affected by the 2008 financial and COVID-19 crises in the Indian electricity sector. The modelling framework is schematically shown in Figure 1.

Within the modelling framework of Figure 1, this paper will explore five specific macroeconomic models, with specific hypotheses in regard to economic shocks. While the growth model, production theory, and trade stochasticity has been explored in the literature, this study adds inflation stochasticity to examine the specific aforementioned research gap. The 3E nexus comes with expected behaviours, and, from a macroeconomic perspective, the following hypothesis can be adopted, extending the purview of the EKC:

In business cycle movements, GDP and energy use/emissions tend to move in the same direction due to increased money supply during the growth phase, which is where interest rates tend to decrease. Similarly, this causes inflation to rise, which ultimately leads to the reversal of the cycle into the recession phase. As a result, the first hypothesis can be written as follows:

Hypothesis 1. *The decoupling of emissions from economic growth is independent of the linkage to inflation emissions.*

The second consideration is one of the key foci of this study, which is economic shocks. During economic shocks, economic growth contracts, and the rebound in the economy results from production increase [18]. The incentive for decarbonization in this rebound phase (or even after a crisis) has not been explained in the past literature; therefore, we can build the second hypothesis, based on the interaction of emissions with the GDP, capital, and TROP.

Hypothesis 2. *The decoupling of emissions is inversely related between pre- and post-phases of shocks.*

The third hypothesis is built around the higher-order behaviour of economic movements and the information contained in a model representing the 3E nexus. While the growth model captures the long-run behaviour of GDP-CO_2 causalities, it does not necessarily represent the impulse response as real-world behaviour [5,49,55]. We can assume that a stochastic model will represent a real-world impulse response much better than a deterministic growth model.

Hypothesis 3. *Decoupling emerges from stochasticity instead of linear relationships with the macroeconomy.*

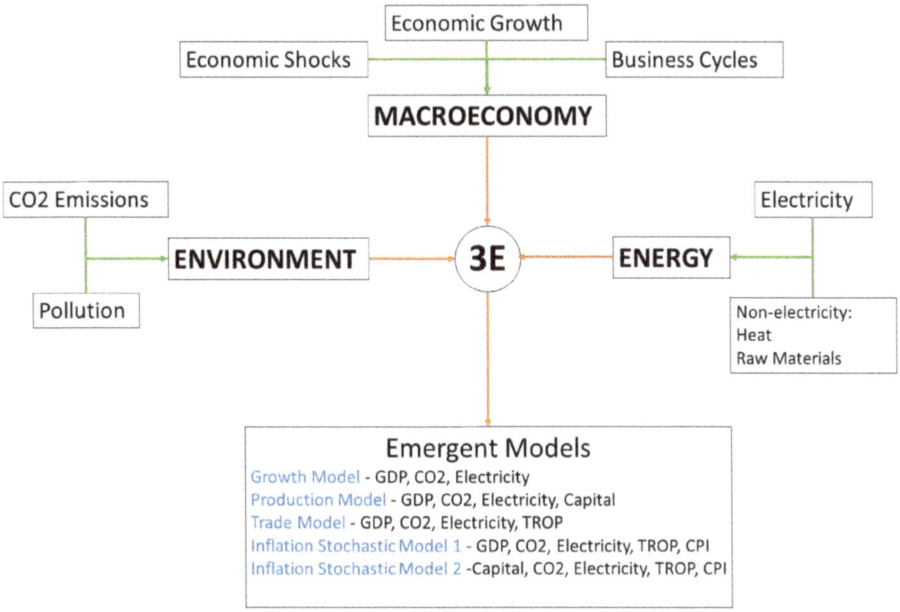

Figure 1. Modelling framework and theory schematic for analysing decoupling in the 3E (economy–electricity–emissions) nexus in India during economic shocks.

3. Materials and Methods

3.1. Modelling Specifications

The models are constructed as per Figure 1, using a long-term cointegration approach, which provides a more concrete econometric model than Vector Auto-Regression (VAR) [66,67]. As seen above, all the models are multivariate, with the dependent variable being CO_2 emissions in each case, representing the macroeconomic effects on decoupling. To account for the higher-order effects and integrating the economic shock rebound endogenously into the 3E nexus system, quarterly data are used from 1996 to 2020, encompassing both the 2008 financial crisis and the pre-crisis period for COVID-19. It must be noted that the Indian macro data are properly reported from 1996 onwards as well. All the series are first normalized to base year 2015 = 100 and then transformed into a logarithmic form to eliminate heteroskedasticity in time-series modelling [15]. Vector Error Correction Models (VECMs) incorporating long-run cointegration are quite prevalent in econometric studies [10,57], allowing for the three hypotheses to be testable. Over short samples, such as in our macroeconomic analysis, VECMs have been shown to have complex dynamic issues, such as multiple lags, difference lags, etc. [22,37,65,66,68]. This issue is dealt with by the robustness analysis method, where we use the approximate entropy method [69] to account for the amount of information present in the explanatory variables. Moreover, we compartmentalize the simulation whenever overfitting occurs in the model to make the VECM achieve a dynamic equilibrium. Both of these are novel approaches to econometric VECM modelling. The 5 models to assess the decoupling situations are given in Equations (1)–(4).

$$\mathrm{ln}\Delta C_t = f(\mathrm{ln}\Delta GDP_t, \mathrm{ln}\Delta EI_t) \tag{1}$$

$$\mathrm{ln}\Delta C_t = f(\mathrm{ln}\Delta GDP_t, \mathrm{ln}\Delta K_t, \mathrm{ln}\Delta EI_t) \tag{2}$$

$$\mathrm{ln}\Delta C_t = f(\mathrm{ln}\Delta GDP_t, \mathrm{ln}TROP_t, \mathrm{ln}\Delta EI_t) \tag{3}$$

$$\mathrm{ln}\Delta C_t = f(\mathrm{ln}\Delta GDP_t, \mathrm{ln}\Delta TROP_t, \mathrm{ln}\Delta CPI_t, \mathrm{ln}\Delta EI_t) \tag{4a}$$

$$\mathrm{ln}\Delta C_t = f(\mathrm{ln}\Delta K_t, \mathrm{ln}\Delta CPI_t, \mathrm{ln}\Delta TROP_t, \mathrm{ln}\Delta EI_t) \tag{4b}$$

Each of the 5 models (Figure 1) are trained over growth and recession periods from 1996Q2 to 2008Q4, and thereafter, an impulse is exogenously applied to GDP and K (where applicable) in 2009Q1, replicating the 2008 financial crisis. Such an impulse is more than the 2x standard deviation, which will allow us to test Hypothesis 3 specifically. The model is trained again from 2012Q1 to 2020Q3 to capture the pre-COVID-19 and COVID-19 impacts on the 3E nexus.

3.2. Data Specifications

The time series data from 1996Q1 to 2020Q4 that are used in this study are described in Table 1, along with the data sources.

Table 1. Modelling variables and abbreviations, data sources, and units.

Variable	Units of Measurement	Data Source
GDP	Constant 2015 US$	[70]
K (capital)	Constant 2015 US$	[70]
El (electricity generation)	Exa Joules (EJ)	[71]
TROP (trade openness)	Quarterly (%)	[70]
CPI (consumer price index)	Ratio	[70]
C (CO_2 emissions)	Mega Tons (MT)	[71]

Figure 2 graphs the data in the sampling period, and Table 2 gives the descriptive statistics of the data. El (electricity generation) and C (CO_2 emissions) are available as annual data and were disaggregated to quarterly data by the Denton-Cholette method [72]. TROP is calculated as the ratio of total trade (imports + exports) to GDP for each quarter. From Figure 2, it can be clearly delineated that there exist distinct higher-order phenomena in the data, even after normalizing and taking the logarithm, specifically for GDP, K (capital), CPI (inflation), and C (CO_2 emissions). It is also seen that the COVID-19 shock is much larger than the 2008 financial crisis in India. From Table 2, it is evident that inflation, emissions, and electricity-use move in a positive direction for India from the skewness and kurtosis values, implying that there is an absence of absolute decoupling in the Indian electricity sector.

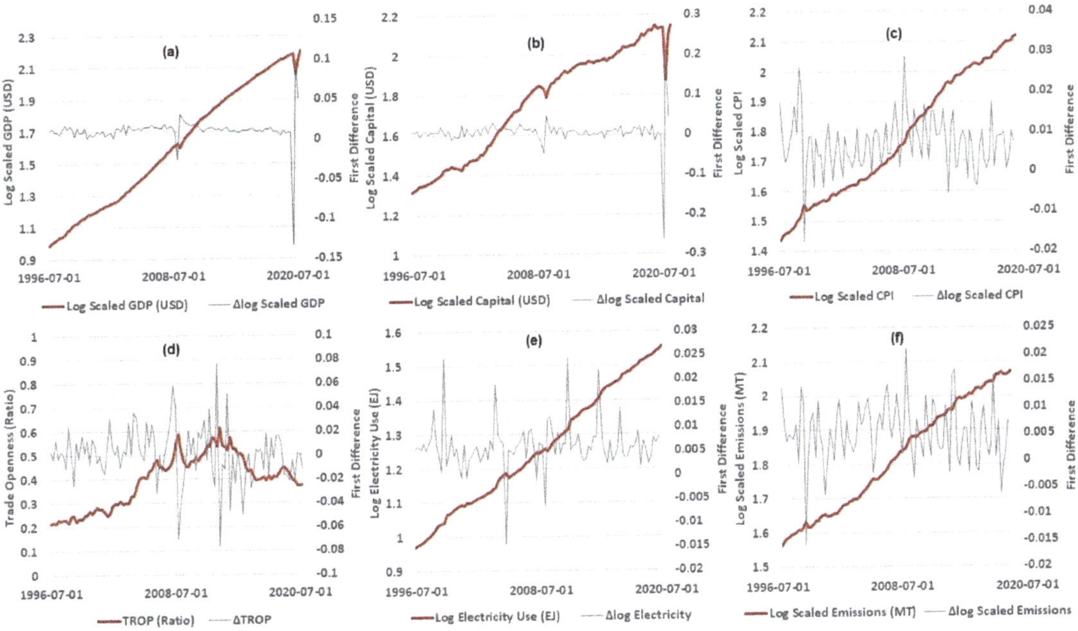

Figure 2. Graphical representation of the normalized data of (**a**) GDP, (**b**) K, (**c**) CPI, (**d**) TROP, (**e**) E-El, (**f**) C {in red—left *y*-axis}, and their first differences {in black—right *y*-axis} (modelling variables from Table 2), along with the units. The system is seen to be resilient against an economic shock (the 2008 financial crisis).

Table 2. Descriptive statistics of the modelling variables from Figure 1.

Variable	Mean	Median	Std. Dev.	Skewness	Kurtosis
ln GDP	1.611	1.602	0.379	−0.001	−1.396
ln K	1.765	1.833	0.264	−0.268	−1.371
ln CPI	1.776	1.743	0.206	0.135	−1.400
TROP	0.395	0.409	0.109	−0.119	−1.083
ln El	1.258	1.245	0.172	0.116	−1.214
ln C	1.821	1.818	0.160	0.049	−1.430
Δln GDP	0.012	0.013	0.019	−3.873	42.97
Δln K	0.008	0.010	0.037	−1.847	39.68
Δln CPI	0.007	0.008	0.007	0.065	2.091
ΔTROP	0.001	0.002	0.023	−0.191	2.266
Δln El	0.006	0.006	0.005	0.670	6.865
Δln C	0.005	0.005	0.006	−0.246	1.468

Note: Δ represents the first differences.

3.3. Unit Root Tests

Before performing the econometric modelling and specifying the model, unit root tests need to be performed on each time series to assess the stationarity and integration order of the variables. Stationarity is absolutely required to remove the statistical uncertainty of spurious regressions [73]. The Augmented Dickey-Fuller (ADF) test has been used in a form (Equation (5)), (ii) a form with intercept (Equation (6)), and (iii) a form with intercept and trend (Equation (7)). However, the ADF test has been reported to give biased results [44]; therefore, Kwiatkowski's KPSS test (reverse hypothesis of the ADF test) is used [74]. The non-inclusion of structural breaks in time series unit root testing by the ADF and KPSS tests has been argued to be spurious by econometricians [22,58,59]. As a result, the Zivot–Andrews structural break unit root test is used in this paper [75] to account for the changes in the time series regime. The Zivot–Andrews test is conducted in three forms in this paper: (i) one-time break in variables at the level form (Equation (8)), (ii) one-time break in the slope of the trend component (Equation (9)), and (iii) one-time break both in the intercept and trend function of the variables to be used for empirical analysis (Equation (10)) [56,75].

$$\Delta Y_t = \mu Y_{t-1} + \sum_{i=1}^{k} \delta_i \Delta Y_{t-i} + \varepsilon_t \quad (5)$$

$$\Delta Y_t = \alpha_0 + \mu Y_{t-1} + \sum_{i=1}^{k} \delta_i \Delta Y_{t-i} + \varepsilon_t \quad (6)$$

$$\Delta Y_t = \alpha_0 + \beta_0 t + \mu Y_{t-1} + \sum_{i=1}^{k} \delta_i \Delta Y_{t-i} + \varepsilon_t \quad (7)$$

where Y represents a time series, t is the time period sampling interval, α_0 is the intercept, β_0 is the coefficient for the time trend, μ is the coefficient of the lagged value of the time series at the level, δ is the coefficient of the lagged value of the time series at first difference, k is the optimal lag length, and ε_t is the random walk error term. The null hypothesis, $\mu = 0$, is agreed when there is no unit root against the alternate hypothesis of $\mu < 0$, when there is a unit root present.

$$\Delta Y_t = a_0 + b_0 t + a_0 Y_{t-1} + b_0 DU_t + \sum_{i=1}^{k} \delta_i \Delta Y_{t-i} + \varepsilon_t \quad (8)$$

$$\Delta Y_t = b_0 + c_0 t + b_0 Y_{t-1} + c_0 DT_t + \sum_{i=1}^{k} \delta_i \Delta Y_{t-i} + \varepsilon_t \quad (9)$$

$$\Delta Y_t = c_0 + c_0 t + c_0 Y_{t-1} + d_0 DU_t + d_0 DT_t + \sum_{i=1}^{k} \delta_i \Delta Y_{t-i} + \varepsilon_t \quad (10)$$

where DU_t is a dummy variable representing the existence of a mean shift with the time break, while DT_t shows that there is a trend shift with the time break. Equation (11) shows the conditions for the hypothesis confirmation of unit root presence.

$$DU_t = \begin{cases} 1 \text{ if } t > TB \\ 0 \text{ if } t < TB \end{cases} \text{ and } DT_t = \begin{cases} t - TB \text{ if } t > TB \\ 0 \text{ if } t < TB \end{cases} \quad (11)$$

The null hypothesis of the unit root break date is $c_0 = 0$, which indicates that the series is not stationary with a trend of not having information about the structural break point, while the $c_0 < 0$ hypothesis implies that the variable is found to be trend-stationary, with one unknown time break. The Zivot-Andrews unit root tests all points as a potential break-point and estimates through regression for all possible break points successively. For both the ADF and Zivot-Andrews tests, the optimum lag order is selected based on the established Akaike Information Criterion (AIC) [76] and Bayesian Information Criterion (BIC) [77].

3.4. *VECM Dynamic Cointegration Modelling*

The ARDL bounds test for cointegration has been used extensively in past studies to identify the causalities in macroeconomic nexuses [8,15,36,44]. While a distributed lag-model has numerous advantages of information preservation, it has the potential of breakdown when an impulse of greater than 2x standard deviation is applied (analogy for an economic shock) [78]. Therefore, this study requires a model that can not only maintain the long-run dynamics of a 3E nexus but also approximate the rebound from the economic shock. An adaptive Vector Error Correction Model (a-VECM) is introduced in this paper, which performs a least error-path Monte Carlo simulation on the unrestricted VAR of the 5 models (Equations (1)–(5)), where the adaptive nature is to freely switch between the 5 Johansen cointegration methods (Table 3) [66] according to the least error of the subsequent stage. Equations (12)–(16) show the a-VECM for the 5 models of Equations (1)–(5).

$$\Delta C_t = \alpha_0 + ect_C C_{t-1} + ect_{GDP} GDP_{t-1} + ect_{El} El_{t-1} + \sum_{i=1}^{n} \alpha_i \Delta C_{t-i} + \sum_{j=1}^{n} \alpha_j \Delta GDP_{t-j} + \sum_{k=1}^{n} \alpha_k \Delta El_{t-k} + \varepsilon_t \quad (12)$$

$$\Delta C_t = \alpha_0 + ect_C C_{t-1} + ect_{GDP} GDP_{t-1} + ect_K K_{t-1} + ect_{El} El_{t-1} + \sum_{i=1}^{n} \alpha_i \Delta C_{t-i} + \sum_{j=1}^{n} \alpha_j \Delta GDP_{t-j} + \sum_{k=1}^{n} \alpha_k \Delta K_{t-k} + \sum_{l=1}^{n} \alpha_l \Delta El_{t-1} + \varepsilon_t \quad (13)$$

$$\Delta C_t = \alpha_0 + ect_C C_{t-1} + ect_{GDP} GDP_{t-1} + ect_{TROP} TROP_{t-1} + ect_{El} El_{t-1} + \sum_{i=1}^{n} \alpha_i \Delta C_{t-i} + \sum_{j=1}^{n} \alpha_j \Delta GDP_{t-j} + \sum_{k=1}^{n} \alpha_k \Delta TROP_{t-k} + \sum_{l=1}^{n} \alpha_l \Delta El_{t-1} + \varepsilon_t \quad (14)$$

$$\Delta C_t = \alpha_0 + ect_C C_{t-1} + ect_{GDP} GDP_{t-1} + ect_{TROP} TROP_{t-1} + ect_{CPI} CPI_{t-1} + ect_{El} El_{t-1} + \sum_{i=1}^{n} \alpha_i \Delta C_{t-i} + \sum_{j=1}^{n} \alpha_j \Delta GDP_{t-j} + \sum_{k=1}^{n} \alpha_k \Delta TROP_{t-k} + \sum_{l=1}^{n} \alpha_l \Delta CPI_{t-1} + \sum_{m=1}^{n} \alpha_m \Delta El_{t-m} + \varepsilon_t \quad (15)$$

$$\Delta C_t = \alpha_0 + ect_C C_{t-1} + ect_K K_{t-1} + ect_{TROP} TROP_{t-1} + ect_{CPI} CPI_{t-1} + ect_{El} El_{t-1} + \sum_{i=1}^{n} \alpha_i \Delta C_{t-i} + \sum_{j=1}^{n} \alpha_j \Delta K_{t-j} + \sum_{k=1}^{n} \alpha_k \Delta TROP_{t-k} + \sum_{l=1}^{n} \alpha_l \Delta CPI_{t-1} + \sum_{m=1}^{n} \alpha_m \Delta El_{t-m} + \varepsilon_t \quad (16)$$

where Δ is the first difference operator with the optimal lags for the differenced terms being determined by the AIC and BIC. The coefficients of the differenced terms form the short-run analysis, while the ect (error correction term) forms the long-run analysis. This selection of the ect is made dynamic by the a-VECM method, which enables the approximation of higher-order behaviour and the return to equilibrium after an exogenous shock. This study employs several diagnostic tests on the residuals of the models, starting with the test for normality by the Jarque-Bera (JB) goodness-of-fit test [79]. In the JB test, if the statistic is far from zero, the null hypothesis of normal distribution must be rejected. The Gaussian autoregressive conditional heteroscedasticity (G-ARCH) test is used to test whether the residuals have constant variance (null hypothesis accepts constant variance) [80]. The Ljung-Box Q (LBQ) test is used to test whether the autocorrelations of the sample of residuals is different from zero [81]. The null hypothesis of the LBQ test refers to the finding that the data are independently distributed.

Table 3. The five methods to determine cointegration in the models by Johansen test [66].

Model	Error Correction Term (ECT)	Cointegrated Series	Data
H2	$AB'y_{t-1}$	No intercept, no trend	No trend
H1*	$A(B'y_{t-1} + c_0)$	Intercept, no trend	No trend
H1	$A(B'y_{t-1} + c_0) + c_1$	Intercept, no trend	Linear trend
H*	$A(B'y_{t-1} + c_0 + d_0t) + c_1$	Intercept, linear trend	Linear trend
H	$A(B'y_{t-1} + c_0 + d_0t)+c_1 + d_1t$	Intercept, linear trend	Quadratic trend

3.5. Robustness Analysis and Approximate Entropy

The final part of the methodology is associated with checking the robustness of estimations using Equations (12)–(16) (the 5 models) pre- and post-crisis. Since sufficient macroeconomic data are available for the 2008 financial crisis, the robustness analysis cannot be applied to the COVID-19 crisis. Mean percentage errors of reproduction and forecasting are evaluated for the 5 models, along with the R-factors during the economic phases from 1996 to 2012 (as defined in Equation (17)).

$$R = \frac{\Sigma||Y_{real}| - |Y_{est}||}{\Sigma|Y_{real}|} \tag{17}$$

where Y_{real} and Y_{est} are the real recorded and model estimate values for the dependent factors from Equations (12)–(16).

Apart from the above, the information contained in the estimated models must be measured to understand which model can most resemble the real world. For this, the principle of maximum entropy is applied to the estimated data of each model [82]. Entropy is a measure of disorder in a system (chaos), and the more chaotic or unpredictable the measurements of a system is, the more information it contains [83]. Thus, chaos in a system (time series) proves to reveal more information about the energy–economy–emissions nexus, which can be seen as the closest application of Occam's Razor (the simplest explanation is the best one, but the explanation needs to be complete) [84]. While the principle of maximum entropy is usually applied to statistical thermodynamics, mechanics, physiology [83], etc., this is the first time (to the best of our knowledge) socio-economic VECM analysis is subject to it. The measure of chaos of the estimated time series data by the models is provided by calculating the approximate entropy of each of the estimates. Each of the estimated time series data in each model is equally spaced in time as follows:

$$U(1), U(2), \ldots, U(N) \tag{18}$$

where N is the raw data values. We define m as a length of run-time data ($0 \leq m \leq N$) and r as a real, positive number specifying a filtering level tolerance for accepting matches. The following sequence of vectors is then formed:

$$Y(1), Y(2), \ldots, Y(N - m - 1) \tag{19}$$

which in an m-dimensional real space defined by (after obtaining the results from Equations (12)–(16))

$$Y(i) = [\Delta C_t(i), \Delta C_t(i+1), \ldots, \Delta C_t(i+m-1)] \tag{20}$$

the above vector sequence is used for each magnitude of i as

$$C_i^m(r) = \frac{(r - d|Y(i), Y(j)|)}{N - m + 1} \tag{21}$$

The functional magnitude of the m-dimensional space is defined as

$$\psi^m(r) = \frac{1}{N - m + 1} \sum_{i=1}^{N-m+1} \log(C_i^m(r)) \tag{22}$$

In the final step, Equation (23) represents the final expression for Approximate Entropy (ApEn) calculation.

$$ApEn(m, r, N) = \psi^m(r) - \psi^{m+1}(r) \tag{23}$$

where $m \geq 1$, and ApEn(0, r, N)(u) = $-C^1(r)$.

The specific econometric methods play critical roles in uncovering the higher-order dynamics of decoupling in a macroeconomic 3E nexus. The Zivot-Andrews test not only exemplifies regime shifts and structural reorientation but demarcates point of an exogenous shock in the time series of stationary data. The a-VECM seamlessly controls the long-run cointegration of key macroeconomic dynamics that are not reoriented in a regime shift but adapts to a different form of cointegration, while maintaining the directionality of feedbacks. This is an advantage over ARDL, which reorients all cointegrations in an economic shock. Multiple cointegrated relationships tend to be independent of each other, whether one reorients or not across a shock. A control system that can capture the maximum number of these independent relationships will contain more information, which is where the ApEn is a practical indicator for the same. Thus, from a methodological perspective, this paper proposes the above specific framework for analysing macroeconomic decarbonization that incorporates business cycle movements and the rebound effects of a shock.

4. Results

4.1. Unit Root Tests Results

Table 4 shows the results of the unit root tests of the ADF and KPSS tests, while Table 5 shows the results of the Zivot-Andrews structural break unit root test. Using the unit root tests, we determine whether the variables are stationary at their first difference and what the order of integration of the explanatory and dependent variables is.

Table 4. The results of the unit root tests (ADF and KPSS tests).

Variable	At Level		At First Difference	
	ADF	KPSS	ADF	KPSS
ln GDP	−3.326 (2) ^{c}	1.284	−82.12 (0) ^{a,*}	0.279
ln K	−3.093 (2) ^{c}	1.454	−90.95 (0) ^{a,*}	0.187
ln CPI	−3.041 (1) ^{c}	1.562	−78.69 (0) ^{a,*}	0.171
TROP	−3.742 (1) ^{c}	1.681	−96.20 (0) ^{b,*}	0.214
ln El	−13.31 (1) ^{c}	1.629	−115.7 (0) ^{a,*}	0.053
ln C	−7.081 (0) ^{c}	1.183	−185.4 (1) ^{b,*}	0.099

Note: (): optimum lags for the ADF test; *: significant at 1% level; ^{a}: intercept and trend are 0; ^{b}: only trend is zero; ^{c}: intercept and trend are non-zero.

Table 5. The results of the unit root tests (Zivot-Andrews structural break test).

Variable	At Level		At First Difference	
	t-Statistic	Breaks	t-Statistic	Breaks
ln GDP	−2.456 (2) ^{c}	2004Q2	−12.12 (0) *	2009Q1
ln K	−4.012 (2) ^{c}	2005Q1	−13.82 (0) *	2009Q1
ln CPI	−2.785 (1) ^{c}	2008Q3	−12.58 (0) *	2009Q2
TROP	−3.247 (1) ^{c}	2020Q2	−11.74 (0) *	2010Q2
ln El	−4.831 (1) ^{c}	2009Q2	−17.65 (0) *	2009Q3
ln C	−4.858 (0) ^{c}	2004Q4	−16.25 (1) *	2009Q1

Note: (): optimum lags for the ADF test; *: significant at 1% level; ^{c}: not significant.

With the KPSS statistic being under 1.0, implying stationarity, it is seen that all the variables of the five models are stationary at their first differences. While there is a non-uniformity of trends and intercepts for the variables, the ADF test confirms that all the modelling variables are integrated at I(1). To mitigate the bias of ADF unit roots in small samples without structural breaks, the Zivot-Andrews test is utilized.

From Table 5, it is very interesting to observe that the structural breaks at level are scattered across the time interval, with no presence of stationarity. However, the variables show a marked break around the 2008 financial crisis at the first difference, when stationarity is detected (except for TROP, which is quite stochastic in nature). With the integration order confirmed at I(1), the a-VECM approach can be applied due to uniformity among the explanatory and dependent variables. Moreover, the presence of the structural break confirms that there would be a regime shift in all the models, which is where the least errors approximating model can be thought to represent the 3E nexus of India with higher-order phenomena.

4.2. Model Lag Order Determination and Cointegrations

Table 6 shows the optimal lag for each of the models using the AIC and BIC of the Unrestricted Vector Auto-Regression (UVAR) model. It has been argued in the literature that the AIC and BIC are quite reliable for small samples, compared to other lag-length selection criteria [85]. It can be seen that the simple model and the inflation stochastic models have the minimum number of lags, while the maximum lags are seen in the trade stochastic model, at six. This is quite counter-intuitive, as compared to the results of [4,12,44] for other developing economies, showing that the economic shocks have a major impact on international trade, rather than domestic economic growth.

Table 6. Lag length selection for the models of Equations (12)–(16) based on the AIC and BIC.

Lags	C = f(GDP,EI)			C = f(GDP,K,EI)			C = f(GDP,TROP,EI)			C = f(GDP,TROP,CPI,EI)			C = f(K,TROP,CPI,EI)		
	LL	AIC	BIC	LL	AIC	BIC	LL	AIC	BIC	LL	AIC	BIC	LL	AIC	BIC
0	355.56	−705.12	−699.32	484.92	−967.83	−954.10	475.47	−942.95	−935.22	638.70	−1267.4	−1257.7	600.56	−1191.1	−1181.5
1	577.96	−1131.9	−1109.0	750.43	−1460.9	−1422.6	714.18	−1388.4	−1350.1	923.86	−1787.7	−1730.4	875.48	−1691.0	−1633.6
2	588.32 *	−1134.7 *	−1109.9 *	750.92	−1433.9	−1365.7	714.93	1357.9	−1289.8	982.57	−1855.1	−1751.1	926.33	−1742.7 *	−1638.6 *
3	583.09	−1116.2	−1080.0	751.37	−1456.7	−1338.4	721.05	−1338.1	−1240.8	999.76 *	−1839.5 *	−1789.8 *	940.26 *	−1720.5	−1570.8
4	579.52	−1081.0	−1008.9	752.25 *	−1498.5 *	−1401.7 *	718.76	−1301.5	−1175.7	995.67	−1793.7	−1599.5	938.15	−1666.3	−1472.0
5	578.42	−1060.8	−973.06	747.76	−1327.5	−1173.9	720.59	−1273.2	−1119.6						
6							834.60 *	−1373.2 *	−1416.1 *						

*: Denotes selection of the lag order.

Cointegration determination by Johansen's test is the subsequent step in determining the interlinkages among macroeconomic variables towards electricity-related CO_2 emissions across two economic shocks in India. The two statistics used for verifying the existence of long-run cointegrations are trace and maximum eigen statistics. In the simple and production models (Equations (12) and (13)), two cointegrating relationships are found at the 5% significance level. The trade model (Equation (14)) shows a surprising result with four cointegrating relationships (at the 5% level), with the $R \leq 1$ and $R \leq 2$ hypotheses showing no evidence of cointegration. The inflation stochastic model (Equation (15)) has the maximum number of cointegrations at five (5% significance level), with significance for the lesser cointegrations at a 1% significance level. Table 7 shows the complete Johansen cointegration results before the 2008 financial crisis, while Table 8 shows it post-financial crisis recovery (from 2012Q2 to 2020Q3).

Table 7. The Johansen long-run cointegration test for the five models of Figure 1 from 1996Q2 to 2008Q3.

Rank	C = f(GDP,EI)		C = f(GDP,K,EI)		C = f(GDP,TROP,EI)		C = f(GDP,TROP,CPI,EI)		C = f(K,TROP,CPI,EI)	
	Trace	Max Eigen	Trace	Max Eigen	Trace	Max Eigen	Trace	Max Eigen	Trace	Max Eigen
$R = 0$	65.45 *	40.72 *	52.92 *	28.29 **	69.79 *	43.71 *	101.2 *	40.80 *	124.5 *	54.75 *
$R \leq 1$	24.73 **	17.48 **	30.63 **	21.55 **	26.08	15.67	60.41 *	22.77	69.75 *	36.32 *
$R \leq 2$	7.255	7.255	13.07	12.95 ***	10.41	6.169	37.64 *	17.12 ***	33.44 **	20.76 ***
$R \leq 3$			0.121	0.121	4.242 **	4.242 **	20.52 *	14.75 **	12.68	10.49
$R \leq 4$							5.778 **	5.778 **	2.190	2.190

*: Significant at the 1% level. **: Significant at the 5% level. ***: Significant at the 10% level.

Table 8. The Johansen long-run cointegration test for the five models of Figure 1 from 2012Q2 to 2020Q3.

Rank	C = f(GDP,EI)		C = f(GDP,K,EI)		C = f(GDP,TROP,EI)		C = f(GDP,TROP,CPI,EI)		C = f(K,TROP,CPI,EI)	
	Trace	Max Eigen	Trace	Max Eigen	Trace	Max Eigen	Trace	Max Eigen	Trace	Max Eigen
$R = 0$	24.84	16.45	108.7 *	54.26 *	95.09 *	42.18 *	903.7 *	751.1 *	149.9 *	58.69 *
$R \leq 1$	8.386	5.042	54.50 *	34.95 *	52.91 *	31.89 *	152.5 *	76.76 *	91.20 *	45.17 *
$R \leq 2$	3.344 **	3.344 **	19.55 **	19.27 *	21.07 *	14.88 **	75.77 *	35.46 *	46.00 *	32.66 *
$R \leq 3$			0.275	0.275	6.139 **	6.139 **	40.37 *	23.14 *	13.37	11.56
$R \leq 4$							17.18 *	17.18 *	1.813	1.813

*: Significant at the 1% level. **: Significant at the 5% level.

A higher rank is found to be significant for the simple growth and production models in the post-crisis period despite a lower sample size. For the stochastic models, the cointegrations are more significant in the post-crisis period, implying that the macroeconomic indicators were more closely linked in recent years than before 2008. We can partly confirm Hypothesis 2 of this research from this result, as long-run relationships imply that decoupling patterns changed with respect to capital, inflation and trade movements significantly across the economic shock. Therefore, we can provide a fresh perspective on the results of [3,4], wherein the macroeconomics of electricity generation is more FF-intensive in India, with RE generation merely being stochastic noise in macroeconomic movements since the introduction of the Paris Agreement in 2015.

4.3. a-VECM Dynamic Cointegration Results

Table 9 shows the long- and short-run results for the dependencies of emissions on macroeconomic variables from 1996Q2 to 2008Q3, while Table 10 shows the same for the 2012Q1 to 2020Q3 period. The adaptive nature of the a-VECM enables the change in the cointegration model type from pre-2008 crisis to post-2008 crisis, which gives a distinct advantage over ARDL, wherein the robustness of the modelling control systems can be checked across growth and recession phases and the rebound from the 2008 financial crisis in forecasting.

Table 9. The a-VECM long- and short-run analysis for the five models from 1996Q2 to 2008Q3.

Variables	C = f(GDP,EI)	C = f(GDP,K,EI)	C = f(GDP,TROP,EI)	C = f(GDP,TROP,CPI,EI)	C = f(K,TROP,CPI,EI)
Long-run Results:					
Constant	0.141 *	−0.043 *	0.334 *	0.131	0.043 **
$\ln GDP_{t-1}$	0.065 *	0.051 *	0.215 *	0.067	
$\ln K_{t-1}$		−0.077 *			0.012 ***
$\ln CPI_{t-1}$				−0.136	−0.075
$TROP_{t-1}$			−0.062 **	0.001	0.022
$\ln EI_{t-1}$	0.040 *	0.224 *	0.352 *	0.374 *	0.472 *
$\ln C_{t-1}$	−0.171 *	−0.171 *	−0.703 *	−0.372 *	−0.441 *
Short-run Results:					
$\Delta\ln GDP_{t-1}$	−0.201 *	0.525 *	−0.377 **	0.075	
$\Delta\ln K_{t-1}$		0.037			0.054
$\Delta\ln CPI_{t-1}$				−0.632 *	−0.324 *
$\Delta TROP_{t-1}$			0.104 *	0.103 **	0.089 *
$\Delta\ln EI_{t-1}$	−0.029	0.283 *	0.300 **	−0.189	−0.281 *
$\Delta\ln C_{t-1}$	0.182 ***	−0.221 **	0.404 *	0.939 *	0.602 *
Diagnostic Tests:					
LL	565.96	734.94	722.15	996.87	909.61
AIC	−1099.9	−1349.9	−1228.3	−1803.7	−1719.2
BIC	−1069.7	−1238.9	−1033.2	−1626.0	−1624.6
$\chi 2$ Normal (JB)	1.342 #	0.173 #	0.662 #	0.208 #	0.788 #
$\chi 2$ Corr (LBQ)	60.49 [b],***	15.99 [a],#	28.92 [a],***	18.98 [a],#	31.40 [a],***
$\chi 2$ ARCH	2.645 #	6.761 **	1.037 #	1.085 #	0.920 #

*: Significant at the 1% level. **: Significant at the 5% level. ***: Significant at the 10% level. #: Significant above the 10% level. [a]: 20 lags involved in the Monte Carlo Auto-correlation test. [b]: 35 lags involved in the Monte Carlo Auto-correlation test.

Table 10. The a-VECM long- and short-run analysis for the five models from 2012Q2 to 2020Q3.

Variables	C = f(GDP,EI)	C = f(GDP,K,EI)	C = f(GDP,TROP,EI)	C = f(GDP,TROP,CPI,EI)	C = f(K,TROP,CPI,EI)
Long-run Results:					
Constant	0.427 *	−0.006	−0.034 *	NA	−0.009 *
$\ln GDP_{t-1}$	0.225 *	0.335 *	0.097 *	NA	
$\ln K_{t-1}$		−0.266 *			−0.065 *
$\ln CPI_{t-1}$				NA	0.137 *
$TROP_{t-1}$			0.312 *	NA	0.162 *
$\ln EI_{t-1}$	−0.220 *	−0.274 *	0.497 *	NA	0.093 *
$\ln C_{t-1}$	−0.216 *	0.190 *	−0.652 *	NA	−0.193 *

Table 10. *Cont.*

Variables	C = f(GDP,EI)	C = f(GDP,K,EI)	C = f(GDP,TROP,EI)	C = f(GDP,TROP,CPI,EI)	C = f(K,TROP,CPI,EI)
Short-Run Results:					
$\Delta lnGDP_{t-1}$	1.206 **	2.594 *	1.642 *	NA	
ΔlnK_{t-1}		−0.421 *			0.136 ***
$\Delta lnCPI_{t-1}$				NA	−0.874 *
$\Delta TROP_{t-1}$			−0.325 *	NA	0.006
$\Delta lnEI_{t-1}$	0.458 *	0.441 *	−0.023	NA	0.057
ΔlnC_{t-1}	0.260	0.177	0.472 *	NA	0.642 *
Diagnostic tests:					
LL	325.91	455.19	426.73	NA	541.31
AIC	−609.82	−806.39	−749.45	NA	−972.61
BIC	−583.40	−743.01	−686.07	NA	−903.42
χ2 Normal (JB)	0.103 #	1.588 #	0.662 #	NA	2.201 #
χ2 Corr (LBQ)	15.13 a,#	16.81 a,#	28.92 a,***	NA	12.89 a,#
χ2 ARCH	0.409 #	0.004 #	1.037 #	NA	0.348 #

*: Significant at the 1% level. **: Significant at the 5% level. ***: Significant at the 10% level. #: Significant above the 10% level. a: 20 lags involved in the Monte Carlo Auto-correlation test.

The a-VECM results show a significant regime shift pre- and post-2008 financial crisis in the macroeconomics of decoupling in the Indian electricity sector. It is found from Table 10 that trade stochastic inflation Model 4 is unstable in the post-financial crisis period, thereby enabling us to disregard the model as a measurement method of electricity-related decoupling in India. The long-run results (derived from the adaptive long-run cointegration) shows the marginal impacts of macroeconomic indicators in different 3E nexus systems, which bear weight for policy implication post-COVID-19 crisis for decarbonizing the power generation sector in India. In the simple growth and production models, GDP-CO_2 coupling is seen in the long-run, which intensifies from the pre-crisis to the post-crisis phase. However, decoupling is actually detected in terms of capital (K) formation, where the EKC can be confirmed as the long-run coefficient of K in Model 2 increases from −0.077 to −0.266. This means that for a 1% increase in capital, electricity-related emissions are reduced by 0.26%, which can be regarded as a 'relative decoupling' phase [3,58]. All these results are significant at the 1% level. However, a reverse long-run impact of TROP is seen in Model 3, where relative decoupling with C pre-2008 turns to coupling at 0.312 (significant at the 1% level) post-2011. Herein, we can confirm Hypothesis 2 of this study, where TROP couples with emissions after the financial crisis, whereas in the trade model, it was slightly decoupled (−0.062). This highlights significant issues in Indian power generation macroeconomics, where the rebound from a crisis is achieved by FF imports, hampering decoupling efforts. This is in opposition to previously established results for Asian economies, which proclaimed that TROP advancement leads to accelerated decoupling [63,86].

Hypothesis 2 with regard to TROP direction change towards C in the different economic regimes can further be consolidated by the long-run relationship between electricity generation and emissions. When considering economic and capital growth (models 1 and 2), we see a change in EI->C from a positive to a negative association. This implies that capital and assets towards RE can macroeconomically reduce the emission intensity of electricity use (confirming the results of [5,49,51]). However, the long-run EI->C shows high coupling (0.497 at 1% significance) in the post-2011 regime, when TROP is considered in the 3E nexus (Model 3 of Table 10), confirming that Indian electricity has increased dependency on imported FF after a crisis and in recent times. In trade stochastic Model 5, it might be thought that EI->C coupling reduction from 0.472 to 0.093 (1% significant level) may indicate a stochastic reduction in electricity-related emissions, but the post-2011 regime shows that the CPI is positively coupled with C at 0.137 (1% significance level), whereas the CPI->C relation was insignificant in the pre-2008 crisis regime. Thus, it can be concluded that long-run inflation is cointegrated with CO_2 emissions, showing the existence of a macroeconomic death spiral from FF imports to inflation to emissions. This is one example of a higher-order macroeconomic phenomenon that is detected by the novel Model 5, not enumerated in any previous decoupling study for India.

The short-run dynamics reveal the quarterly and immediate impacts on decoupling. In all the 3E systems, the GDP becomes strongly coupled with C in the post-crisis period compared to the pre-2008 period. For example, in the production model, a 1% change in GDP growth will result in a 2.594% increase in the subsequent quarter's CO_2 emissions, significant at the 1% level. However, in both economic regimes, CPI->C shows short-run decoupling, with a 1% change in the CPI, reducing CO_2 emissions by 0.874% in the post-2011 period (1% significance). In both the production Model 2 and trade Model 3 short-runs, TROP and K promote decoupling, which can be seen as critical power generation policy directions. In the short-run, we can confirm Hypothesis 1 of this study, that inflation and economic growth have similar impacts on decarbonization in the Indian electricity sector, but in inflation stochastic Model 2, GDP emissions and inflation emissions are independent of each other in the short-run. This is because the monetary policy of India does not follow business cycles, as financial budgeting sees massive changes annually in India [4,14,87], which hampers decoupling efforts. On the other hand, from stochastic models 4 and 5, it can be confirmed that inflation promotes decoupling in an immediate sense (quarterly) but actually increases emissions in the long-run due to dependency on FF imports. This result agrees with the review of decoupling nexus studies by [44] for developing economies.

The diagnostic tests show that the residual terms in all the models for dependent variable C are normally distributed in both the regimes. Some evidence of serial autocorrelation is seen for models 1, 3, and 5 in the pre-financial crisis regime, and for Model 3 in the post-crisis regime, it is seen at the 10% significance level. Production Model 2 shows strong evidence of autoregressive conditional heteroskedasticity at the 5% significance level in the pre-crisis regime. Figure 3 shows the CUSUM test figures, and Figure 4 shows the CUSUM-sq tests figures for the pre-crisis period for all the models. The CUSUM plots are all within the critical bounds, while the CUSUM-sq plots for models 2, 3, and 4 exceed the critical bounds. For Model 5, the stochasticity pushes the CUSUM-sq plot of C towards the critical limit but does not exceed it. Thus, it can be concluded from the diagnostic tests that stochastic Model 5 is the least-error statistical representation of a higher-order 3E nexus for the Indian electricity sector from the period of 1996 to 2020.

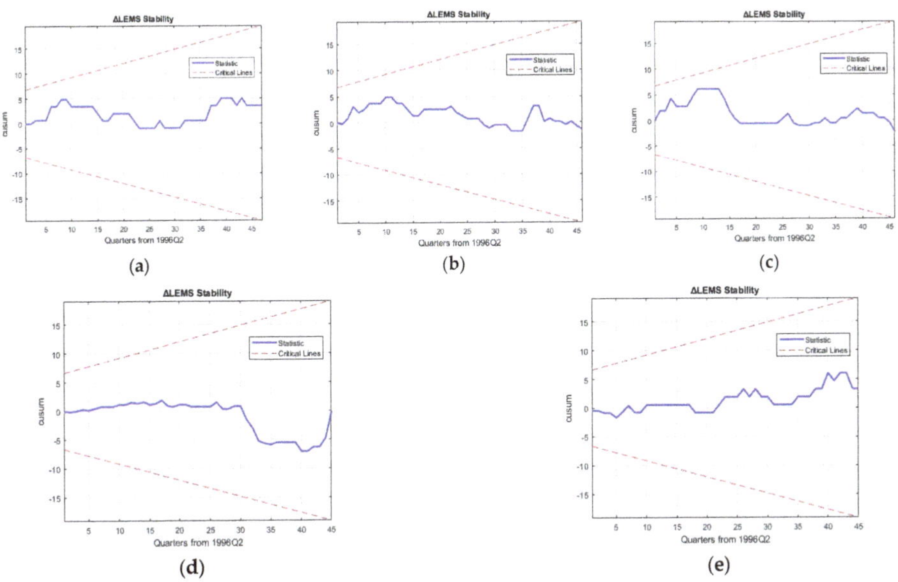

Figure 3. Plot of cumulative sum (CUSUM) of residuals for dependent variable C in the a-VECM modelling for the Indian 3E nexus from 1996Q2 to 2008Q4 for (**a**) Model 1 (Equation (12)), (**b**) Model 2 (Equation (13)), (**c**) Model 3 (Equation (14)), (**d**) Model 4 (Equation (15)), and (**e**) Model 5 (Equation (16)).

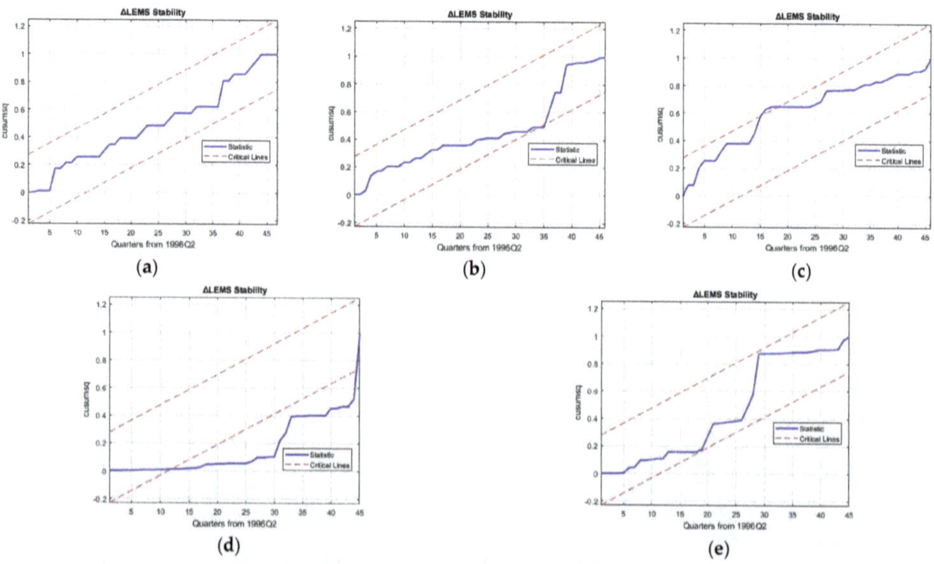

Figure 4. Plot of cumulative sum of squares (CUSUM-sq) of residuals for dependent variable C in the a-VECM modelling for the Indian 3E nexus from 1996Q2 to 2008Q4 for (**a**) Model 1 (Equation (12)), (**b**) Model 2 (Equation (13)), (**c**) Model 3 (Equation (14)), (**d**) Model 4 (Equation (15)), and (**e**) Model 5 (Equation (16)).

5. Robustness Discussion and Policy Implications

Robustness of the 3E nexus models is required to hold up against the reproduction and forecasting of business cycle movements of macroeconomic variables and rebound from the 2008 financial crisis to ascertain the results of decoupling in the electricity sector. Figures 5–9 show reproduction from the 1996Q2 to 2008Q3 training period and forecasts from the 2009Q1 to 2011Q4 testing period for the five a-VECMs. Exogenous shock is applied to the GDP and K in the 2008Q4 period. The detailed coefficients of all the a-VECMs are given in Appendices A–E, along with the residual diagnostic tests for all the variables. Table 11 shows the mean aggregated percentage reproduction error (MAPRE), mean aggregated percentage forecasting error (MAPFE) and approximate entropy (ApEn) for the five models, which are the key robustness tests of this study. Specifically, ApEn is a measure of the amount of information contained in the modelled values, giving an indication of the independence of the variables in the 3E nexus considered.

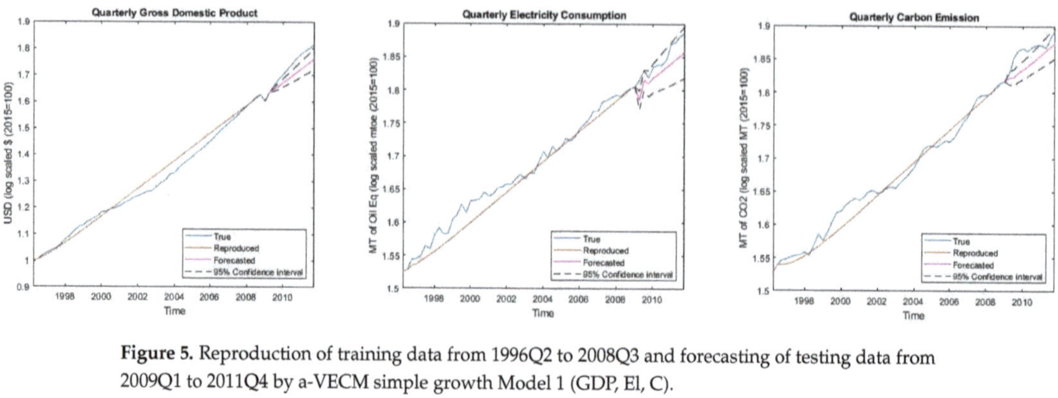

Figure 5. Reproduction of training data from 1996Q2 to 2008Q3 and forecasting of testing data from 2009Q1 to 2011Q4 by a-VECM simple growth Model 1 (GDP, El, C).

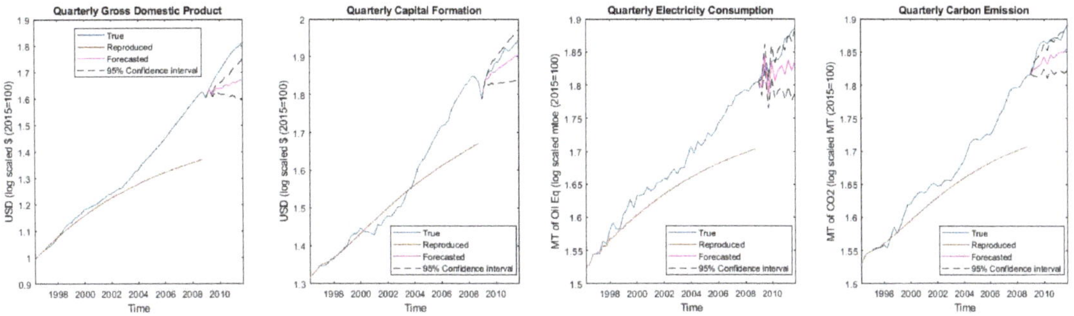

Figure 6. Reproduction of training data from 1996Q2 to 2008Q3 and forecasting of testing data from 2009Q1 to 2011Q4 by a-VECM production Model 2 (GDP, K, El, C).

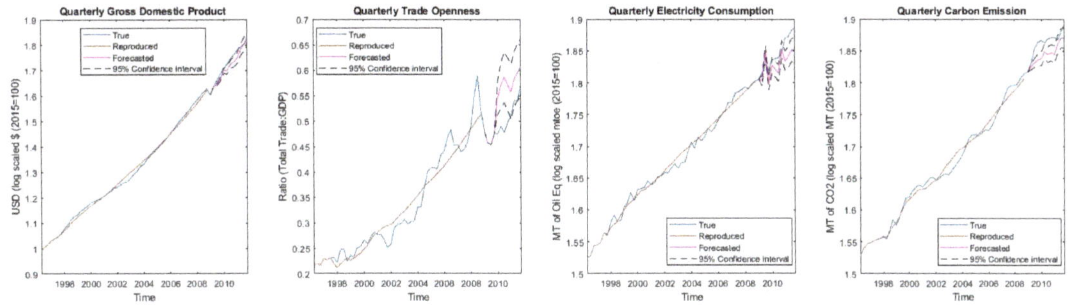

Figure 7. Reproduction of training data from 1996Q2 to 2008Q3 and forecasting of testing data from 2009Q1 to 2011Q4 by a-VECM trade Model 3 (GDP, TROP, El, C).

Figure 8. Reproduction of training data from 1996Q2 to 2008Q3 and forecasting of testing data from 2009Q1 to 2011Q4 by a-VECM trade stochastic Model 4 (GDP, CPI, TROP, El, C).

Figure 9. Reproduction of training data from 1996Q2 to 2008Q3 and forecasting of testing data from 2009Q1 to 2011Q4 by a-VECM inflation stochastic Model 5 (K, CPI, TROP, El, C).

Table 11. MAPRE, MAPFE, and approximate entropy for Models 1 to 5 for training data of 1996Q2 to 2008Q4 and forecasting data from 2009Q1 to 2011Q4.

Model	MAPRE (%)	MAFRE (%)	ApEn
1 (GDP, El, C)	1.00	1.46	0.015
2 (GDP, K, El, C)	3.20	2.30	0.137
3 (GDP, TROP, El, C)	1.93	3.06	0.471
4 (GDP, CPI, TROP, El, C)	4.05	2.07	0.342
5 (K, CPI, TROP, El, C)	2.90	1.72	0.441

Confirming the results, Model 4 (trade stochastic 3E system) is seen to oscillate out-of-bounds in Figure 8, and Model 2 (production 3E system) is seen to reproduce large errors in Figure 6. As a result, the discussion on robustness will be limited to models 1, 3, and 5. While MAPRE of the simple growth 3E nexus is seen to be the best at an aggregated level for all the variables, ApEn is considerably less at 0.015. This implies that the amount of information in the modelling system is significantly less; it is exactly in Figure 5 where it is seen that higher-dimensional business cycle movements are not captured by this nexus. As a result, the causalities examined by previous research using the trivariate nexus of GDP-El-C can be inferred to not be representative of real-world movement [4,88,89]. Moreover, the rebound effect post-2008 crisis is also directionally opposite in the forecast for Model 1, although the forecasting error is the smallest.

The advocacy that international trade is highly impactful on nexus modelling by the authors of [90] is highly agreeable with the results of Model 3 (Figure 7). The cyclic movements of El, C, and GDP itself is captured by the inclusion of TROP in both growth and recession periods, with a MAPRE of just 1.93%. The GDP–TROP–El–C nexus can indeed be considered an important 3E nexus policy tool with the highest entropy value of 0.47. The reason for high levels of information content in this model is that the Indian electricity sector depends on coal generation, which mainly depends on international trade. The results of [47,57,90] for the economy and electricity linkage of Italy, Ghana, and Indonesia is, thus, in agreement with the reproducibility and information quantity (entropy) of Model 3. However, the downside of the trade model is its forecasting error, which is the largest among all the considered 3E nexus systems at 3.06%, showing that the forecasting capability is limited for this model in the face of an economic shock.

Finally, inflation stochastic Model 5 has a slightly larger MAPRE than the trade Model 3 (Table 11) but still captures the cyclic movements of the variables, as seen in Figure 9. This is because the CPI (inflation) is associated with inherent stochasticity, which affects the predictive accuracy of any modelling system. Model 5 maintains a similar ApEn value as Model 3, indicating a high content of information in the modelled variables towards representing the 3E nexus for India. Where Model 5 is superior to Model 3 is in forecasting accuracy after the application of the 2008Q4 shock, which is greatly improved. Therefore, a stochastic model internalizing inflation and TROP is capable of capturing higher-dimensional business cycle movement, as well as the rebound effect of an exogenous shock better than growth, production, and trade models, confirming Hypothesis 3 of this research.

To ascertain Hypothesis 3, we utilize Equation (17) to calculate the R factors in four identified economic phases, as shown in Table 12, with respect to models 1, 3, and 5.

Table 12. Summary of the R factors (normalized) in two growth phases, one recession phase, and the recovery phase post-2008 financial crisis for models 1, 3 and 5 (models 2 and 4 contain large errors of reproducing 3E nexus data).

Economic Phase	Model	GDP	K	TROP	CPI	El	C
Growth Phase 1 (1996Q1–2000Q1)	1 (GDP, El, C)	0.010				0.011	0.006
	3 (GDP, TROP, El, C)	0.006		0.036		0.002	0.001
	5 (K, CPI, TROP, El, C)		0.009	0.021	0.036	0.012	0.016
Recession Phase 1 (2000Q2–2003Q3)	1 (GDP, El, C)	0.021				0.008	0.008
	3 (GDP, TROP, El, C)	0.009		0.066		0.004	0.005
	5 (K, CPI, TROP, El, C)		0.045	0.125	0.009	0.008	0.014
Growth Phase 2 (2003Q4–2008Q4)	1 (GDP, El, C)	0.015				0.004	0.004
	3 (GDP, TROP, El, C)	0.005		0.060		0.003	0.004
	5 (K, CPI, TROP, El, C)		0.010	0.055	0.012	0.005	0.009
Post-crisis Recovery (2008Q2–2011Q4)	1 (GDP, El, C)	0.021				0.010	0.010
	3 (GDP, TROP, El, C)	0.008		0.083		0.011	0.009
	5 (K, CPI, TROP, El, C)		0.007	0.024	0.026	0.004	0.018

An R factor greater than 0.1 shows insignificance and lesser than 0.01 shows high significance.

R factors reveal some interesting dynamics of the capability of the models to represent higher-order business cycle movements and crisis rebound effects in different regimes. As seen in Figures 5–9, the most significant R factors are reported by all the models in the first growth phase of the business cycle, at the start of the modelling period. It is in the recession phase where we see fewer significant R factors greater than 0.01 for all the modelling variables, which, again, shows improvement for models 3 and 5 in the second growth phase. Thus, it can be concluded that cointegration modelling is limited in capturing economic movement from a growth to recession phase, but it accurately predicts the opposite movement from a recession to growth phase. This is because cointegration models tend to correct errors in the direction of the overall trend in the training data, which, in this case, is an incremental direction, as opposed to the recession phase. This adds valuable insight to existing structural cointegration models [9,37,38,68]. Where Model 5 contains more significant R factors than Model 3 is in the post-crisis recovery phase, which can be attributed to the stochastic nature of the CPI and TROP being internalized to more accurately approximate the rebound from an exogenous shock to a 3E nexus system [63], affirming Hypothesis 3.

Several policy implications for the current status and future of decoupling in the Indian electricity sector can be derived from this study. Firstly, unlike previous studies in the literature where the EKC is measured solely against economic growth (GDP) [8,44,58,59], the dimensions of EKC interpretation are extended here. From 1996 to 2020, which encompasses two distinct economic periods—the pre-2008 and post-2008 financial crisis periods and multiple business cycle movements—the GDP and electricity-related CO_2 emissions are absolutely not decoupled. However, decoupling is observed through production and inflation stochastic models with respect to capital formation over a long-run cointegrated relationship. Moreover, trade openness and inflation promote decoupling in the short-run, while TROP shows evidence of coupling with emissions in the long-run. Therefore, it can be confirmed that the incentive of decoupling lies in boosting the capability of the manufacturing sector and ensuring resiliency in domestic RE markets, which builds the equity of entities invested in reducing the emission intensity of power generation [3]. Capital can also be accelerated by promoting innovation in the RE sector, which is exactly what policy programs like 'make in India' [91] and 'surya ghar muft Bijli yojana' (free electricity from rooftop solar panels) [92] aim to promote. Not only can capital formation be ensured by domestic production but also stabilized domestic markets can ensure foreign direct investment (FDI), which also boosts capital [14].

On the other hand, inflation has been rising constantly with the fast-growing GDP of India, and while short-run decoupling was seen in the post-crisis period with the CPI, it ultimately inhibited decarbonization of the electricity sector. The novel policy pathway that is uncovered in Model 5 unveils the negative feedback loop involving TROP, CPI, and emissions. Looking at Tables A23 and A25, when inflation peaks and a recession phase begins, GDP growth is ensured through FF imports, which boosts TROP and ultimately reduces the CPI. However, this causes an increase in CO_2 emissions from power generation, which is where we see a long-run positive association between the CPI and TROP and between TROP and C. This resembles a death spiral with the positive effect of capital reducing CO_2 emissions during recession phases. This loop was also magnified during the rebound from the 2008 economic shock, which can also be thought to have been replicated during the COVID-19 shock, as the macroeconomic dynamics in the 3E nexus are similar to those before the 2008 financial crisis (as seen in Tables 9 and 10). This is coupled with the fact that economic shocks destabilize RE stock markets [65], which further nullifies the positive effect of capital on decoupling, magnifying the negative feedback in the process.

6. Conclusions

This study explores the macroeconomic dynamics of decoupling in the Indian electricity sector through long-term cointegration models from 1996Q2 to 2020Q3 across the economic shocks of the 2008 financial crisis and the COVID-19 pandemic. Five models covering aspects of economic growth, Cobb-Douglas production, trade, and inflation stochasticity within the economy–electricity–emissions (3E) nexus were tested for their robustness to reproduce higher-order business cycle movements and to forecast the rebound from an economic shock. Robustness was analysed using a novel information-theory-based approximate entropy (ApEn) approach to the 3E systems and R factors in economic phases.

The key findings of the study are as follows:

1. An inflation stochastic 3E system can detect higher-order behaviour in decoupling dynamics better than linear growth nexuses by containing more information in the coefficients (ApEn value of 0.441 as opposed to 0.015 of a simple growth model).
2. Cointegration models can approximate decoupling and 3E dynamics more accurately in growth phases of business cycles than recession phases, with the inflation stochastic model again being superior to the other 3E systems.
3. AN EKC hypothesis for the Indian electricity sector is non-existent from the period of 1996–2020 with respect to the GDP but decouples from capital growth on either side of the 2008 financial crisis.
4. Inflation and TROP decoupling directions are inversely related in economic regimes across economic shocks, showing recoupling in the post-crisis regime. These factors are responsible for the rebound of the economy, at the expense of decarbonization incentives in the Indian 3E nexus.
5. GDP and CO_2 emissions are not directly causal. This study found the existence of a negative feedback loop from the CPI to TROP to CO_2 emissions that inhibits the positive effect of decoupling by capital growth.

The following policy pathway can be proposed for the fast-growing, developing economy of India from the empirical evidence that was gathered in this study to ensure continued progress towards 'net-zero' targets in the post-COVID-19 regime:

a. Risk hedging of RE investments and energy sustainability markets must be practiced during a crisis, such that the recoupling of emissions with increasing inflation in a post-crisis demand surge can be prevented.
b. Macroeconomic decoupling of power generation should be tied to capital formation, rather than GDP growth incentives. Capital growth, both domestic and FDI, can accelerate RE infrastructure and introduce innovations in capacity factors, such as giga-scale solar projects and thorium-based heavy-water nuclear power reactors.

c. Economic rebound should not be fostered by increased FF imports in the post-crisis period but by investments in energy transition technologies like electric vehicles and solar manufacturing, which ultimately creates capital that enables decoupling.
d. Post-crisis reliance on FF imports can be reduced by removing all forms of subsidies on oil and gas sectors, such that a quick turnaround of the economy does not occur at the expense of decarbonization efforts in the electricity sector.

In terms of the limitations of this study, the tested cointegration models were not able to reproduce economic movements that lay in recession phases. Future studies should consider stochastic time series that exhibit complimentary trends during growth and recession periods, such that 3E systems are able to minimize the reproduction errors of recession phases. A second limitation is that the impact of inflation on overall energy use needs to be delineated to introduce comprehensive decoupling pathways in the face of economic shocks. More expansive theoretical bases, involving factors that were not considered in the modelling in this paper or using more critical methodological approaches, such as data assimilation, should be explored. Thirdly, the financial policies of every major developing nation are vastly different, implying that the macroeconomic dynamics of inflation and CO_2 emissions may be quite different than those of India. There are several macroeconomic dynamics that are required to be addressed in future studies for developing the progression of economies' decoupling processes. How can social engineering be performed to artificially control inflation with a rebound effect in crisis regimes and prevent it from affecting decarbonization initiatives? The interplay of labour dynamics, wage equity, and social justice as concerns CPI-TROP-CO_2 emissions needs to be explored to determine the role of human and social capital in decoupling dynamics.

It is indeed easy to foster economic growth from imported FFs and subsidies on oil use in a post-COVID-19 recovery scenario, but this study clearly shows that such actions will inhibit decoupling in the long-run. Such short-run economic growth will ultimately lead to uneconomic growth in the pathway to achieve net-zero targets.

Supplementary Materials: The following supporting information can be downloaded at https://doi.org/10.17632/vyxs2c4nfw.1 (accessed on 26 May 2024) (contains codes and data for replication of the results). The Appendices A–E containing the detailed results of the modelling are included as Supplementary Materials also.

Author Contributions: Conceptualization, S.B. and K.I.; methodology, S.B. and K.I.; software, S.B.; validation, S.B., T.O., H.O. and K.I.; formal analysis, S.B.; investigation, S.B.; resources, S.B.; data curation, S.B.; writing—original draft preparation, S.B.; writing—review and editing, T.O., H.O. and K.I.; visualization, S.B.; supervision, K.I; project administration, H.O.; funding acquisition, S.B. All authors have read and agreed to the published version of the manuscript.

Funding: This work was supported by JST SPRING, Grant Number JPMJSP2110.

Data Availability Statement: The original contributions presented in the study are included in the article/Supplementary Materials. Further inquiries can be directed to the first author (Soumya Basu).

Acknowledgments: The authors are grateful to the data repository agencies in India for enabling the explorative analysis that was performed in this study. The authors are also grateful to the Japan Science and Technology Agency for their support of this work.

Conflicts of Interest: The authors declare no conflicts of interest.

Appendix A. Model 1

Table A1. Lag length selection for Model 1 UVAR estimation.

Lags	Log Likelihood	AIC	BIC
0	355.56	−705.12	−699.32
1	577.96	−1131.9	−1109.0

Table A1. *Cont.*

Lags	Log Likelihood	AIC	BIC
2	588.32 *	−1134.7 *	−1109.9 *
3	583.09	−1116.2	−1080.0
4	579.52	−1081.0	−1008.9
5	578.42	−1060.8	−973.06

*: Denotes selection of the lag order.

Table A2. Johansen Cointegration test results for Model 1 (using H1* model) pre-2008.

Rank	Trace Statistic	Maximum Eigen Stat.	Eigen Value
R = 0	65.45 *	40.72 *	0.564
R ≤ 1	24.73 **	17.48 **	0.300
R ≤ 2	7.255	7.255	0.138

*, **: Denotes significance at the 1% and 5% levels, respectively.

Table A3. Results of the VECM long- and short-run estimations of Model 1 pre-2008.

Independent Variable	Dependent Variable		
	ΔLGDP	ΔLELEC	ΔLEMS
Constant	0.266 *	−0.041	0.141 *
LGDP(−1)	0.123 *	−0.019	0.065 *
LELEC(−1)	0.076 *	−0.012	0.040 *
LEMS(−1)	−0.323 *	0.050	−0.171 *
ΔLGDP(−1)	−0.123	0.793 *	−0.201 *
ΔLELEC(−1)	−0.142 **	−0.406 *	−0.029
ΔLEMS(−1)	0.082	−0.025	0.182 ***
Log-Likelihood		565.96	
AIC		−1099.9	
BIC		−1069.7	
Diagnostic Tests			
JB	0.730 #	0.799 #	1.342 #
Q (LBQ)	26.16 a,#	26.05 b,#	60.49 b,***
ARCH	0.306 #	17.76 *	2.645 #

*: Significant at the 1% level. **: Significant at the 5% level. ***: Significant at the 10% level. #: Significant above the 10% level. a: 20 lags involved in the Monte Carlo Auto-correlation test. b: 45 lags involved in the Monte Carlo Auto-correlation test.

Table A4. Johansen Cointegration test results for Model 1 (using H1* model) post-2011.

Rank	Trace Statistic	Maximum Eigen Stat.	Eigen Value
R = 0	24.84	16.45	0.468
R ≤ 1	8.386	5.042	0.176
R ≤ 2	3.344 **	3.344 **	0.120

**: Denotes significance at the 1% and 5% levels, respectively.

Table A5. Results of the VECM long- and short-run estimations of Model 1 post-2011.

Independent Variable	Dependent Variable		
	ΔLGDP	ΔLELEC	ΔLEMS
Constant	0.048 *	0.225 *	0.427 *
LGDP(−1)	−0.027 *	0.073 *	0.225 *
LELEC(−1)	0.033 *	−0.065 *	−0.220 *
LEMS(−1)	−0.026 *	−0.122 *	−0.216 *
ΔLGDP(−1)	0.285	−0.217	−1.840 *
ΔLELEC(−1)	−0.063 ***	−0.051	0.458 *
ΔLEMS(−1)	0.091 ***	−0.086	0.260
ΔLGDP(−2)	0.002	1.165	1.206 **
ΔLELEC(−2)	−0.084 **	−0.125	0.085
ΔLEMS(−2)	−0.102 **	−0.025	0.066
Log-Likelihood		325.91	
AIC		−609.82	
BIC		−583.40	
Diagnostic Tests			
JB	2.934 ***	9.972 *	0.103 #
Q (LBQ)	15.35 a,#	8.196 a,#	15.13 a,#
ARCH	0.002 #	0.005 #	0.409 #

*: Significant at the 1% level. **: Significant at the 5% level. ***: Significant at the 10% level. #: Significant above the 10% level. a: 20 lags involved in the Monte Carlo Auto-correlation test.

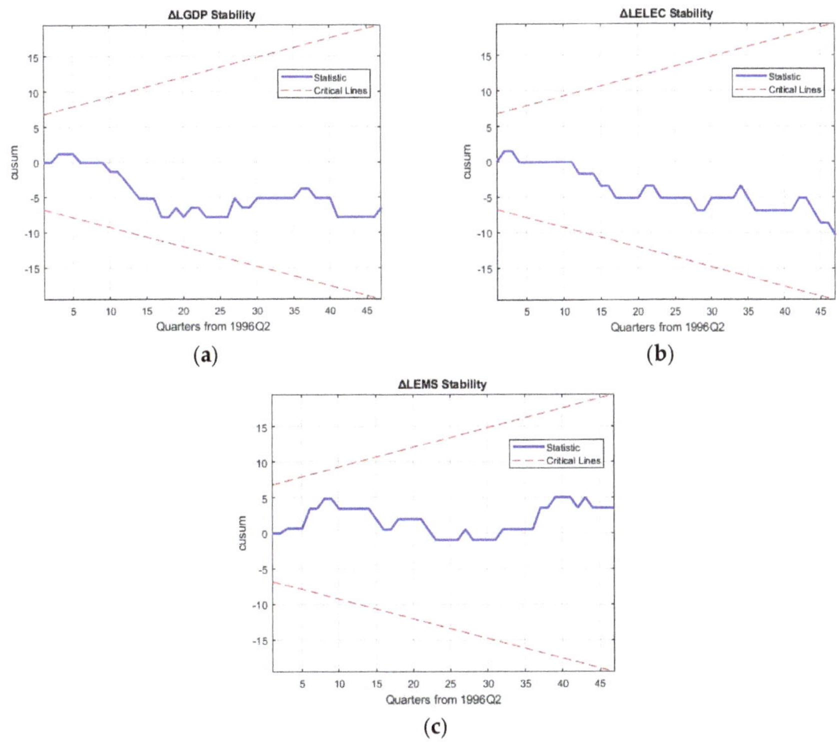

Figure A1. Plot of cumulative sum of parameters for Model 1 for India from 1996Q2 to 2008Q4 for (**a**) Gross Domestic Product, (**b**) Electricity Consumption, and (**c**) Electricity-related CO_2 Emissions.

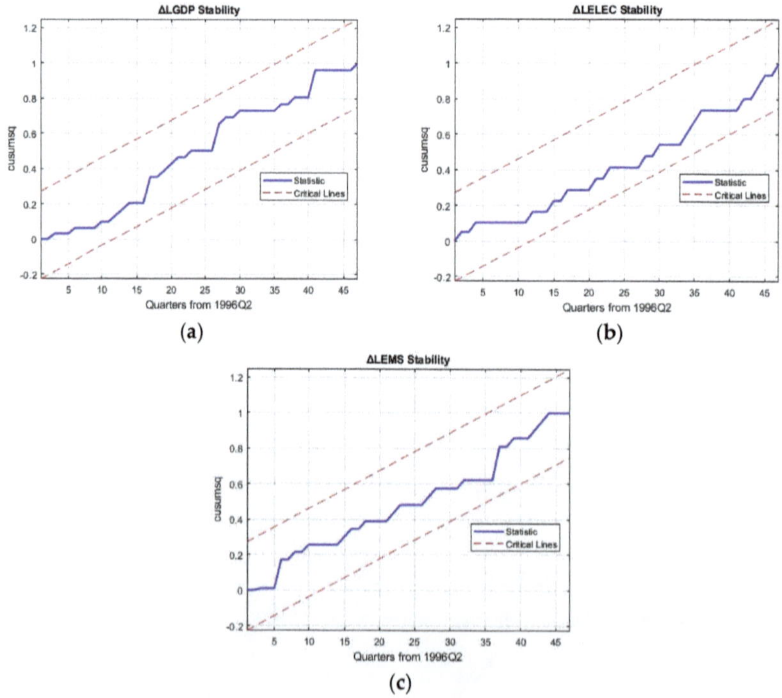

(**a**) (**b**) (**c**)

Figure A2. Plot of cumulative sum of squares of parameters for Model 1 for India from 1996Q2 to 2008Q4 for (**a**) Gross Domestic Product, (**b**) Electricity Consumption, (**c**) and Electricity-related CO_2 Emissions.

Appendix B. Model 2

Table A6. Lag length selection for Model 2 UVAR estimation.

Lags	Log Likelihood	AIC	BIC
0	484.92	−967.83	−954.10
1	750.43	−1460.9	−1422.6
2	750.92	−1433.9	−1365.7
3	751.37	−1456.7	−1338.4
4	752.25 *	−1498.5 *	−1401.7 *
5	747.76	−1327.5	−1173.9

*: Denotes selection of the lag order.

Table A7. Johansen Cointegration test results for Model 2 (using H1 model) pre-2008.

Rank	Trace Statistic	Maximum Eigen Stat.	Eigen Value
R = 0	52.92 *	28.29 **	0.378
R ≤ 1	30.63 **	21.55 **	0.312
R ≤ 2	13.07	12.95 ***	0.241
R ≤ 3	0.121	0.121	0.003

*, **, ***: Denotes significance at the 1%, 5%, and 10% levels, respectively.

Table A8. Results of the VECM long- and short-run estimations of Model 2 pre-2008.

Independent Variable	Dependent Variable			
	ΔLGDP	ΔLCAP	ΔLELEC	ΔLEMS
Constant	0.004	−0.064 *	0.025 ***	−0.043 *
LGDP(−1)	0.0001	0.0853 *	−0.028	0.0506 *
LCAP(−1)	−0.0002	−0.1292 *	0.0424	−0.0766 *
LELEC(−1)	0.0006	0.377 *	−0.1237	0.2236 *
LEMS(−1)	−0.0005	−0.2876 *	0.0944	−0.1706 *
ΔLGDP(−1)	0.085	0.236	0.54 **	0.052
ΔLCAP(−1)	0.073	0.070	−0.006	0.074
ΔLELEC(−1)	−0.041	−0.202	−0.187	0.170 ***
ΔLEMS(−1)	−0.189 ***	−0.029	0.212	0.135
ΔLGDP(−2)	0.377 *	0.242	−0.883*	0.054
ΔLCAP(−2)	0.075	0.062	0.034	0.037
ΔLELEC(−2)	0.200 **	0.067	−0.32 **	0.283 *
ΔLEMS(−2)	−0.011	0.084	0.038	−0.221 **
ΔLGDP(−3)	0.152	0.135	0.705 *	0.525 *
ΔLCAP(−3)	−0.004	0.376 **	−0.106	−0.05
ΔLELEC(−3)	0.001	−0.062	−0.254 ***	0.183 ***
ΔLEMS(−3)	−0.100	−0.038	−0.073	0.133
Log-Likelihood	734.94			
AIC	−1349.9			
BIC	−1238.9			
Diagnostic Tests				
JB	0.489 #	4.483 ***	2.232 #	0.173 #
Q (LBQ)	48.96 b,***	16.48 a,#	40.74 b,#	15.99 a,#
ARCH	2.467 #	9.245 *	0.378 #	6.761 **

*: Significant at the 1% level. **: Significant at the 5% level. ***: Significant at the 10% level. #: Significant above the 10% level. [a]: 20 lags involved in the Monte Carlo Auto-correlation test. [b]: 35 lags involved in the Monte Carlo Auto-correlation test.

Table A9. Johansen Cointegration test results for Model 2 (using H1 model) post-2011.

Rank	Trace Statistic	Maximum Eigen Stat.	Eigen Value
R = 0	108.7 *	54.26 *	0.886
R ≤ 1	54.50 *	34.95 *	0.753
R ≤ 2	19.55 **	19.27 *	0.537
R ≤ 3	0.275	0.275	0.011

*, **: Denotes significance at the 1% and 5% levels, respectively.

Table A10. Results of the VECM long- and short-run estimations of Model 2 post-2011.

Independent Variable	Dependent Variable			
	ΔLGDP	ΔLCAP	ΔLELEC	ΔLEMS
Constant	0.469 *	0.548 *	1.103 *	−0.006
LGDP(−1)	0.091 *	0.213 *	1.203 *	0.335 *

Table A10. *Cont.*

Independent Variable	Dependent Variable			
	ΔLGDP	ΔLCAP	ΔLELEC	ΔLEMS
LCAP(−1)	−0.071 *	−0.168 *	−0.950 *	−0.266 *
LELEC(−1)	−0.101 *	−0.204 *	−1.050 *	−0.274 *
LEMS(−1)	−0.137 *	−0.088 *	0.220 *	0.190 *
ΔLGDP(−1)	−0.564 *	−1.500 ***	−0.864 **	−0.520
ΔLCAP(−1)	0.135 *	0.036	0.878 *	0.057
ΔLELEC(−1)	−0.078 *	−0.042	0.555 *	0.441 *
ΔLEMS(−1)	0.313 *	0.794 *	−0.809 *	0.177
ΔLGDP(−2)	−0.735 *	−0.938	0.303	2.594 *
ΔLCAP(−2)	0.295 *	0.535 *	0.586 *	−0.013
ΔLELEC(−2)	−0.128 *	−0.612 *	0.232 *	0.004
ΔLEMS(−2)	−0.021	−0.158	−0.015	−0.213
ΔLGDP(−3)	−0.742 *	−1.758 ***	5.402 *	1.389 ***
ΔLCAP(−3)	0.289 *	0.357 **	−0.609 *	−0.421 *
ΔLELEC(−3)	−0.041	0.050	0.127	0.024
ΔLEMS(−3)	0.020	−0.026	−0.047	−0.026
Log-Likelihood	455.19			
AIC	−806.39			
BIC	−743.01			
Diagnostic Tests				
JB	0.750 #	0.994#	0.734 #	1.588 #
Q (LBQ)	11.24 a,#	16.17 a,#	26.77 b,***	16.81 a,#
ARCH	0.757 #	1.798 #	0.112 #	0.004 #

*: Significant at the 1% level. **: Significant at the 5% level. ***: Significant at the 10% level. #: Significant above the 10% level. a: 20 lags involved in the Monte Carlo Auto-correlation test. b: 35 lags involved in the Monte Carlo Auto-correlation test.

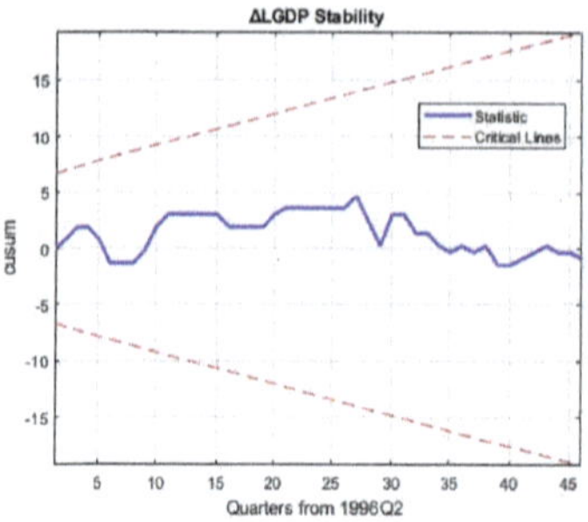

(a)

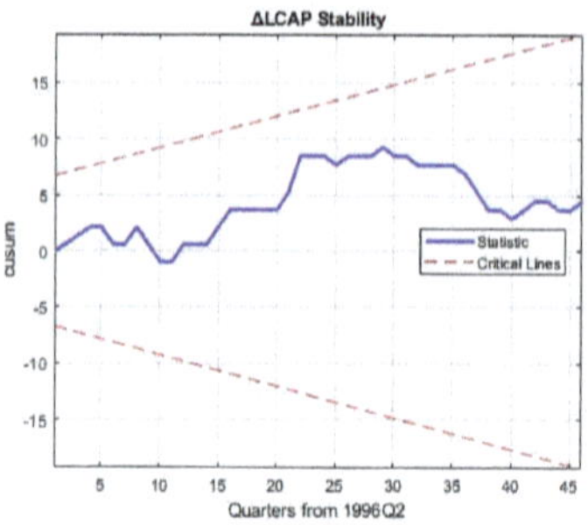

(b)

Figure A3. *Cont.*

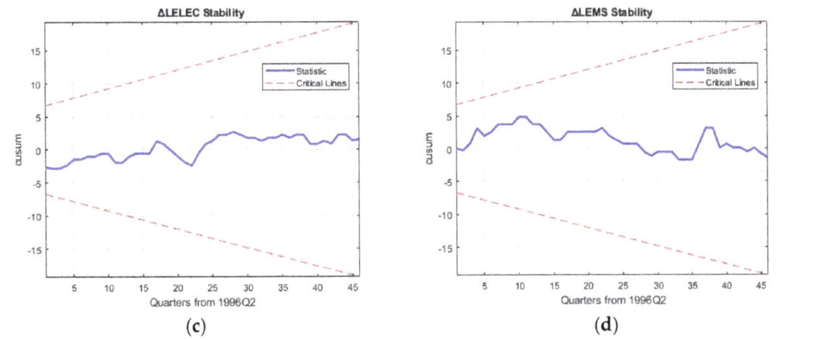

Figure A3. Plot of cumulative sum of parameters for Model 2 for India from 1996Q2 to 2008Q4 for (**a**) Gross Domestic Product, (**b**) Capital formation, (**c**) Electricity Consumption, (**d**) and Electricity-related CO_2 Emissions.

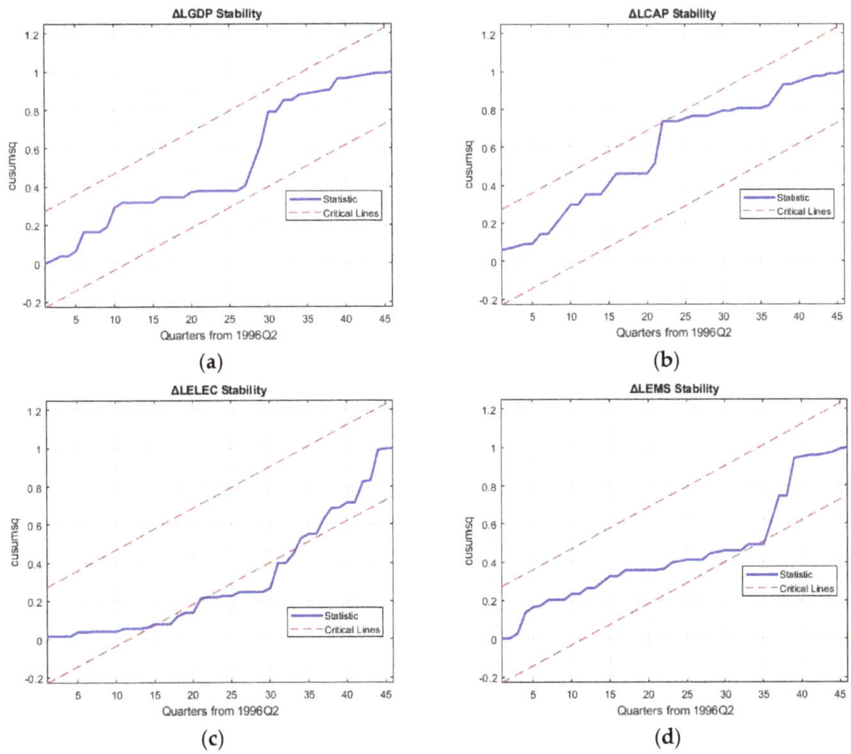

Figure A4. Plot of cumulative sum of squares of parameters for Model 2 for India from 1996Q2 to 2008Q4 for (**a**) Gross Domestic Product, (**b**) Capital formation, (**c**) Electricity Consumption, and (**d**) Electricity-related CO_2 Emissions.

Appendix C. Model 3

Table A11. Lag length selection for Model 3 UVAR estimation.

Lags	Log Likelihood	AIC	BIC
0	475.47	−942.95	−935.22
1	714.18	−1388.4	−1350.1
2	714.93	1357.9	−1289.8
3	721.05	−1338.1	−1240.8
4	718.76	−1301.5	−1175.7
5	720.59	−1273.2	−1119.6
6	834.60 *	−1373.2 *	−1416.1 *

*: Denotes selection of the lag order.

Table A12. Johansen Cointegration test results for Model 3 (using H1 model) pre-2008.

Rank	Trace Statistic	Maximum Eigen Stat.	Eigen Value
R = 0	69.79 *	43.71 *	0.621
R ≤ 1	26.08	15.67	0.294
R ≤ 2	10.41	6.169	0.128
R ≤ 3	4.242 **	4.242 **	0.090

*, **: Denotes significance at the 1% and 5% levels, respectively.

Table A13. Results of the VECM long- and short-run estimations of Model 3 pre-2008.

Independent Variable	Dependent Variable			
	ΔLGDP	ΔTROP	ΔLELEC	ΔLEMS
Constant	0.125	−1.123 *	0.039	0.334 *
LGDP(−1)	0.041	−0.045	−0.088	0.215 *
TROP (−1)	0.028	−0.604 *	0.118 **	−0.062 **
LELEC(−1)	−0.095	0.145	0.095	0.352 *
LEMS(−1)	−0.012	0.645	−0.070	−0.703 *
ΔLGDP(−1)	0.060	1.744 **	0.775 *	−0.377 **
ΔTROP (−1)	0.029	0.224	−0.109 ***	0.104 *
ΔLELEC(−1)	0.112	0.106	−0.676 *	0.072
ΔLEMS(−1)	−0.040	−1.755 *	0.461 **	0.494 *
ΔLGDP(−2)	0.227	1.327 ***	−0.474 **	−0.342 **
ΔTROP (−2)	−0.071 **	0.019	−0.096 ***	0.011
ΔLELEC(−2)	0.263 **	0.229	−0.831 *	0.267 **
ΔLEMS(−2)	−0.176	−1.884 **	0.297	0.036
ΔLGDP(-3)	0.054	0.961	0.552 **	0.002
ΔTROP (-3)	0.026	0.160	−0.064	−0.004
ΔLELEC(-3)	0.195	1.018	−0.807 *	0.241 ***
ΔLEMS(-3)	0.167	0.498	0.242	−0.016
ΔLGDP(-4)	−0.425 *	−0.045	0.038	−0.116
ΔTROP (-4)	−0.031	−0.404 **	−0.053	0.133 *

Table A13. *Cont.*

Independent Variable	Dependent Variable			
	ΔLGDP	ΔTROP	ΔLELEC	ΔLEMS
ΔLELEC(-4)	0.113	0.690	−0.641 *	0.300 **
ΔLEMS(-4)	−0.230 **	−0.536	−0.211	0.404 *
ΔLGDP(-5)	0.189	1.784 **	0.332	−0.482 *
ΔTROP (-5)	0.059 ***	0.019	−0.033 ***	−0.007 *
ΔLELEC(-5)	−0.011	0.250	−0.289	0.342
ΔLEMS(-5)	0.054	0.764	−0.123	0.111
Log-Likelihood	722.15			
AIC	−1228.3			
BIC	−1033.2			
Diagnostic Tests				
JB	0.877 #	0.814 #	4.771 ***	0.662 #
Q (LBQ)	18.74 a,#	12.40 a,#	23.42 a,#	28.92 a,***
ARCH	0.110 #	5.973 ***	0.281 #	1.037 #

*: Significant at the 1% level. **: Significant at the 5% level. ***: Significant at the 10% level. #: Significant above the 10% level. a: 20 lags involved in the Monte Carlo Auto-correlation test.

Table A14. Johansen Cointegration test results for Model 3 (using H1 model) post-2011.

Rank	Trace Statistic	Maximum Eigen Stat.	Eigen Value
R = 0	95.09 *	42.18 *	0.815
R ≤ 1	52.91 *	31.89 *	0.721
R ≤ 2	21.07 *	14.88 **	0.449
R ≤ 3	6.139 **	6.139 **	0.218

*, **: Denotes significance at the 1% and 5% levels, respectively.

Table A15. Results of the VECM long- and short-run estimations of Model 3 post-2011.

Independent Variable	Dependent Variable			
	ΔLGDP	ΔTROP	ΔLELEC	ΔLEMS
Constant	0.500 *	0.718 *	0.469 *	−0.034 *
LGDP(−1)	0.137 *	1.552 *	0.459 *	0.097 *
TROP (−1)	−0.213 *	0.226 *	−0.213 *	0.312 *
LELEC(−1)	−0.525 *	−1.533 *	−0.976 *	0.497 *
LEMS(−1)	0.197 *	−0.471 *	0.329 *	−0.652 *
ΔLGDP(−1)	−0.311 **	0.265	−1.404	−0.809
ΔTROP (−1)	0.184 *	−1.041 *	−0.065	−0.325 *
ΔLELEC(−1)	0.347 *	0.239	0.645 *	−0.023
ΔLEMS(−1)	0.102 **	0.310	−0.295	0.472 *
ΔLGDP(−2)	−0.232	−2.873	0.892	1.642 *
ΔTROP (−2)	0.124 *	−0.984 *	−0.123	−0.257 *
ΔLELEC(−2)	0.153 *	0.352	0.117	−0.157
ΔLEMS(−2)	−0.117 **	0.341	0.272	−0.054

Table A15. *Cont.*

Independent Variable	Dependent Variable			
	ΔLGDP	**ΔTROP**	**ΔLELEC**	**ΔLEMS**
ΔLGDP(-3)	−0.551 *	6.018 *	0.972	−0.567
ΔTROP (-3)	0.048 *	−0.854 *	0.086	−0.067
ΔLELEC(-3)	0.128 *	0.182	0.126	0.020
ΔLEMS(-3)	0.130 *	−0.501	−0.434 ***	−0.004
Log-Likelihood	426.73			
AIC	−749.45			
BIC	−686.07			
Diagnostic Tests				
JB	1.119 #	0.719 #	0.908 #	0.215 #
Q (LBQ)	20.96 a,#	23.04 a,#	17.41 a,#	23.81 a,#
ARCH	2.478 #	3.894 **	0.283 #	3.962 **

*: Significant at the 1% level. **: Significant at the 5% level. ***: Significant at the 10% level. #: Significant above the 10% level. [a]: 20 lags involved in the Monte Carlo Auto-correlation test.

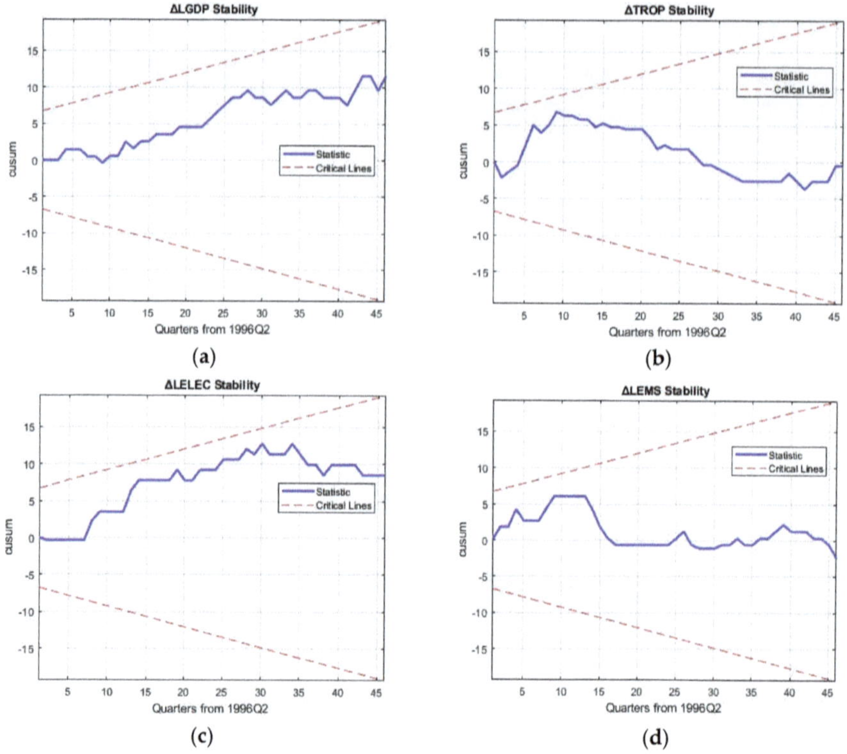

Figure A5. Plot of cumulative sum of parameters for Model 3 for India from 1996Q2 to 2008Q4 for (**a**) Gross Domestic Product, (**b**) Trade Openness, (**c**) Electricity Consumption, and (**d**) Electricity-related CO_2 Emissions.

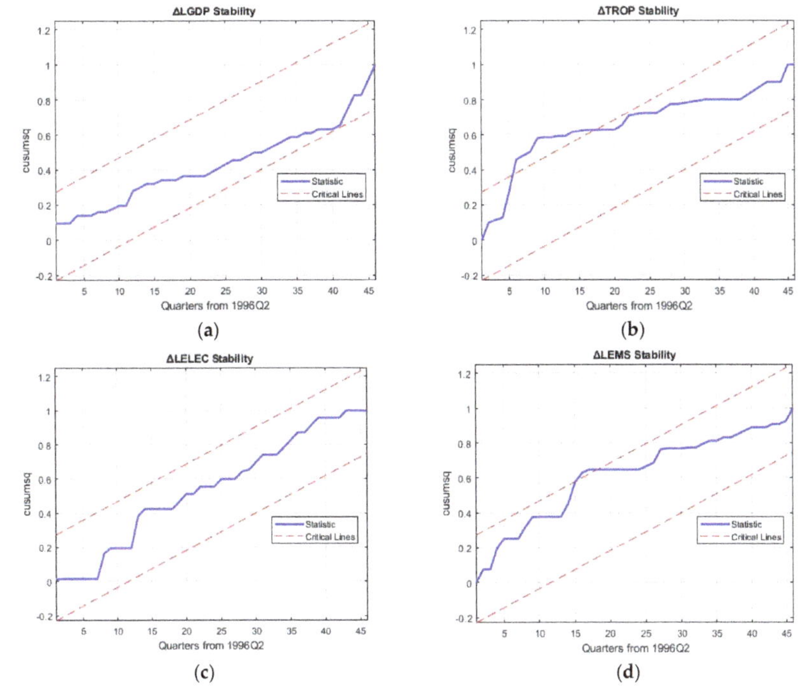

Figure A6. Plot of cumulative sum of squares of parameters for Model 3 for India from 1996Q2 to 2008Q4 for (**a**) Gross Domestic Product, (**b**) Trade Openness, (**c**) Electricity Consumption, and (**d**) Electricity-related CO_2 Emissions.

Appendix D. Model 4

Table A16. Lag length selection for Model 4 UVAR estimation.

Lags	Log Likelihood	AIC	BIC
0	638.70	−1267.4	−1257.7
1	923.86	−1787.7	−1730.4
2	982.57	−1855.1	−1751.1
3	999.76 *	−1839.5 *	−1789.8 *
4	1001.9	−1793.7	−1599.5

*: Denotes selection of the lag order (the BIC is preferred over the AIC in this case).

Table A17. Johansen Cointegration test results for Model 4 (using H1 model) pre-2008.

Rank	Trace Statistic	Maximum Eigen Stat.	Eigen Value
R = 0	101.2 *	40.80 *	0.573
R ≤ 1	60.41 *	22.77	0.378
R ≤ 2	37.64 *	17.12 ***	0.300
R ≤ 3	20.52 *	14.75 **	0.265
R ≤ 4	5.778 **	5.778 **	0.113

*, **, ***: Denotes significance at the 1%, 5%, and 10% levels, respectively.

Table A18. Results of the VECM long- and short-run estimations of Model 4 pre-2008.

Independent Variable	Dependent Variable				
	ΔLGDP	ΔLCPI	ΔTROP	ΔLELEC	ΔLEMS
Constant	0.170	0.328 ***	1.657 **	0.762 *	0.131
LGDP(−1)	0.069	0.191 **	1.188 *	0.288 **	0.067
LCPI(−1)	−0.120	−0.399 *	−0.996 **	0.222	−0.136
TROP (−1)	0.012	−0.062	−0.781 *	0.012	0.001
LELEC(−1)	0.236 ***	0.470 *	−0.244	−0.750 *	0.374 *
LEMS(−1)	−0.274 *	−0.416 *	−0.562	−0.137	−0.372 *
ΔLGDP(−1)	−0.045	−0.004	−0.112	0.411	0.075
ΔLCPI(−1)	0.210	0.529 **	1.616 ***	−0.220	−0.632 *
ΔTROP (−1)	0.008	0.123 *	0.325 ***	−0.107 ***	0.103 **
ΔLELEC(−1)	−0.301 **	−0.281 ***	−0.511	−0.045	−0.189
ΔLEMS(−1)	−0.194	−0.325	−2.148 ***	0.035	0.939 *
ΔLGDP(−2)	0.318 **	−0.242	1.267 ***	−0.605 *	−0.077
ΔLCPI(−2)	−0.173	−0.092	−1.890 ***	−0.450	0.281
ΔTROP (−2)	−0.030	0.027	0.354 ***	−0.023	−0.036
ΔLELEC(−2)	0.042	−0.044	−0.331	−0.013	0.039
ΔLEMS(−2)	0.316	−0.093	2.253 ***	0.477	−0.509 ***
Log-Likelihood			996.87		
AIC			−1803.7		
BIC			−1626.0		
Diagnostic Tests					
JB	0.828 #	0.262#	1.511 #	1.492 #	0.208 #
Q (LBQ)	13.70 a,#	16.27 a,#	16.09 a,#	17.71 a,#	18.98 a,#
ARCH	1.074 #	0.209#	3.224 #	1.354 #	1.085 #

*: Significant at the 1% level. **: Significant at the 5% level. ***: Significant at the 10% level. #: Significant above the 10% level. a: 20 lags involved in the Monte Carlo Auto-correlation test.

Table A19. Johansen Cointegration test results for Model 4 (using H1 model) post-2011.

Rank	Trace Statistic	Maximum Eigen Stat.	Eigen Value
R = 0	903.7 *	751.1 *	1.000
R ≤ 1	152.5 *	76.76 *	0.957
R ≤ 2	75.77 *	35.46 *	0.758
R ≤ 3	40.37 *	23.14 *	0.604
R ≤ 4	17.18 *	17.18 *	0.497

*: Denotes significance at the 1% level.

Table A20. Results of the VECM long- and short-run estimations of Model 4 post-2011.

Independent Variable	Dependent Variable				
	ΔLGDP	**ΔLCPI**	**ΔTROP**	**ΔLELEC**	**ΔLEMS**
Constant	NA	NA	NA	NA	NA
LGDP(−1)	NA	NA	NA	NA	NA
LCPI(−1)	NA	NA	NA	NA	NA
TROP (−1)	NA	NA	NA	NA	NA
LELEC(−1)	NA	NA	NA	NA	NA
LEMS(−1)	NA	NA	NA	NA	NA
ΔLGDP(−1)	NA	NA	NA	NA	NA
ΔLCPI(−1)	NA	NA	NA	NA	NA
ΔTROP (−1)	NA	NA	NA	NA	NA
ΔLELEC(−1)	NA	NA	NA	NA	NA
ΔLEMS(−1)	NA	NA	NA	NA	NA
ΔLGDP(−2)	NA	NA	NA	NA	NA
ΔLCPI(−2)	NA	NA	NA	NA	NA
ΔTROP (−2)	NA	NA	NA	NA	NA
ΔLELEC(−2)	NA	NA	NA	NA	NA
ΔLEMS(−2)	NA	NA	NA	NA	NA
Log−Likelihood			NA		
AIC			NA		
BIC			NA		
Diagnostic Tests					
JB	NA	NA	NA	NA	NA
Q (LBQ)	NA	NA	NA	NA	NA
ARCH	NA	NA	NA	NA	NA

Note: The model was not stable, so every value is 'NA' in this time period.

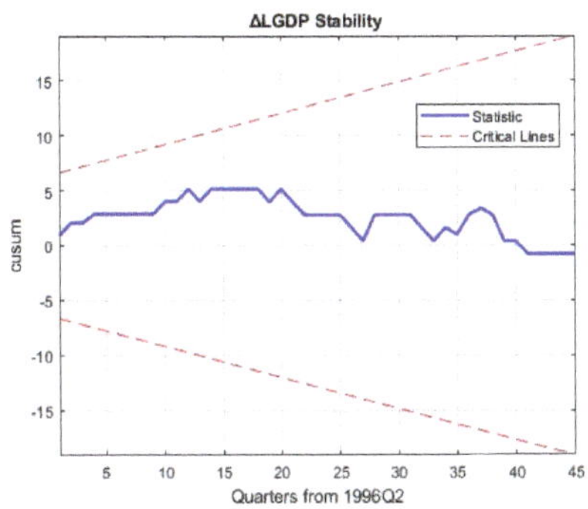

(a)

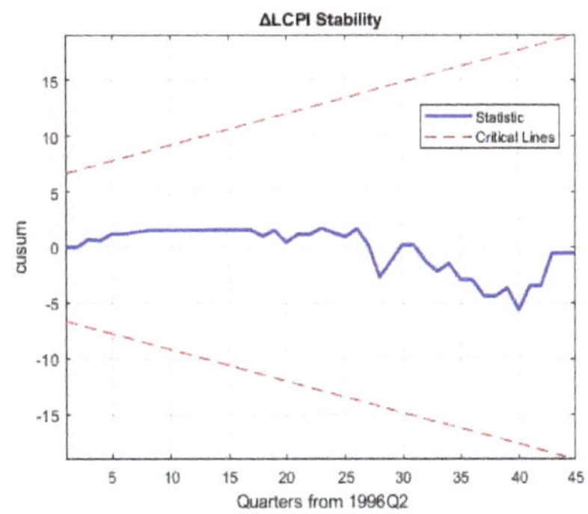

(b)

Figure A7. *Cont.*

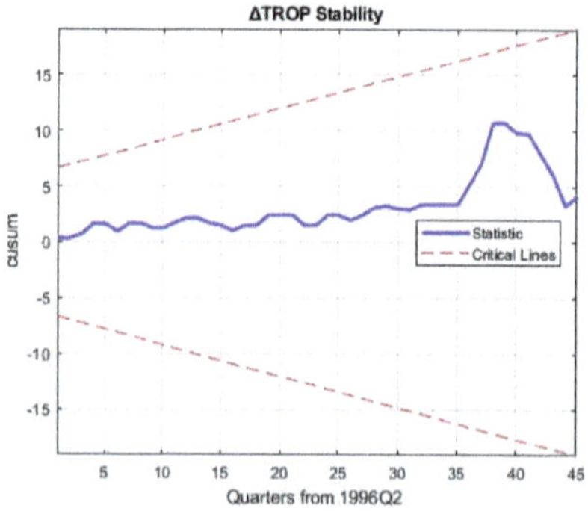

(c)

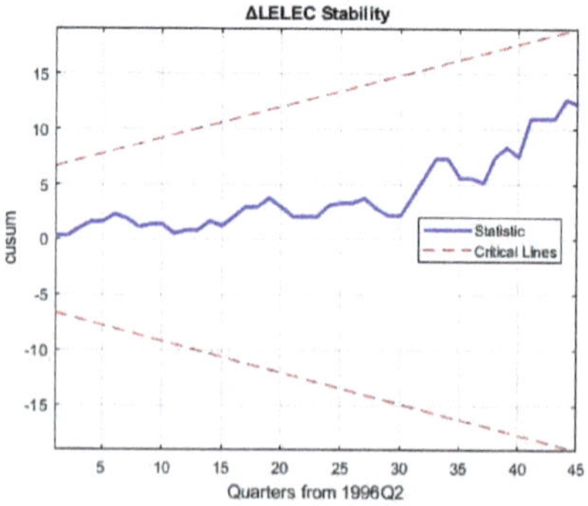

(d)

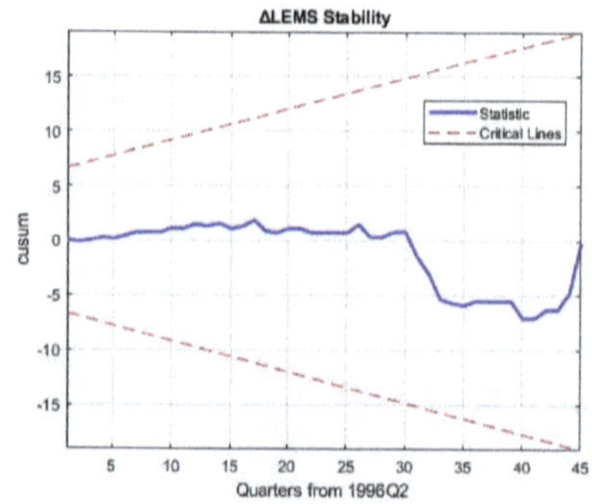

(e)

Figure A7. Plot of cumulative sum of parameters for Model 4 for India from 1996Q2 to 2008Q4 for (**a**) Gross Domestic Product, (**b**) Consumer Price Index, (**c**) Trade Openness, (**d**) Electricity Consumption, and (**e**) Electricity-related CO_2 Emissions.

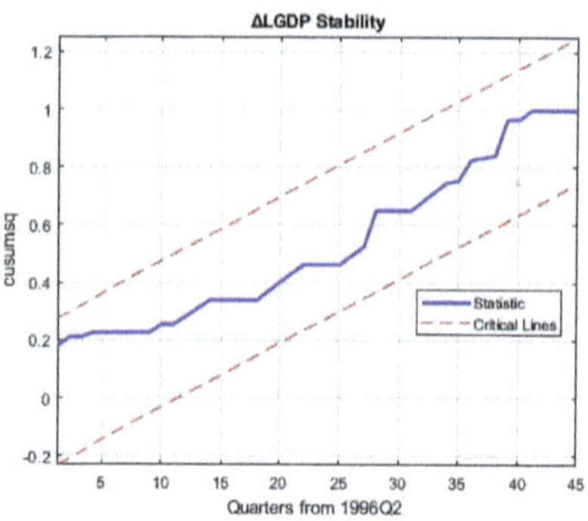

(a)

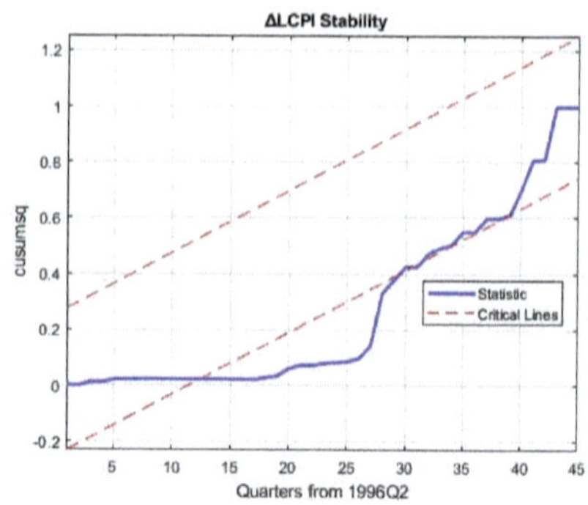

(b)

Figure A8. *Cont.*

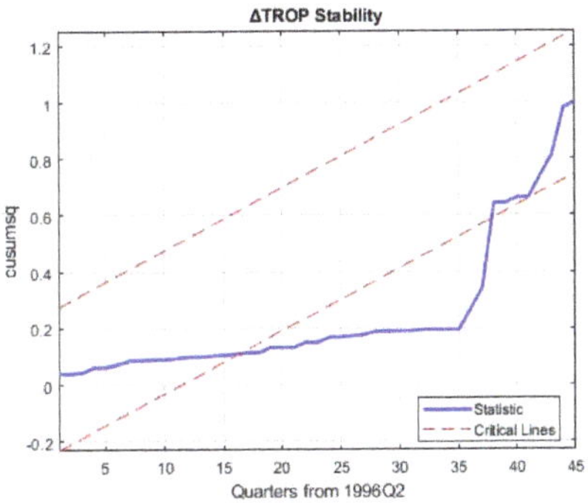

(c)

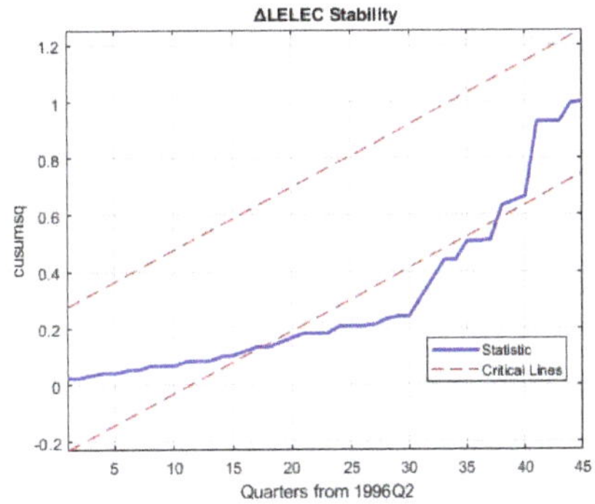

(d)

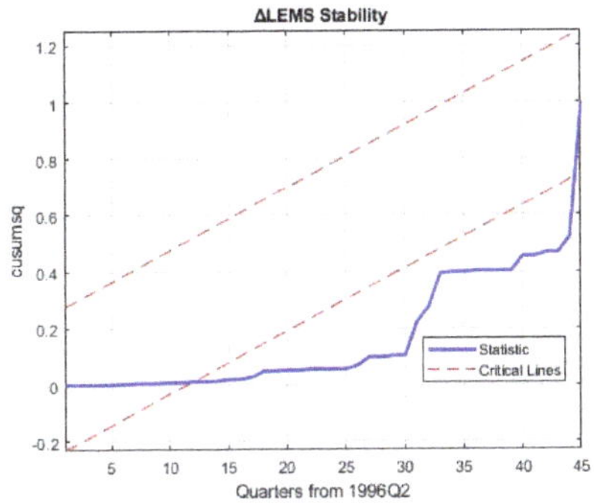

(e)

Figure A8. Plot of cumulative sum of squares of parameters for Model 4 for India from 1996Q2 to 2008Q4 for (**a**) Gross Domestic Product, (**b**) Consumer Price Index, (**c**) Trade Openness, (**d**) Electricity Consumption, and (**e**) Electricity-related CO_2 Emissions.

Appendix E. Model 5

Table A21. Lag length selection for Model 5 UVAR estimation.

Lags	Log Likelihood	AIC	BIC
0	600.56	−1191.1	−1181.5
1	875.48	−1691.0	−1633.6
2	926.33 *	−1742.7 *	−1638.6 *
3	940.26	−1720.5	−1570.8
4	938.15	−1666.3	−1472.0

*: Denotes selection of the lag order.

Table A22. Johansen Cointegration test results for Model 5 (using H1 model) pre-2008.

Rank	Trace Statistic	Maximum Eigen Stat.	Eigen Value
R = 0	124.5 *	54.75 *	0.673
R ≤ 1	69.75 *	36.32 *	0.523
R ≤ 2	33.44 **	20.76 ***	0.345
R ≤ 3	12.68	10.49	0.193
R ≤ 4	2.190	2.190	0.044

*, **, ***: Denotes significance at the 1%, 5%, and 10% levels, respectively.

Table A23. Results of the VECM long- and short-run estimations of Model 5 pre-2008.

Independent Variable	Dependent Variable				
	ΔLCAP	ΔLCPI	ΔTROP	ΔLELEC	ΔLEMS
Constant	0.001	−0.037	−0.340 *	0.082 **	0.043 **
LCAP(−1)	0.017	0.036 *	0.101 *	−0.031 *	0.012 ***
LCPI(−1)	−0.125	−0.267 *	−0.805 *	0.242 *	−0.075
TROP (−1)	−0.014	−0.045 **	−0.287 *	0.065 **	0.022
LELEC(−1)	0.350 ***	0.62 *	0.573	−0.353 **	0.472 *
LEMS(−1)	−0.238	−0.365 *	0.369	0.093	−0.441 *
ΔLCAP(−1)	0.040	0.001	0.174	0.114	0.054
ΔLCPI(−1)	−0.280	0.611 *	0.422	−0.314	−0.324 *
ΔTROP (−1)	−0.081	0.129 *	0.169	−0.119 **	0.089 *
ΔLELEC(−1)	−0.208	−0.375 *	−0.393	−0.259 ***	−0.281 *
ΔLEMS(−1)	0.345	−0.457 *	−0.604	0.193	0.602 *
Log-Likelihood			909.61		
AIC			−1719.2		
BIC			−1624.6		
Diagnostic Tests					
JB	1.103 #	0.284 #	0.200 #	0.697 #	0.788 #
Q (LBQ)	49.20 b,#	37.36 a,**	12.92 a,#	26.69 a,#	31.40 a,***
ARCH	1.658 #	2.213 #	6.661 **	0.245 #	0.920 #

*: Significant at the 1% level. **: Significant at the 5% level. ***: Significant at the 10% level. #: Significant above the 10% level. a: 20 lags involved in the Monte Carlo Auto-correlation test. b: 35 lags involved in the Monte Carlo Auto-correlation test.

Table A24. Johansen Cointegration test results for Model 5 (using H1 model) post-2011.

Rank	Trace Statistic	Maximum Eigen Stat.	Eigen Value
R = 0	149.9 *	58.69 *	0.895
R ≤ 1	91.20 *	45.17 *	0.824
R ≤ 2	46.00 *	32.66 *	0.715
R ≤ 3	13.37	11.56	0.359
R ≤ 4	1.813	1.813	0.067

*: Denotes significance at 1% level.

Table A25. Results of the VECM long- and short-run estimations of Model 5 post-2011.

Independent Variable	Dependent Variable				
	ΔLCAP	ΔLCPI	ΔTROP	ΔLELEC	ΔLEMS
Constant	0.187 *	0.098 *	−0.416 *	−0.084 *	−0.009 *
LCAP(−1)	0.122 *	−0.037 *	−0.035 *	−0.101 *	−0.065 *
LCPI(−1)	−1.242 *	−0.568 *	2.694 *	0.663 *	0.137 *
TROP (−1)	−0.100 *	0.228 *	−0.460 *	0.158 *	0.162 *
LELEC(−1)	0.705 *	0.632 *	−2.295 *	−0.258 *	0.093 *
LEMS(−1)	0.344 *	−0.124 *	−0.050 *	−0.292 *	−0.193 *

Table A25. *Cont.*

Independent Variable	Dependent Variable				
	ΔLCAP	ΔLCPI	ΔTROP	ΔLELEC	ΔLEMS
ΔLCAP(−1)	−0.610 *	0.098 ***	0.207	0.327 **	0.136 ***
ΔLCPI(−1)	−0.497 *	0.163	2.527 *	−0.892 *	−0.874 *
ΔTROP (−1)	−0.201 *	−0.139 *	0.140	−0.137 *	0.006
ΔLELEC(−1)	−0.464 *	−0.263 *	1.551 *	−0.297 **	0.057
ΔLEMS(−1)	2.105 *	−0.024	−2.421 **	0.641 **	0.642 *
ΔLCAP(−2)	−0.005	0.413 *	0.784 ***	0.682 *	0.468 *
ΔLCPI(−2)	2.990 *	−0.235 ***	−5.899 *	−0.811 **	−0.673 *
ΔTROP (−2)	−0.242 *	−0.076 *	0.426 *	−0.146 *	0.025
ΔLELEC(−2)	−0.749 *	−0.238 *	0.035	−0.406 *	−0.126
ΔLEMS(−2)	−2.180 *	−0.471 *	3.654 *	0.070	−0.195
Log-Likelihood			541.31		
AIC			−972.61		
BIC			−903.42		
Diagnostic Tests					
JB	2.144 #	5.459 **	6.302 **	0.229 #	2.201 #
Q (LBQ)	23.74 a,#	14.92 a,#	14.85 a,#	29.62 b,#	12.89 a,#
ARCH	0.803 #	0.470 #	0 #	4.700 **	0.348 #

*: Significant at the 1% level. **: Significant at the 5% level. ***: Significant at the 10% level. #: Significant above the 10% level. [a]: 20 lags involved in the Monte Carlo Auto-correlation test. [b]: 35 lags involved in the Monte Carlo Auto-correlation test.

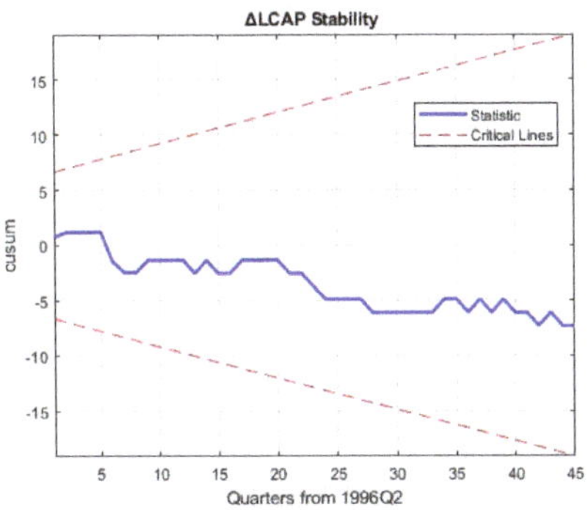

(a)

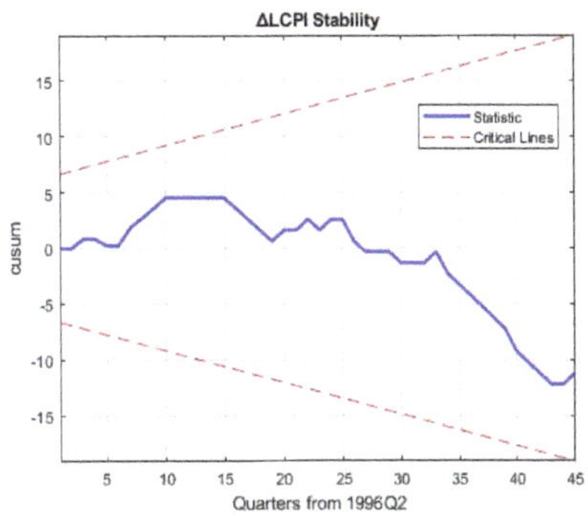

(b)

Figure A9. *Cont.*

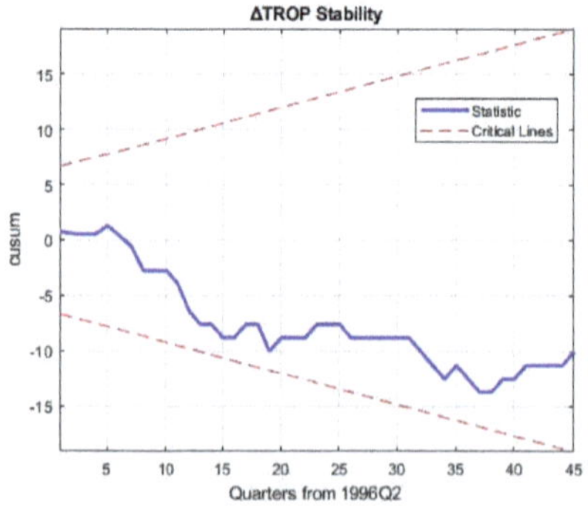

(c)

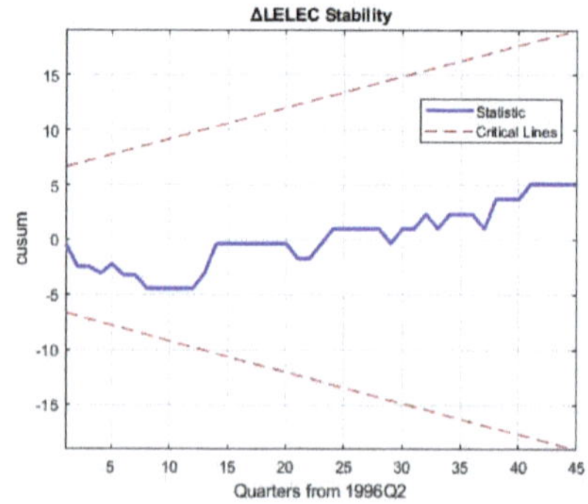

(d)

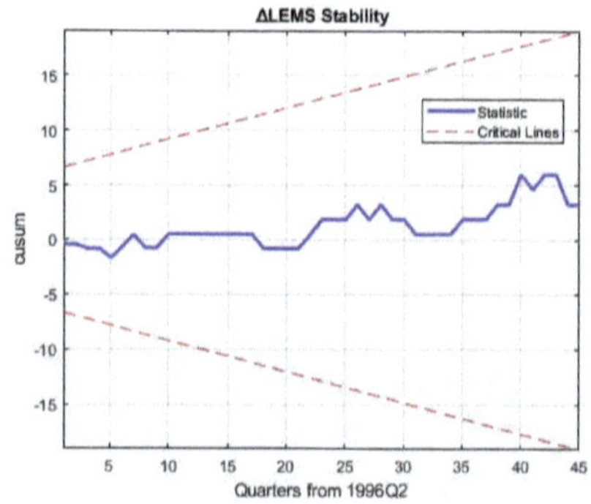

(e)

Figure A9. Plot of cumulative sum of parameters for Model 5 for India from 1996Q2 to 2008Q4 for (**a**) Capital formation, (**b**) Consumer Price Index, (**c**) Trade Openness, (**d**) Electricity Consumption, and (**e**) Electricity-related CO_2 Emissions.

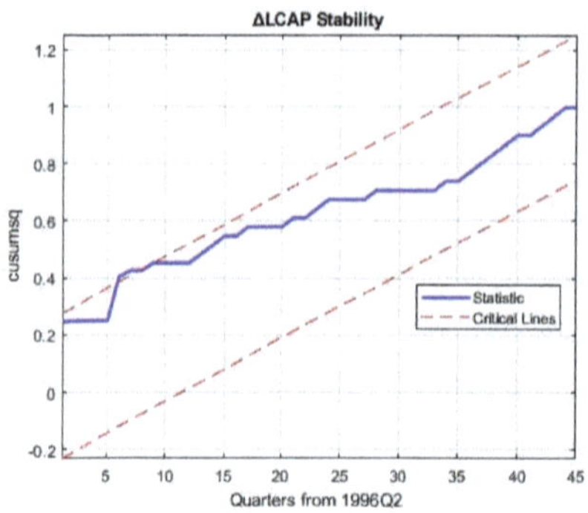

(a)

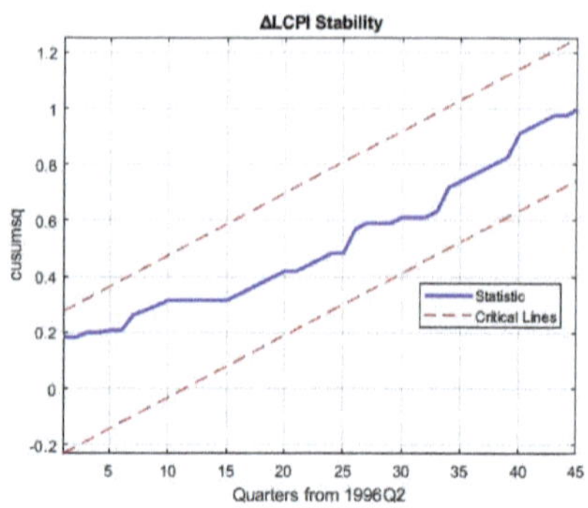

(b)

Figure A10. *Cont.*

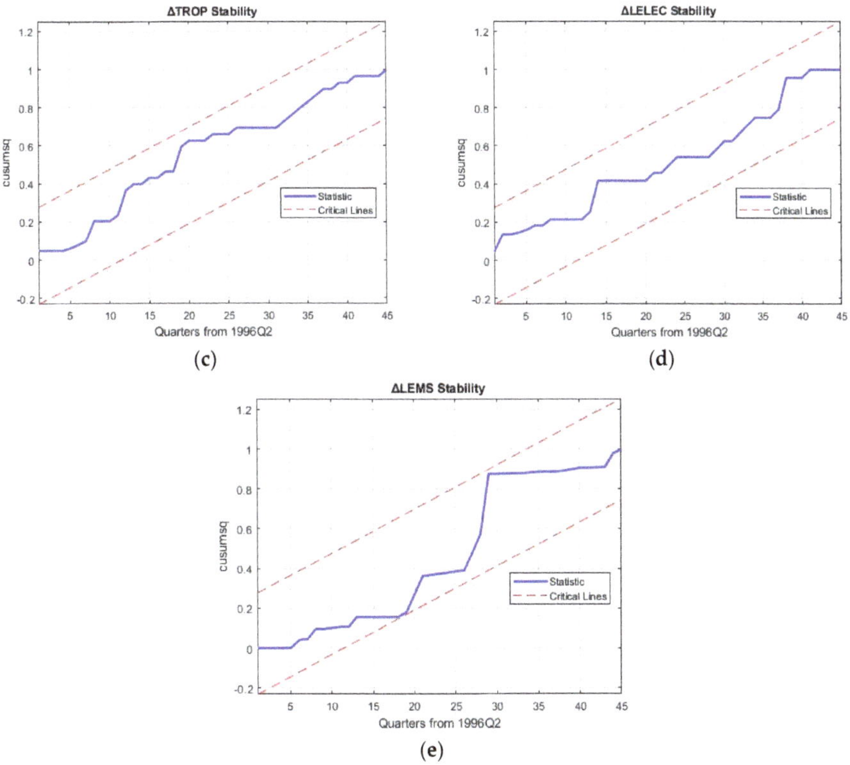

Figure A10. Plot of cumulative sum of squares of parameters for Model 5 for India from 1996Q2 to 2008Q4 for (**a**) Capital formation, (**b**) Consumer Price Index, (**c**) Trade Openness, (**d**) Electricity Consumption, and (**e**) Electricity-related CO_2 Emissions.

References

1. Ibrahim, H.A.; Ayomoh, M.K.; Bansal, R.C.; Gitau, M.N.; Yadavalli, V.S.S.; Naidoo, R. Sustainability of power generation for developing economies: A systematic review of power sources mix. *Energy Strateg. Rev.* **2023**, *47*, 101085. [CrossRef]
2. Fu, Q.; Liu, L.; Wang, H. Role of fossil fuels resources on high-quality economic development: Evidence from China. *Resour. Policy* **2023**, *86*, 104126. [CrossRef]
3. Basu, S.; Ogawa, T.; Ishihara, K.N. The methods and factors of decoupling energy usage and economic growth. In *Waste-to-Energy Approaches Towards Zero Waste*; Elsevier: Amsterdam, The Netherlands, 2022; pp. 269–313. [CrossRef]
4. Paul, S.; Bhattacharya, R.N. Causality between energy consumption and economic growth in India: A note on conflicting results. *Energy Econ.* **2004**, *26*, 977–983. [CrossRef]
5. Jamel, L.; Derbali, A. Do energy consumption and economic growth lead to environmental degradation? Evidence from Asian economies. *Cogent Econ. Financ.* **2016**, *4*, 1170653. [CrossRef]
6. Dong, F.; Li, J.; Wang, Y.; Zhang, X.; Zhang, S.; Zhang, S. Drivers of the decoupling indicator between the economic growth and energy-related CO_2 in China: A revisit from the perspectives of decomposition and spatiotemporal heterogeneity. *Sci. Total Environ.* **2019**, *685*, 631–658. [CrossRef] [PubMed]
7. Holdren, J.P. Population and the energy problem. *Popul. Environ.* **1991**, *12*, 231–255. [CrossRef]
8. Marques, A.C.; Fuinhas, J.A.; Leal, P.A. The impact of economic growth on CO_2 emissions in Australia: The environmental Kuznets curve and the decoupling index. *Environ. Sci. Pollut. Res.* **2018**, *25*, 27283–27296. [CrossRef] [PubMed]
9. Mikayilov, J.I.; Hasanov, F.J.; Galeotti, M. Decoupling of CO_2 emissions and GDP: A time-varying cointegration approach. *Ecol. Indic.* **2018**, *95*, 615–628. [CrossRef]
10. Mohamed, M.M.A.; Liu, P.; Nie, G. Causality between Technological Innovation and Economic Growth: Evidence from the Economies of Developing Countries. *Sustainability* **2022**, *14*, 3586. [CrossRef]
11. Zeng, J. Reexamining the interaction between innovation and capital accumulation. *J. Macroecon.* **2003**, *25*, 541–560. [CrossRef]
12. Işik, C.; Kasımatı, E.; Ongan, S. Analyzing the causalities between economic growth, financial development, international trade, tourism expenditure and/on the CO_2 emissions in Greece. *Energy Sources Part B Econ. Plan. Policy* **2017**, *12*, 665–673. [CrossRef]

13. Shahbaz, M.; Khan, S.; Tahir, M.I. The dynamic links between energy consumption, economic growth, financial development and trade in China: Fresh evidence from multivariate framework analysis. *Energy Econ.* **2013**, *40*, 8–21. [CrossRef]
14. Nepal, R.; Paija, N.; Tyagi, B.; Harvie, C. Energy security, economic growth and environmental sustainability in India: Does FDI and trade openness play a role? *J. Environ. Manag.* **2021**, *281*, 111886. [CrossRef]
15. Abbasi, K.R.; Shahbaz, M.; Jiao, Z.; Tufail, M. How energy consumption, industrial growth, urbanization, and CO_2 emissions affect economic growth in Pakistan? A novel dynamic ARDL simulations approach. *Energy* **2021**, *221*, 119793. [CrossRef]
16. King, R.G.; Watson, M.W. Money, Prices, Interest Rates and the Business Cycle. *Rev. Econ. Stat.* **1996**, *78*, 35. [CrossRef]
17. Neumeyer, P.A.; Perri, F. Business cycles in emerging economies: The role of interest rates. *J. Monet. Econ.* **2005**, *52*, 345–380. [CrossRef]
18. Zhang, P.; Zhao, X.; Sun, L.; Zuo, J.; Wei, W.; Liu, X.; Peng, X.; Shan, Y.; Li, S.; Ge, L.; et al. What can we learn from the 2008 financial crisis for global power decarbonization after COVID-19? *Fundam. Res.* **2023**, *in press*. [CrossRef]
19. Heffron, R.J.; Sheehan, J. Rethinking international taxation and energy policy post COVID-19 and the financial crisis for developing countries. *J. Energy Nat. Resour. Law* **2020**, *38*, 465–473. [CrossRef]
20. Piłatowska, M.; Włodarczyk, A. Decoupling Economic Growth From Carbon Dioxide Emissions in the EU Countries. *Montenegrin J. Econ.* **2018**, *14*, 7–26. [CrossRef]
21. Pao, H.-T.; Chen, C.-C. Decoupling strategies: CO_2 emissions, energy resources, and economic growth in the Group of Twenty. *J. Clean. Prod.* **2019**, *206*, 907–919. [CrossRef]
22. Shahbaz, M.; Lean, H.H.; Farooq, A. Natural gas consumption and economic growth in Pakistan. *Renew. Sustain. Energy Rev.* **2013**, *18*, 87–94. [CrossRef]
23. Andrijevic, M.; Schleussner, C.-F.; Gidden, M.J.; McCollum, D.L.; Rogelj, J. COVID-19 recovery funds dwarf clean energy investment needs. *Science* **2020**, *370*, 298–300. [CrossRef] [PubMed]
24. Andreoni, V. The energy metabolism of countries: Energy efficiency and use in the period that followed the global financial crisis. *Energy Policy* **2020**, *139*, 111304. [CrossRef]
25. World Bank. Poverty Rate by Country. 2024. Available online: https://worldpopulationreview.com/country-rankings/poverty-rate-by-country (accessed on 1 May 2024).
26. IEA. World Energy Outlook, 2020. International Energy Agency. Available online: https://www.iea.org/reports/world-energy-outlook-2020 (accessed on 24 September 2023).
27. Basu, S.; Ogawa, T.; Okumura, H.; Ishihara, K.N. Assessing the geospatial nature of location-dependent costs in installation of solar photovoltaic plants. *Energy Rep.* **2021**, *7*, 4882–4894. [CrossRef]
28. International Energy Agency. India Energy Outlook. 2021. Available online: https://www.iea.org/reports/india-energy-outlook-2021 (accessed on 24 September 2023).
29. Worldometer. GDP by Country. 2023. Available online: https://www.worldometers.info/gdp/gdp-by-country/ (accessed on 24 September 2023).
30. World Bank. GDP (Current US$), India. World Development Indicators. Available online: https://data.worldbank.org/indicator/NY.GDP.MKTP.CD?locations=IN (accessed on 20 January 2024).
31. World Bank. Inflation, Consumer Prices (Annual %), India. World Development Indicators. Available online: https://data.worldbank.org/indicator/FP.CPI.TOTL.ZG?locations=IN (accessed on 20 January 2024).
32. Worldometer. CO_2 Emissions by Country. 2023. Available online: https://www.worldometers.info/co2-emissions/co2-emissions-by-country/ (accessed on 24 September 2023).
33. Li, L.; Willett, T.D.; Zhang, N. The Effects of the Global Financial Crisis on China's Financial Market and Macroeconomy. *Econ. Res. Int.* **2012**, *2012*, 961694. [CrossRef]
34. IEA. *Global Energy Demand to Plunge This Year as a Result of the Biggest Shock since the Second World War*; IEA: Paris, France, 2020.
35. Vogel, J.; Hickel, J. Is green growth happening? An empirical analysis of achieved versus Paris-compliant CO_2–GDP decoupling in high-income countries. *Lancet Planet. Health* **2023**, *7*, e759–e769. [CrossRef] [PubMed]
36. Acaravci, A.; Ozturk, I. On the relationship between energy consumption, CO_2 emissions and economic growth in Europe. *Energy* **2010**, *35*, 5412–5420. [CrossRef]
37. Ghosh, S. Electricity supply, employment and real GDP in India: Evidence from cointegration and Granger-causality tests. *Energy Policy* **2009**, *37*, 2926–2929. [CrossRef]
38. Ben Jebli, M.; Ben Youssef, S. Renewable energy consumption and agriculture: Evidence for cointegration and Granger causality for Tunisian economy. *Int. J. Sustain. Dev. World Ecol.* **2017**, *24*, 149–158. [CrossRef]
39. Lee, C.-C.; Chang, C.-P. The impact of energy consumption on economic growth: Evidence from linear and nonlinear models in Taiwan. *Energy* **2007**, *32*, 2282–2294. [CrossRef]
40. Ho, C.-Y.; Siu, K.W. A dynamic equilibrium of electricity consumption and GDP in Hong Kong: An empirical investigation. *Energy Policy* **2007**, *35*, 2507–2513. [CrossRef]
41. Yu, E.S.H.; Hwang, B.-K. The relationship between energy and GNP. *Energy Econ.* **1984**, *6*, 186–190. [CrossRef]
42. Payne, J.E. Survey of the international evidence on the causal relationship between energy consumption and growth. *J. Econ. Stud.* **2010**, *37*, 53–95. [CrossRef]
43. Payne, J.E. A survey of the electricity consumption-growth literature. *Appl. Energy* **2010**, *87*, 723–731. [CrossRef]
44. Ozturk, I. A literature survey on energy-growth nexus. *Energy Policy* **2010**, *38*, 340–349. [CrossRef]

45. Apergis, N.; Payne, J.E. CO_2 emissions, energy usage, and output in Central America. *Energy Policy* **2009**, *37*, 3282–3286. [CrossRef]
46. Adedoyin, F.F.; Zakari, A. Energy consumption, economic expansion, and CO_2 emission in the UK: The role of economic policy uncertainty. *Sci. Total Environ.* **2020**, *738*, 140014. [CrossRef] [PubMed]
47. Cerdeira Bento, J.P.; Moutinho, V. CO_2 emissions, non-renewable and renewable electricity production, economic growth, and international trade in Italy. *Renew. Sustain. Energy Rev.* **2016**, *55*, 142–155. [CrossRef]
48. Yu, E.S.H.; Jin, J.C. Cointegration tests of energy consumption, income, and employment. *Resour. Energy* **1992**, *14*, 259–266. [CrossRef]
49. Lean, H.H.; Smyth, R. On the dynamics of aggregate output, electricity consumption and exports in Malaysia: Evidence from multivariate Granger causality tests. *Appl. Energy* **2010**, *87*, 1963–1971. [CrossRef]
50. Stern, D.I. Energy and economic growth in the USA. *Energy Econ.* **1993**, *15*, 137–150. [CrossRef]
51. Ghali, K.H.; El-Sakka, M.I.T. Energy use and output growth in Canada: A multivariate cointegration analysis. *Energy Econ.* **2004**, *26*, 225–238. [CrossRef]
52. Lee, C.-C.; Chang, C.-P.; Chen, P.-F. Energy-income causality in OECD countries revisited: The key role of capital stock. *Energy Econ.* **2008**, *30*, 2359–2373. [CrossRef]
53. Stern, D.I. A multivariate cointegration analysis of the role of energy in the US macroeconomy. *Energy Econ.* **2000**, *22*, 267–283. [CrossRef]
54. Shahbaz, M.; Zeshan, M.; Afza, T. Is energy consumption effective to spur economic growth in Pakistan? New evidence from bounds test to level relationships and Granger causality tests. *Econ. Model.* **2012**, *29*, 2310–2319. [CrossRef]
55. Narayan, P.K.; Smyth, R. Electricity consumption, employment and real income in Australia evidence from multivariate Granger causality tests. *Energy Policy* **2005**, *33*, 1109–1116. [CrossRef]
56. Raza Abbasi, K.; Hussain, K.; Abbas, J.; Fatai Adedoyin, F.; Ahmed Shaikh, P.; Yousaf, H.; Muhammad, F. Analyzing the role of industrial sector's electricity consumption, prices, and GDP: A modified empirical evidence from Pakistan. *AIMS Energy* **2021**, *9*, 29–49. [CrossRef]
57. Abokyi, E.; Appiah-Konadu, P.; Sikayena, I.; Oteng-Abayie, E.F. Consumption of Electricity and Industrial Growth in the Case of Ghana. *J. Energy* **2018**, *2018*, 8924835. [CrossRef]
58. Tiwari, A.K.; Shahbaz, M.; Adnan Hye, Q.M. The environmental Kuznets curve and the role of coal consumption in India: Cointegration and causality analysis in an open economy. *Renew. Sustain. Energy Rev.* **2013**, *18*, 519–527. [CrossRef]
59. Shahbaz, M.; Khraief, N.; Uddin, G.S.; Ozturk, I. Environmental Kuznets curve in an open economy: A bounds testing and causality analysis for Tunisia. *Renew. Sustain. Energy Rev.* **2014**, *34*, 325–336. [CrossRef]
60. Kasman, A.; Duman, Y.S. CO_2 emissions, economic growth, energy consumption, trade and urbanization in new EU member and candidate countries: A panel data analysis. *Econ. Model.* **2015**, *44*, 97–103. [CrossRef]
61. Lee, C.-C.; Wang, F.; Lou, R.; Wang, K. How does green finance drive the decarbonization of the economy? Empirical evidence from China. *Renew. Energy* **2023**, *204*, 671–684. [CrossRef]
62. Bersalli, G.; Tröndle, T.; Heckmann, L.; Lilliestam, J. Economic crises as critical junctures for policy and structural changes towards decarbonization—The cases of Spain and Germany. *Clim. Policy* **2024**, *24*, 410–427. [CrossRef]
63. Nguyen, V.M.H.; Ho, T.H.; Nguyen, L.H.; Pham, A.T.H. The Impact of Trade Openness on Economic Stability in Asian Countries. *Sustainability* **2023**, *15*, 11736. [CrossRef]
64. Adebayo, T.S.; AbdulKareem, H.K.K.; Bilal; Kirikkaleli, D.; Shah, M.I.; Abbas, S. CO_2 behavior amidst the COVID-19 pandemic in the United Kingdom: The role of renewable and non-renewable energy development. *Renew. Energy* **2022**, *189*, 492–501. [CrossRef] [PubMed]
65. Basu, S.; Ishihara, K.N. Multivariate time-frequency interactions of renewable and non-renewable energy markets with macroeconomic factors in India. *Energy Syst.* **2023**, 1–37. [CrossRef]
66. Johansen, S. Estimation and Hypothesis Testing of Cointegration Vectors in Gaussian Vector Autoregressive Models. *Econometrica* **1991**, *59*, 1551. [CrossRef]
67. Yussuf, Y.C. Cointegration test for the long-run economic relationships of East Africa community: Evidence from a meta-analysis. *Asian J. Econ. Bank.* **2022**, *6*, 314–336. [CrossRef]
68. Jordan, S.; Philips, A.Q. Cointegration Testing and Dynamic Simulations of Autoregressive Distributed Lag Models. *Stata J. Promot. Commun. Stat. Stata* **2018**, *18*, 902–923. [CrossRef]
69. Delgado-Bonal, A.; Marshak, A. Approximate Entropy and Sample Entropy: A Comprehensive Tutorial. *Entropy* **2019**, *21*, 541. [CrossRef] [PubMed]
70. Federal Bank of St. Louis (FRED). Economic Data for India. Available online: https://fred.stlouisfed.org/tags/series?t=india (accessed on 1 April 2024).
71. British Petroleum. BP Energy Outlook. 2023. Available online: https://www.bp.com/content/dam/bp/business-sites/en/global/corporate/xlsx/energy-economics/energy-outlook/bp-energy-outlook-2023-chart-data-pack.xlsx (accessed on 24 September 2023).
72. Sax, C.; Steiner, P. Temporal Disaggregation of Time Series. *R J.* **2013**, *5*, 80. [CrossRef]
73. Dickey, D.A.; Fuller, W.A. Distribution of the Estimators for Autoregressive Time Series With a Unit Root. *J. Am. Stat. Assoc.* **1979**, *74*, 427. [CrossRef]

74. Kwiatkowski, D.; Phillips, P.C.B.; Schmidt, P.; Shin, Y. Testing the null hypothesis of stationarity against the alternative of a unit root. *J. Econom.* **1992**, *54*, 159–178. [CrossRef]
75. Zivot, E.; Andrews, D.W.K. Further Evidence on the Great Crash, the Oil-Price Shock, and the Unit-Root Hypothesis. *J. Bus. Econ. Stat.* **2002**, *20*, 25–44. [CrossRef]
76. Akaike, H. Fitting autoregressive models for prediction. *Ann. Inst. Stat. Math.* **1969**, *21*, 243–247. [CrossRef]
77. Schwarz, G. Estimating the Dimension of a Model. *Ann. Stat.* **1978**, *6*, 461–464. [CrossRef]
78. Sam, C.Y.; McNown, R.; Goh, S.K. An augmented autoregressive distributed lag bounds test for cointegration. *Econ. Model.* **2019**, *80*, 130–141. [CrossRef]
79. Jarque, C.M.; Bera, A.K. A Test for Normality of Observations and Regression Residuals. *Int. Stat. Rev./Rev. Int. Stat.* **1987**, *55*, 163. [CrossRef]
80. Engle, R.F. Autoregressive Conditional Heteroscedasticity with Estimates of the Variance of United Kingdom Inflation. *Econometrica* **1982**, *50*, 987. [CrossRef]
81. Ljung, G.M.; Box, G.E.P. On a Measure of Lack of Fit in Time Series Models. *Biometrika* **1978**, *65*, 297. [CrossRef]
82. Zhao, N.; Lin, W.T. A copula entropy approach to correlation measurement at the country level. *Appl. Math. Comput.* **2011**, *218*, 628–642. [CrossRef]
83. Olbrys, J.; Majewska, E. Approximate entropy and sample entropy algorithms in financial time series analyses. *Procedia Comput. Sci.* **2022**, *207*, 255–264. [CrossRef]
84. Gill, R.D. Schrödinger's Cat Meets Occam's Razor. *Entropy* **2022**, *24*, 1586. [CrossRef]
85. Shan, T.A.; Rather, S.A.; Dar, H.A. Revisiting openness-growth nexus: Panel data estimates for selected developing Asian countries. *Transnatl. Corp. Rev.* **2024**, *16*, 200051. [CrossRef]
86. Nasreen, S.; Anwar, S. Causal relationship between trade openness, economic growth and energy consumption: A panel data analysis of Asian countries. *Energy Policy* **2014**, *69*, 82–91. [CrossRef]
87. International Energy Agency. Energy-Intensive economic Growth, Compounded by Unfavourable Weather, Pushed Emissions up in China and India. 2023. Available online: https://www.iea.org/reports/CO2-emissions-in-2023/ (accessed on 24 September 2023).
88. Sultan, Z.A.; Alkhateeb, T.T.Y. Energy consumption and economic growth: The evidence from India. *Int. J. Energy Econ. Policy* **2019**, *9*, 142–147. [CrossRef]
89. Ang, J.B. CO_2 emissions, energy consumption, and output in France. *Energy Policy* **2007**, *35*, 4772–4778. [CrossRef]
90. Shahbaz, M.; Hye, Q.M.A.; Tiwari, A.K.; Leitão, N.C. Economic growth, energy consumption, financial development, international trade and CO_2 emissions in Indonesia. *Renew. Sustain. Energy Rev.* **2013**, *25*, 109–121. [CrossRef]
91. Bhambu, M.K. Make in India and Future of Renewable Energy. *SSRN Electron. J.* **2015**, *9*, 142–147. [CrossRef]
92. KhareSaxena, A.; Saxena, S.; Sudhakar, K. Solar energy policy of India: An overview. *CSEE J. Power Energy Syst.* **2020**. [CrossRef]

energies

MDPI

Article

Heat Pumps for Germany—Additional Pressure on the Supply–Demand Equilibrium and How to Cope with Hydrogen

Andreas von Döllen [1] and Stephan Schlüter [2,*]

1 Wintershall Dea, Friedrich-Ebert-Straße 160, 34119 Kassel, Germany; andreas.von-doellen@wintershalldea.com
2 Institute of Energy Engineering and Energy Economics, Ulm University of Applied Sciences, 89075 Ulm, Germany
* Correspondence: stephan.schlueter@thu.de

Abstract: In the context of the German Energiewende, the current government intends to install six million heat pumps by 2030. Replacing gas heating by power has significant implications on the infrastructure. One of the biggest advantages of using gas is the existing storage portfolio. It has not been clarified yet how power demand should be structured on an annual level—especially since power storage is already a problem and solar power is widely promoted to fuel heat pumps, despite having an inverse profile. In this article, three different solutions, namely, hydrogen, batteries, and carbon capture and storage, are discussed with respect to resources, energy, and financial demand. It shows that relying solely on batteries or hydrogen is not solving the structuring problem. A combination of all existing technologies (including fossil fuels) is required to structure the newly generated electricity demand.

Keywords: hydrogen; heat pumps; renewable energy; battery storage; carbon capture and storage

1. Introduction

Global climate change manifests itself in Germany in various forms, such as excessive heat, draughts, and extreme weather phenomena, such as hailstorms, for example. Past governments have been aware of Germany's contribution to this process and both designed and implemented various strategies to bring Germany on a road toward carbon-neutrality. One major pillar is the German hydrogen (H_2) strategy 2030 [1] that aims to reduce carbon emissions, mainly in the fields of industry and transport. Von [2] showed that this strategy would imply a significant carbon bow wave, as well as bow waves in financial and resource demand terms. Based on an economic model, the authors derived the necessity to include turquoise hydrogen, i.e., hydrogen based on natural gas, as well. However, one of their fundamental assumptions is that the supply of natural gas to Germany is ample and gas is cheap. The beginning of the conflict in Ukraine in February 2022, as well as the explosion of the North Stream 1 pipeline, altered this assumption. Not only did the price of natural gas change but also—due to increased liquefied natural gas transport—the carbon emission levels rose. Moreover, in order to reduce the German dependence on natural gas, the German government launched a campaign for a considerable transformation of the German heating sector away from the predominant gas heating toward heat pumps in 2023. From a financial point of view and considering only the heat pump installation, this looks like a daring move, imposing substantial investment costs either on the individual households, real estate investors, or community (in case of subsidies). Switching from gas to electricity naturally increases the demand for the latter, which has to be of renewable form. The consequence has to be a net increase in renewable power sources (wind or solar), in addition to the already existing Energiewende efforts. The shift from gas heating to electricity-based heat pumps also implies the requirement of structuring supply, e.g., via inter-seasonal storage [3]. Gas and oil can be easily stored. Thus, continuous production is possible, and supply can be structured according to demand via the storage and distribution infrastructure (pipelines also contribute a small amount of flexibility). Currently, there is no solution for large seasonal storage of electricity. However, in contrast to gas and oil, in which production is almost flat or even increased in the winter, renewable power is anticyclical, meaning that most is produced during summer (solar power) and the shoulder months, i.e., especially from September to November and March to May (wind power). Hence, structuring issues increase considerably, and an inter-seasonal storage possibility is needed. Batteries are for short-term and intraday structuring (solar power vs. charging cars and supplying household baseload during the nighttime). Technologies such as thermal energy storage and flow batteries offer a longer time horizon of several days up to a couple of weeks. However, most of these concepts are still in the development stage and not scalable within the next five years [4,5]. The potentials of water storage, such as pumped hydro storage, are also limited considering environmental effects and would require substantial planning. In addition, Refs. [4,6] argue that this kind of storage is not suitable for bridging the inter-seasonal gap. The predominately discussed solution is hydrogen generation [2] and storage [5]. For storing hydrogen, there are two approaches, i.e., using the existing gas grid or investing in a separate infrastructure. The first one faces several technical hurdles; the second alternative requires large investments and considerable time.

In this article, we estimate the costs of a nationwide introduction of heat pumps and calculate the residual energy demand, including the required installed power capacity and storage demand. We also estimate the implications for the hydrogen and battery demand. Carbon capture and storage (CCS) as an alternative solution to realize carbon-neutral electricity is discussed. In this context, it is shown that solar power is not an adequate source of heat pump electricity due to its anticyclical annual period, although this is exactly what is considerably subsidized on a micro level. An increase in installed heat pumps substitutes natural gas, including the corresponding storage capacity; hence, we boost the already existing problem of structuring the power supply inter-seasonally. Ceteris paribus, this either increases the need for additional fossil power capacity in winter times or causes massive renewable overcapacity to meet the demand (at least statistically). Neither the use of household batteries nor hydrogen is able to close this gap economically. Based on historical data, we conclude that an intelligent mix of hydrogen and fossil power that is decarbonized by CCS is the most promising path (within a reasonable time frame) in order to provide a cost-efficient possibility for seasonal structuring.

The article is structured as follows: In Section 2, we describe the data sources and explain our calculation steps. In Section 3, relevant technical fundamentals are given. In Section 4, the results of our calculations are given, which are discussed in more detail in Section 5. Section 6 concludes the article.

2. Data Sources and Core Assumptions

Our heating and power demand data for the heat pumps is based [7] which also provides historical temperature data. This information allows us to derive a history of heating years (2007–2014). We are aware of the age of the dataset. Nevertheless, transformation in the housing sector is slow due to high costs. Hence, the dataset is sufficient to estimate the effects of various strategies, as tested in Section 4. For wind power, we use the numbers and assumptions from [2], and for solar power, the numbers from [8]. Let us assume 900 full load hours (flh) for solar power at a lifetime of 20 years. Currently, about one-third of German solar power is field-based (consuming steel and concrete) and two-thirds are mounted on roofs [9]. Given a certain installed capacity, we can also simulate historic solar power and wind production based on the aforementioned sources. Combining the data, we can project the past into the future to simulate demand and supply and structure needs for the future. For future renewable energy installation, we allocate the technology based on assumptions made by the German Federal Ministry of Economic Affairs and Climate Action [10]. Here, 32% is assumed to be onshore wind, 9% offshore wind, and 59% installed solar power panels. Regarding energy demand, we use data from the German energy market database [11] between July 2021 and July 2023 and assume it will be more or less constant in the future. For the gas demand and storage injection, we rely on Bundesnetzagentur data. All further required data, such as German onshore/offshore wind and solar power load factors, are taken from the open-source platform Open Power System Data [12]. As a simplified assumption for industrial demand, we use data from 2022/23 [11] and use it as a projection, as well. Given our assumptions on renewables and certain technologies like battery storage or H_2, we calculate the hourly residual demand. This is the demand to be met by waterpower, waste burning, and fossil power generation, which provide most of the remaining power. Since we see a limited growth potential of water and waste power, we assume them to be constant and project all changes in the residual demand to be met by the fossil share. To compare the effect, we aggregate the data on an annual basis.

3. Heat Pumps and Renewable Resource Demand

3.1. Heat Pumps

Heat pumps are used in private households for heating and warm water supply. Their efficiency therefore depends on both the type of heat pump and the type of heating itself (radiator or floor heating). A detailed yet brief overview of the respective energy demand is given by [7]. Here, they discriminate between air-source heat pumps (ASHP), ground-source heat pumps (GSHP), and groundwater-source heat pumps. In most cases, an ASHP is used for retrofitting household heating systems, which therefore will be the considered type in this study. The main trick of heat pumps is to profit from the temperature difference between ambient (or ground) temperature and the cooling liquid within the pump. The smaller this difference, the higher the electricity demand, especially for ASHPs, which is unfortunate as this means that heat pump demand is anticyclical to the annual temperature movement. A measure of heat pump efficiency is the coefficient of performance (COP), which quantifies the relation between the energy output of the heat pump and its energy demand. Ref. [13] compute the electricity demand of heat pumps for England and find a linear relationship between air temperature and the COP of both GSHP and ASHP [7], again, compute the COP based on outside temperature using historical data. In Figure 1, we show their results for the hourly COP (averaged for Germany) in 2014 for ASHPs. Based on these COP values and the actual heat demand, we are able to calculate the corresponding electricity demand. Results are shown in Figure 2. Both in Figures 1 and 2, a predominant annual seasonality is visible, which implies the demand for flexible energy production, as well as annual structuring capacity. In this article, we ceteris paribus focus on heat pumps for households, as this is the prime target of current political measures (especially subsidies). Thereby, we acknowledge that large-scale heat pumps would be an efficient alternative for various heat-intensive industries. However, estimating the potential industry-driven demand would imply a number of additional assumptions. Hence, we restrict this analysis to households, noting that industrial demand would only augment the conclusions drawn in this article.

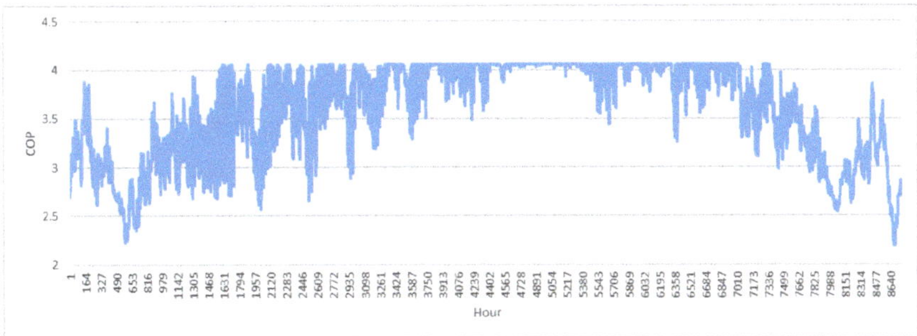

Figure 1. Hourly coefficient of performance (COP) for air-supplied heat pumps in Germany in 2014.

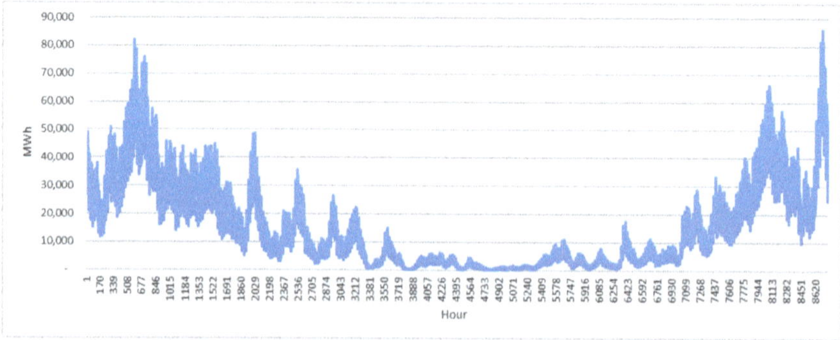

Figure 2. Hourly electricity demand in megawatt hours (MWh) for air-supplied heat pumps (household sector) in Germany in 2014, assuming that the total heat demand is satisfied by this technology.

3.2. *Solar and Wind Power Resource Demand*

Here, we summarize the carbon footprint, resource demand, and costs of installing and using solar power panels. For specific numbers of wind power plants, we refer to [2]. Ref. [8] estimate the carbon footprint of solar panels to be around 13–30 g (g) per produced kilowatt hour (kWh), depending on the module's location in Germany and where it has been produced, i.e., basically China or the European Union (EU). Due to the missing production capacity in the EU and the fact that we assume an installed capacity all across Germany, we work with an average of 22 g/kWh full life cycle emissions. Thus, solar power production and installation cause about twice the emissions of wind turbines, assuming a lifetime of 20 years [2].

Regarding leverage costs, we see a range of 0.025 to 0.04 EUR/kWh for ground-based and 0.04 to 0.08 EUR/kWh for rooftop solar power panels [14]. As mentioned, in Germany, one-third of all solar power plants are ground-based, i.e., mounted on open fields, and two-thirds are mounted on roofs. Hence, we work with 0.04 EUR/kWh, considering the cheapest rooftop level, which equals the worst case for ground-based panels. Regarding resources, we focus on the major ones, as follows: next to steel and concrete, whose production is CO_2-intensive, we have aluminum, copper, silicon, silver, germanium, and gallium. The specific numbers are summarized in Table 1, which is based on [15]. Note that the volumes of concrete and steel, especially, refer to ground-based solar power plants; hence, we scale these numbers by one-third.

Table 1. Resources demand per MW for installed solar power panels.

Type	Ton per MW
Concrete (only open-field solar power)	60.7
Steel (only open-field solar power)	67.9
Aluminum	6.5
Copper	4.6

4. Consequences of a Nationwide Change in Heating Technology in Numbers

We are discussing four different scenarios in four sections, which are listed in Table 2. All are based on the assumption that six million heat pumps are installed. For further core assumptions, see Section 2.

Table 2. The considered scenarios, including the underlying assumptions.

Section	Description	Assumptions
Section 4.1	The current energy mix is fixed until 2030, and residual heat pump demand is satisfied by renewables.	• An average COP of 3.5 is used; • An average household heat demand amounts to about 18 MWh per year; • Heat pump demand is satisfied using only additional renewables; • The share between solar and wind for additional power supply is 59% to 41%.
Section 4.2	Battery storages are used to flatten the daily demand.	• Batteries have an efficiency of 92%; • We use data from 2008 to 2014 for a backtest and assume their validity for predictions.
Section 4.3	Hydrogen is used for storing energy seasonally.	• Neither thermal storage nor flow batteries are an alternative within the next 10 years; • Investment costs amount to 5000 EUR/MWh power input; • Overall electrolyzer efficiency is 70%.
Section 4.4	Carbon capture and storage are used to decarbonize fossil energy supply.	• The missing energy in winter will be in large parts supplied by existing gas turbines; • CCS for gas-to-power will be used to decarbonize energy supply; • Levelized costs of electricity will be around 100 to 170 EUR/t CO_2 for the total CCS chain.

4.1. The Current Energy Mix in 2030 Plus Renewables for Heat Pump Demand

If we average the COP values from Figure 1 (although being not perfect and a rather positive estimate), we yield an annual average COP of 3.5. For the sake of simplicity, we assume that this COP holds for all 6 million heat pumps. According to the Federal Statistical Office of Germany (Destatis), an average household consumes 17,644 kWh of energy, 85% of which is for air/floor heating and warm water [16]. This means we have a heat pump-related energy demand of 15 MWh $\times$ 6,000,000/3.5 = 25.714 TWh. Extending Figure 2 to the period from 2008 to 2014, we compare the heat pump demand to the hourly production of renewable energy. We see, not unexpectedly, that the power demand is highly seasonal, with a peak between November and February, the usual heating months. In Figure 3 (right side), we show the simulated heat pump demand profile vs. the renewable energy production if we assume that the additional demand is installed in a way that demand can be supplied by renewables—on an annual level—even in the worst year of our historic data. To compare the profiles, we show the renewable production as negative numbers in the sense that supply is negative demand. The requirement of structuring is obvious. We also show the residual demand in Figure 3 (left side), which shows that there is a significant overproduction in summer due to the solar power share.

Note that as explained in Section 2, we work with the German government's target for a renewable energy split of 59% solar power, 32% onshore wind, and 9% onshore wind. Historical data [7] show a calculated energy demand of 21,318 MW. This equals 12,450 MW of solar panel power, 6897 MW of onshore wind power, and 1881 MW of offshore wind power. Numbers for individual years are given in Table 3. The first question is whether it is statistically possible to generate the required power via renewables at any point in time. Under ceteris paribus conditions, this also means that sufficient battery capacity for intraday balancing has to be installed since, due to the nature of solar energy, huge overcapacity in wind would be required if we want to meet that condition on an hourly basis. Batteries are discussed in detail in Section 4.2; here, we ignore this issue and only consider the daily aggregated demand.

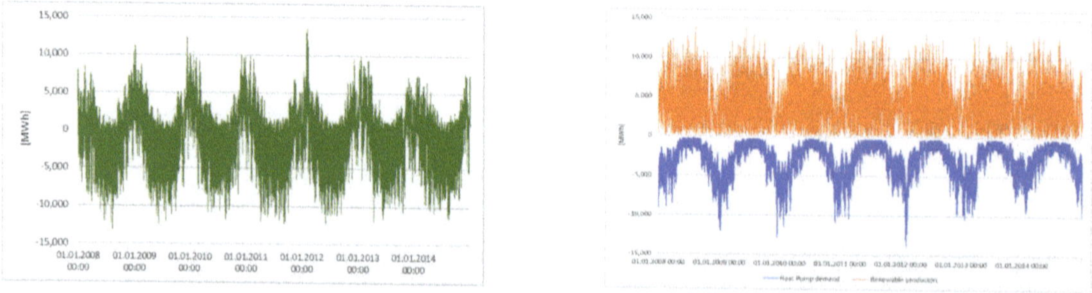

Figure 3. (**Left side**): Heat pump demand minus renewable supply; (**Right side**): Heat pump demand vs. renewable supply.

Table 3. Simulated demand and supply.

	2008	2009	2010	2011	2012	2013	2014
Heat pump demand (TWh)	26.06	25.61	28.54	23.99	25.69	27.58	23.79
Renewable production (TWh)	31.83	30.03	28.58	32.37	30.89	29.39	30.19
Share (%)	81.8%	85.3%	99.9%	74.1%	83.2%	93.9%	78.8%

Based on our backtest dataset, we compute how much surplus capacity is needed to guarantee that on 95% of all days, the demand is met by renewable sources. Following standard risk management logic, we ignore the remaining 5% and assume that fossil production capacity is available for this case. Given the data, we calculate a factor of 3.792, i.e., we have to install almost four times the amount of renewable capacity involving an increased demand for money and rare metals by a factor of four, compared to the flat scenario (25 TWh are produced on a yearly basis).

If we aim for 100% supply security—and note that this still does not guarantee supply, but 100% supply based on our historical data—we need 10.93 times the capacity, so almost factor 11. This result is consistent with the increasing marginal cost view common in economic theory [17].

For a closer look at the current situation, we use the German 2022/23 energy demand and project it into the past. We also use the mix of installed capacity in July 2023, thereby ignoring nuclear power capacity and considering coal as the first step in a reserve capacity. Now, let us consider the additional heat pump-induced demand shown in Figure 4 and all information gathered above as follows: (a) the July 2023 installed wind and solar capacities [11], (b) the simulated factors as described above, and (c) the demand for fueling the installed heat pump. As a consequence, we can derive the residual demand similar to Figure 3, i.e., the required supply from other sources (fossil fuels or water, for example). In Figure 4, we see that due to seasonality, winter demand has increased even more; especially in winter, there is up to 80,000 MWh/h additional demand, while in summer peak times, we have 47,000 MWh/h of unused supply. That is not a new problem and already exists today. However, heat pumps increase the issue by approximately 25%, assuming that their demand is satisfied by the current renewable supply mix.

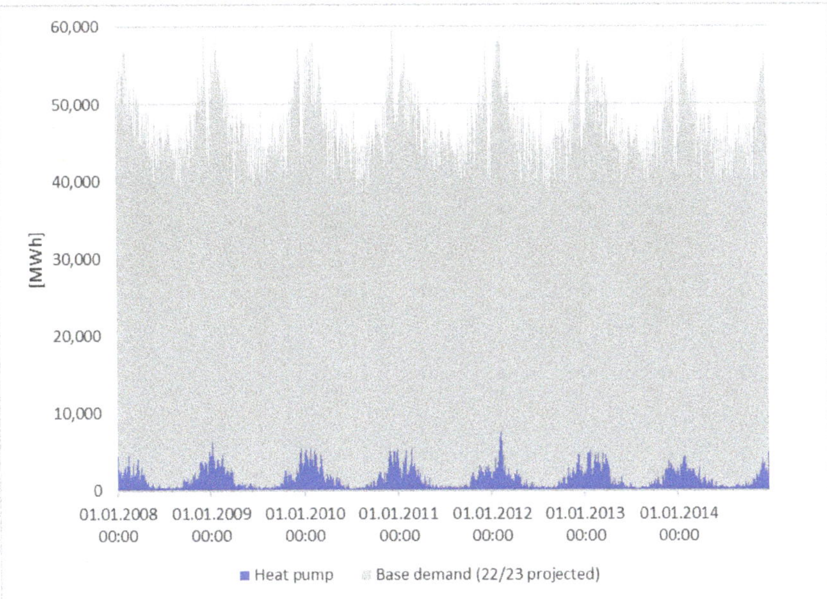

Figure 4. Heat pump demand compared to the total demand.

This leads to our first finding, as follows: a heat pump-based strategy increases the need for seasonal storage, especially since the commodity substitutes, namely, oil and gas, offer this solution. We do not only substitute a commodity for another one but also an infrastructure without having a market-ready solution for the latter one.

Another crucial question is whether and how the existing energy mix can cover the demand peaks; as seen in Figure 5, the maximum is about 78,630 MW. Given the 2023 supply mix [11,18], Germany has a controllable capacity of 85,350 MW (coal, gas, and pump storage). If we assume that coal should be removed from the mix, this number decreases to a 41,364 MW capacity, which is not enough anymore and implies investments into state-of-the-art gas-fired power plants as replacements (see also Section 4.3). Given these numbers and numbers from Table 4 and considering the gap of 37,000 MW, approximately 64% of the coal-fired power plants have to be replaced by pump storage or gas-fired power plants in order to structure demand. In the case of gas-fired power plants, this immediately raises the question of whether this move is consistent with the climate targets. The answer might be CCS and/or H_2, which of course increases the costs significantly (see also Sections 4.3 and 4.4). Following the results from [2], in which the diversification logic is discussed for H_2, an equivalent social cost approach can show that it is very likely economically efficient to split the technologies to fill the above-mentioned gap between pump storage, which is capital-intensive and linked with social cost, and battery/renewable overcapacity. Batteries are thereby used to flatten the daily renewable production.

These derived capacity requirements allow for a view of the implications for resource demand. The capacity numbers for 2023 are summarized in Table 4. If we include the required renewables, we end up with the numbers in Table 5.

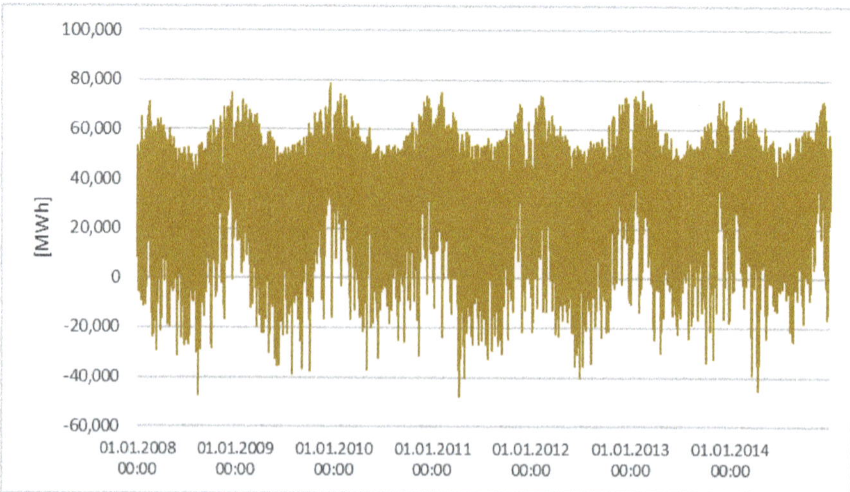

Figure 5. Residual demand in the base model given the July 2023 energy mix.

Table 4. German installed power generation in July 2023. All values in MW [11].

Source	Wind Offshore	Wind Onshore	Solar Power	Water/Biomass	Gas and Coal	Others	Total
amount	7774	54,499	53,302	22,934	75,928	7927	222,364
in %	3.5%	24.5%	24.0%	10.3%	34.2%	3.6%	

Table 5. Adjusted German installed power generation in July 2023. All values in MW [11].

Source	Wind Offshore	Wind Onshore	Solar Power	Water/Biomass	Gas and Coal	Others	Total
amount	9654.94	61,395.76	65,841,57	22,934	75,928	7927	243,681
in %	3.96%	25.20%	27.02%	9.41%	31.16%	3.25%	

Of a demand of 21,317 MW in total, 6897 MW would be onshore wind power capacity, 1881 MW would be offshore wind power, and 12,540 MW would be solar power. Based on the sources and numbers from Sections 2 and 3.2, we obtain a rough idea of the costs and raw material demand (see Table 6), e.g., over 1700 new onshore windmills would be required. The total costs of this transformation would be estimated at 21.69 bn EUR (ceteris paribus).

Table 6. Costs and raw material demand [19].

	#	Bn EUR	Copper (mill t)	Steel (mill t)	Concrete (mill t)	CO_2 (mill t)	Aluminum (mill t)
Onshore wind	1724	7.59	0.01	0.74	2.85	4.33	
Offshore wind	235	5.08	0.01	0.22	0.46	1.43	
Solar power		9.03	0.12	1.84	0.76	4.97	0.18
Total		21.69	0.14	2.8	4.07	10.73	0.18

Regarding wind power, we talk of a required net growth of installed capacity of 1200 MW per year, which is approximately the rate in 2023, ignoring the growth needed for charging electric vehicles, general demand growth, e.g., caused by the debate about de-risking (i.e., battery and chip production in Germany) and the increase in the renewable share in general.

With the numbers stated above, the consequences of an overcapacity in terms of investment and resource demand are shown. Thereby, the fact that the increase in resource demand is likely to increase, their prices are ignored. Effects on the energy demand are ignored as well, although [20] argue that this increase in resource demand will increase global energy demand as well.

Therefore, doubling the renewables will mean that the cost will be multiplied by a factor of two or more. Even if we assume a linear relationship, we talk about EUR 82 billion (bn) (factor 3.79) or EUR 237 bn (factor 10.92), ignoring further costs, e.g., for batteries, as described above. We can conclude that this renewable-only approach is not going to work and that at least for a certain residual part, we need a flexible solution, which in the short term—due to the absence of alternative options—has to be natural gas supported by pumped hydro storage.

If we structure the hourly demand using only the sources of the system, i.e., fossil fuels, we end up with the numbers in Table 7; for all tested years, the fossil share was above 14%. Note that based on the 2023 installed capacity, we assume a capacity of 22,934 MW for biomass and water-based power, which we always use first and consider to be available. Effectively, this is not true and may increase the share of fossil power generation even more. This illustrates that without an effective storage solution, we will still end up with a high fossil share, even if we overshoot the annual demand. Diversification is needed, as well as an efficient seasonal storage approach or a solution to decarbonize the fossil power share (potentially CCS).

Table 7. Demand, installed renewable power, and fossil fuels (TWh).

Year	Demand	Renewables	Residual	Non-Fossil	Fossil	% Fossil
2008	491.6	316.7	174.9	104.5	704	14%
2009	389.7	296.7	193.0	113.1	79.9	16%
2010	492.6	282.1	210.6	125.5	85.1	17%
2011	488.1	320.9	167.2	97.7	69.5	14%
2012	491.2	306.0	185.2	109.5	75.7	15%
2013	491.7	290.5	201.2	119.9	81.4	17%
2014	487.9	297.6	190.3	116.0	74.3	15%
Total	3432.8	2110.5	1322.3	786.2	536.2	16%

4.2. Flattening the Daily Demand Using Batteries

So far, in Germany, we operate on an hourly basis and assume that the daily structuring is performed by the system (i.e., fossil power sources), which is currently the reality. But there is a clear target and agenda in Germany, as well as in other parts of the world, to increase the amount of battery storage significantly. For the case of Texas it has already been shown that this has a positive effect on both energy transition and CO_2 footprint [5]—although the production of batteries is very CO_2- and resource-intensive and induces some bow wave problems. This is especially the case if the timing of constructing heat pumps, batteries, and renewable power capacities is not well matched. For this section, we choose a stepwise approach. First, we derive the costs and resource demand for batteries. Based on the above-described backtest, we can calculate the required capacity for the daily flattening of the heat pump demand (Case 1). In a second step, using the same methodology, we will analyze the case of the 100% renewable portfolio (Case 2). This gives us an insight into possible synergy effects.

Based on [5] we assume that a battery has an efficiency of 92% both for injection and withdrawal. As a consequence, from any produced quantity, after injection and withdrawal, only 84.62% is actually available, which makes our calculations slightly more complicated. Again, we use an easily replicable two-step approach; in the first step, we calculate the flat quantity given the daily production. Doing this, we can calculate the daily energy loss and reduce the flat quantity by the corresponding amount. This is not an exact 1:1 calculation, but since this is close to a theoretical perfect foresight scenario, it is already very idealized. In reality, the steering and optimization of a battery cluster is an important but complicated issue, but this is out of the scope of this paper. If batteries are used myopically, which may be the case for, e.g., household battery storage, there may be a notable social loss with respect to the system's optimal steering. In Table 8, the respective maximum monthly numbers for the years 2008–2014 are given. In the last two columns, we compute the ratio of maximum withdrawal and maximum injection to the installed capacity of renewables in order to see how much battery capacity is needed. Thereby we see that you need between 20% and 40% battery capacity.

Table 8. MW needed to flatten the daily production.

Month	Max Withdrawal [MWh]	Max Injection [MWh]	Capacity Renewables [MWh]	%Withdrawal	%Injection
Oct	7280.48	−5213.35	213,180	34.2%	−24.5%
Noc	6893.65	−4428.99	213,180	32.3%	−20.8%
Dec	4963.94	−5013.97	213,180	23.3%	−23.5%
Jan	6143.15	−4681.08	213,180	28.8%	−22.0%
Feb	7140.60	−4359.76	213,180	33.5%	−20.5%
Mar	7772.43	−4655.55	213,180	36.5%	−21.8%
Apr	8185.28	−5204.35	213,180	38.4%	−24.4%
May	7817.73	−5110.98	213,180	36.7%	−24.0%
Jun	8195.88	−6299.21	213,180	38.4%	−29.5%
Jul	8156.97	−5369.87	213,180	38.3%	−25.2%
Aug	8455.49	−5641.11	213,180	39.7%	−26.5%
Sep	7478.04	−5565.29	213,180	35.1%	−26.1%

In the next step, we simulate the utilization of the storage to show the effectiveness and analyze the change compared to the simulated residual power demand in total. From Figure 6, we see that the assumed storage system is, in general, well-designed and able to provide the necessary structuring, especially in winter. Nevertheless, the loss of energy induces low levels in the nighttime.

The question is now how the power demand–supply balance, in general, is affected. For that, we analyze the peak residual demand, the peak surplus, and the total overproduction. One expected finding is that batteries reduce the overproduction in both peak capacity and non-utilized power, which will be an important factor for the hydrogen analysis in Section 4.3. However, peak demand is only marginally reduced—both regarding total quantity and required peak capacity (Table 9). While in the base case, our simulations show a demand of 78,630 MW, we now have a peak demand for other power generation of 77,876 MW in the simulation, thus only 1% less. Given the nature of the analysis, one may ask whether this is a significant effect after all. In addition, the effective renewable energy consumption decreases slightly, and the residual demand increases, which implies that it is likely that more fossil power generation is necessary. Again, the difference is small, so approximately, one can say that batteries imply no significant quantity effect, which is a somewhat surprising result.

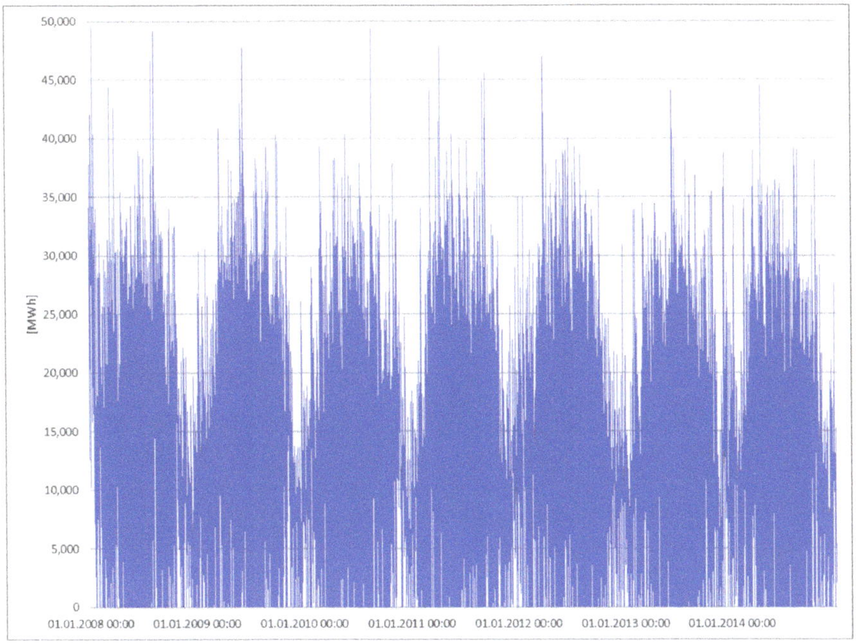

Figure 6. Utilization of the battery storage system given historical demand data (Case 1).

Table 9. Total aggregated demand and supply for the base case.

Total (TWh) 2008–2014	Demand	Renewable	Residual
No batteries	3432.8	1515.0	1917.8
Batteries	3432.8	1506.5	1926.3

Finally, we add the numbers of scaling-up batteries and installing a corresponding number of renewables. We again use the methodology described at the beginning of this chapter, but this time consider the fully modeled renewable production. We end up with almost 49,540 MW of 7 h battery storage and a battery utilization pattern, as described in Figure 7. Now, we talk about an investment of 14.88 bn Euro and a CO_2 bow wave of about 22 to 35 million tons. This already shows a more significant impact with regard to the necessary backup capacity (74,684 MW vs. 78,630 MW). Moreover, the overproduction is reduced; nevertheless, we see a decrease of 3.7% in the effective production of renewable energy (see Table 10).

Table 10. Total aggregated demand and supply in case of flattening the total renewable production.

Total (TWh) 2008–2014	Demand	Renewable	Residual
No batteries	3432.8	1515.0	1917.8
Batteries	3432.8	1461.2	1971.6

As a consequence, we have to state an even strengthened counterintuitive effect of battery storage "wasting" renewable energy. This is mainly due to the aforementioned energy loss of almost 14%, which is completely eating up gross efficiency gains in renewable energy production. If battery storage would work without loss, we indeed would see a boost in production. Note that this conclusion is rather simplified and based on multiple assumptions; hence, it cannot be used as an argument against renewables and batteries. However, it highlights an important issue; a myopic and unstructured investment into batteries may lead to an inefficient infrastructure and may increase the total amount of wasted energy. Research efforts to develop an intelligent and adaptive system control are required to realize the benefits of a battery park.

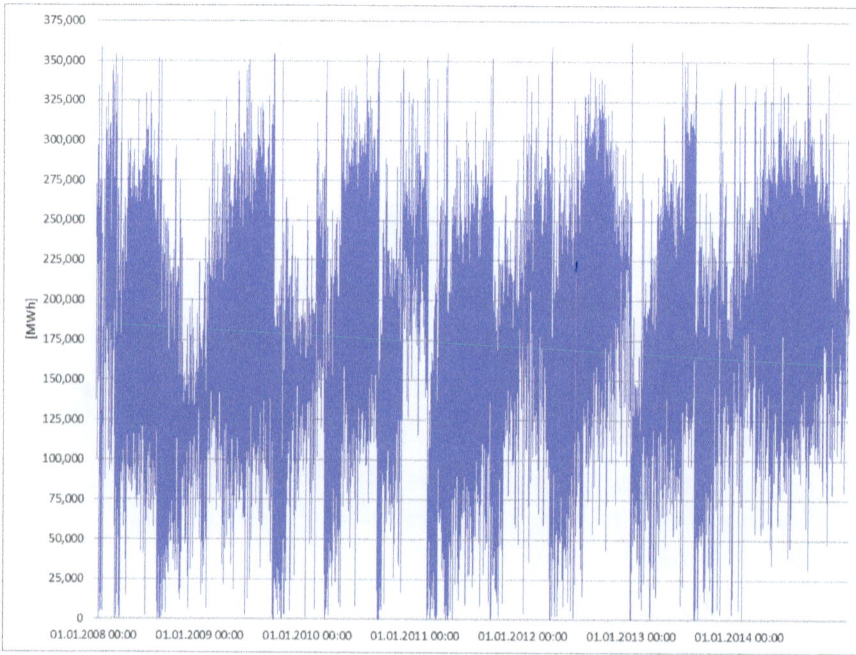

Figure 7. Utilization of the battery storage system in the 100% renewable case (Case 2).

If no residual fossil power generation is available, batteries are advantageous and required, as shown by our calculation of ensuring 100% supply only with renewables. However, in this scenario, the main advantage of batteries is them being the only remaining flexibility in the system. One may reduce the effect and turn it into a positive contribution using a more focused storage control, but this analysis is out of scope for this paper.

4.3. The Hydrogen Strategy Revised—Hydrogen/Natural Gas Mix

An alternative to batteries is to use H_2 as a storage medium. We also include the possibility of decarbonizing the remaining share of power plants (which are needed for H_2 anyway) by utilizing CCS, thereby following the portfolio approach for the transition pathway discussed in von [2]. Moreover, we now know the estimated costs of one alternative, namely, batteries.

The basic idea is as follows: for natural gas, there exists a significant storage and pipeline infrastructure. According to the German (European) grid regulation, up to a 10% hydrogen share is acceptable, despite, of course, potentially inducing some issues for old household connections and some industrial applications. Not every industrial plant is ready to work with a natural gas/hydrogen mixture, which means that a hydrogen removal unit may be necessary, especially if the share exceeds the 10% limit. It is also not proven that the gas network can be operated safely with a higher ratio of hydrogen. Additional investments may be needed. For more information regarding this topic, please refer to [21,22]. Moreover—like biogas—it is a reasonable approach to virtualize parts of the H_2 logistics by injecting it into the natural gas grid and working with H_2 certification, which disconnects the physical and contractual usage [23,24]. This would at least partly relax the situation if we base the analysis on the recent regulation and ignore further costs and efforts to improve the grid for higher hydrogen degrees. Moreover, in the sense of a diversification approach, one may compare the additional cost induced by the 95% case and the 100% case of Section 4.1 with the potential costs of these measures, whereby the costs of the H_2 generation infrastructure have to be included as well.

Note that we also have to consider the positive synergies with the hydrogen strategy, which is beyond the scope of this work. In particular, increasing the wind share of the required capacity of Section 4.1 might contribute to the hydrogen targets; having more wind power increases the likelihood of excess power, which would then be available for hydrogen production.

As a first step, we ignore the storage issue since there are a lot of unsolved questions; large parts of the German storage capacity are depleted gas fields. In addition to the question of how a large hydrogen share will influence the capacity and dynamics of the storage, one must consider that there will be a lot of migration of the hydrogen in the first years (cushion gas) to the permanent gas in the storage, so 1 MW hydrogen injected will be mean less than 1 MW hydrogen withdrawn. In a virtual system, this is manageable, but in general, this needs to be analyzed. Research about how hydrogen can be stored in depleted gas fields still needs to be conducted.

We again consider the renewable power production from our model and aggregate our data to a monthly level. Let us also assume an electrolyzer efficiency of 70% [1] and a machine life of 20 years, which is a typical but nevertheless hydrogen-friendly assumption for leverage cost calculation. Then, we yield the hydrogen levels aggregated in Table 11 in MWh. Since in the summertime, physical consumption is generally flat within the month, a monthly granularity is sufficient for our analysis. Using data from [25] and the website of Trading Hub Europe, Germany's virtual gas trading hub (URL: https://www.tradinghub.eu, accessed on 12 June 2024), we derive the numbers in Table 12.

Table 11. Hydrogen production in MWh with surplus renewable energy.

Year	Month												Total
	Jan	Feb	Mar	Apr	May	Jun	Jul	Aug	Sept	Oct	Nov	Dec	
2008	126	258	889	204	213	473	334	771	93	283	160	1	3805
2009	0	25	95	345	655	1017	388	282	401	426	134	94	3861
2010	6	18	107	490	221	273	177	262	225	83	397	25	2283
2011	27	315	95	1152	659	351	761	535	569	159	152	107	4879
2012	420	107	233	258	362	856	540	325	165	205	7	185	3665
2013	97	17	230	274	18	419	229	112	155	633	28	338	2550
2014	17	31	344	683	224	197	60	499	98	34	0	318	2507

So, on average for the backtest period, 2008–2014, the ratio of hydrogen and gas is comfortably below 5%, but in the low-demand months, in which we have the highest solar intensity (June–August), we are at a peak above 10%. Moreover, in reality, the allocation is an issue, so locally, the degree of H_2 may significantly exceed 10%. This effect could be managed since solar generation, which is the main driver of the surplus power generation, can be built more or less uniformly distributed across Germany. Nevertheless, one should note that the direct effect, i.e., the amount of H_2 stored in the system, is quite low. We talk on average about approximately 23.55 TWh. On the other side of the balance sheet, we have to consider the loss of energy. As stated above, only 70% of the energy is transferred in H_2. The loss of energy in re-converting the H_2 to power is more complicated to estimate and depends on the use case. However, since we assume introducing H_2 into the natural gas infrastructure, highly efficient cases like H_2 cells can be neglected since this necessitates refining the H_2 from the gas mix, which is linked with a significant energy loss. The best assumption is that the H_2 is just burned in a modern power plant. If we assume that in this case, only modern highly efficient turbines are used with 50% efficiency, we end up with 35% of the original energy used in such a storage concept. This means that we can generate 12.28 TWh of power. This is only 1% of the fossil power needed, but more or less the green H_2 target for 2030 in the German hydrogen strategy 2030. This highlights another dimension of the issue. We cannot easily replace fossil power if we want to balance and structure the system in the short run. Reality lies between this scenario and the 95% case scenario stated in the beginning if we want to replace fossil power generation in total. The alternative is to decarbonize a fraction for a period of 20 to 30 years via CCS. We do not consider resource availability issues but highlight that in a more detailed analysis, this is an important external factor for the transition strategy.

Table 12. Hydrogen production in major months with surplus renewable energy.

Month	Ø Gas Consumption 2018–2021 [GWh/day]	Ø H_2 Production 2018–2021 [GWh/day]	Peak Daily H_2 Production [GWh]	% Peak to Demand
Apr	2475	14	139	5.6%
May	2019	10	228	11.3%
Jun	1561	15	242	15.5%
Jul	1528	10	232	15.2%
Aug	1489	12	152	10.2%
Sep	1765	7	238	13.5%

In the next step, we have to evaluate the costs. The use of the gas grid may be considered for free at first (also some investment to be H_2 ready should be expected), but the installation of the required H_2 generation capacity has to be calculated. Ref. [26] estimate the costs for a 70% electrolyzer to be around 500 EUR/kW. In our backtest, we see a peak power surplus of almost 48,000 MW in summer. Fueling respective electrolyzer capacities with this excess power would thus mean investment costs of EUR 16.8 bn. Adding the construction costs of windmills and solar panels (EUR 21.7 bn, see above), we end up at a total sum of about EUR 40 bn. Of course, H_2 production can be increased by installing more renewables that partly replace fossil sources, and given the German target of 80% renewable share compared to the model share of approximately 36%, this would likely be the case. This would boost the overproduction, which can be then utilized for H_2 production; as long as the power would be unused and therefore has a marginal cost of zero, we obtain an expensive but acceptable price, e.g., considering 50,000 MW of installed renewable capacity, given the 500 EUR/kW investment costs for the electrolyzer and calculating an aggregated efficiency of 35% (power $\rightarrow$ H_2 $\rightarrow$ power), the leverage price of this power would be about 744 EUR/MWh since the units will be only utilized 96 h per year (see Table 13). This can be optimized to reduce the capacity for H_2 generation not covering all peak days. Table 13 shows various scenarios depending on the amount of utilizable excess power.

Therefore, we see in all cases a level far below 5000 average operation hours per year (mentioned in the current government's hydrogen strategy), which is below any threshold in which hydrogen as a measure to prevent the loss of power makes sense.

If we install more renewables, it will provide more peak power supply but may also have more days in which we have excess power. Let us now consider the 95% renewable case, in which we scale the installation of each type by 3.792. Results are given in Table 14. We see that we are still below the 5000 average operating hours mentioned in the hydrogen strategy, which is not surprising since, all in all, the renewable production has 1430 usage hours.

Table 13. Costs of the hydrogen, depending on installed capacity.

Electrolysis [MW Installed]	Utilizable Excess Power [TWh]	TWh H_2	Average Operating hours	Costs [bn EUR]	Power Costs [EUR/MWh]
5000	13.7	9.59	391	2.5	182.48
10,000	22.4	15.68	320	5.0	223.21
15,000	27.7	19.39	264	7.5	270.76
20,000	30.7	21.49	219	10.0	325.73
25,000	32.3	22.61	185	12.5	287.00
30,000	33.1	23.17	158	15.0	453.17
35,000	33.5	23.45	137	17.5	522.39
40,000	33.6	23.52	120	20.0	595.24
45,000	33.6	23.52	107	22.5	669.64
50,000	33.6	23.42	96	25.0	744.05

Table 14. Costs of hydrogen-generated power in the 95% renewables case.

Electrolysis (MW Installed)	Utilizable Excess Power [TWh]	TWh H_2	Average Operating Hours	Costs [bn EUR]	Power Costs [EUR/MWh]
10,000	91.6	64.1	1309	5	54.59
20,000	145.3	101.7	1038	10	68.80
30,000	174.1	1219	829	15	86.16
40,000	188.5	132.0	673	20	106.10
50,000	195.3	136.7	558	25	128.01
60,000	198.0	138.6	471	30	151.52
70,000	198.8	139.2	406	35	176.06
80,000	199.0	139.3	355	40	201.01
90,000	199.0	139.3	316	45	226.13

If we assume 10,000 MW as the installed capacity, which would provide the highest degree of usage and a power price level that seems internationally competitive, we could replace merely 4% of the fossil stocks.

In total, we consider H_2 not to be a realistic option, as long as natural gas is available and not extraordinarily expensive. If we now install seasonal overcapacity to reduce fossil fuels as much as possible, it can be analogously shown that a certain capacity is effective, but H_2 is still too expensive to utilize peak power. More or less 20% of the potential excess capacity can be used for H_2. These findings raise doubts about whether H_2 can be a seasonal storage solution—at least with regard to a short- to medium-term time horizon. Finally, we must note that the findings are based on a number of simplifications and assumptions. However, these were mostly beneficial for the H_2 case, so in reality, the costs are likely to be higher.

An alternative approach would be a 100% hydrogen infrastructure, which allows us to install more efficient fuel cell units, increasing efficiency from 50% to 70%. However, infrastructure investments, as well as high fuel cell costs, are likely to erase the gains in efficiency. While the first issue is complex and location-dependent, there is a lot of research about fuel cell technology and cost. High-end fuel cells generating electricity at efficiencies of 70% or higher degrees are estimated to have a range of about 65 USD/MWh to 87 USD/MWh leveraged cost of energy [27]. According to [27], assuming a EUR/USD exchange rate of 1.08, 5000 operating hours per year, and 20 years of operation, the 100% fuel cell options would add costs of between 18.24 and 15.37 EUR/MWh to the tables above. Hence, compared to the results in Table 14, we conclude that this approach would not provide a significant cost advantage and add additional questions about locations, time of availability, etc. without changing our general conclusions. Hence, we refrain from a more detailed discussion of this case.

4.4. Carbon Capture and Storage—A Mid-Term Solution for Structuring

We see in all scenarios described above that fossil power generation is needed in winter in order to structure demand and provide security of supply. Thereby, on average, we still have a share of fossil power generation of around 36%, assuming that we only install new renewables to meet additional demand for heat pumps. Nevertheless, due to the limited contribution of H_2 and battery storage, we have a trade-off between installing a massive overcapacity and accepting a certain share of fossil fuels. Decarbonizing this share requires methods to capture and store carbon, i.e., CCS. This idea is not new [28] but in the last 20 years, CCS has faced considerable opposition both from the scientific [29] and political and emotional sides (especially in Germany and Central Europe). However, with the current EU regulation and investment initiatives in Norway underway, CCS has become a key element in emission reduction policy. That is why we also consider this case. Schmelz et al. [30] see CCS costs for sites in the United States of between 52 and 90 USD/t, depending on the form of storage, whereby we can expect a certain decrease in costs. Thereby, experts estimate costs of 2–14 USD/t for onshore transport in the United States and 15 USD/t CO_2 for (onshore) storage [31,32]; thus, we yield (simplified) a range of 15 to 30 USD/t, or about 14 to 28 EUR/t at the current (2024) exchange rate. According to European project sources, storage cost and transportation are expected to be at 50 to 100 EUR/t for many locations, which corresponds to [32]. Therefore, we can work with a scenario range of 100 to 170 EUR/t CO_2 for the full CCU chain (leverage cost-based) if we want to use decarbonized fossil power. For reasons of simplification, we assume 100% efficiency of the capturing. In reality, this will be 95% to 98%, and we would still have to invest some money in purchasing European emission allowances.

5. Discussion

One major finding of this paper is consistent with the results in the H_2 case of [2]; the solution is not a single technology but a mix of different approaches, such as the gas grid combined with hydrogen injection, CCS, additional renewable capacity, and battery storage in large amounts. Given our calculations, this proves to be the only realistic option for the transformation of the German heating landscape toward heat pumps. This is because we not only have to replace the fuel but also the storage system of the fossil alternatives. Nevertheless, contrary to [2], we see a distinct tendency; as long as gas and coal are available, the best option is to use fossil power generation combined with CCS for structuring and providing the marginal demand. This is simply due to the fact that charging and using batteries consumes too much power; in addition, batteries prove to be only suitable for providing short-term flexibility. The often-promoted perfect fit of heat pumps and solar power for households is not working on an annual basis due to opposed seasonal patterns. Hydrogen, again, is too expensive in our calculation, despite rather favorable assumptions; we do not assume any cost either to adjust the gas grid to inject hydrogen or to utilize the hydrogen-rich natural gas. The installation of battery storage will be linked to massive adjustments as well. Also not considered but important is the water consumption of green hydrogen generation. Some locations will imply the usage of salt water. Desalination also consumes energy and will decrease the already low degree of effective usable energy. We will definitely need a hydrogen and battery infrastructure in the future, especially hydrogen as an alternative if gas becomes limited and expensive. This may be the case for Germany much earlier than for other countries due to political factors. This is—such as all disruptions—both a competitive disadvantage and a chance. Germany will need a strategy that focuses on industry sectors that can also survive under high energy prices.

Another issue is the timing of the transition strategy—especially from an environmental point of view. All measures described in this paper consume time and money, especially the CO_2 transport infrastructure for CCS and the hydrogen infrastructure. Even utilizing the gas grid requires a considerable investment and a clear mid-term plan about where to locate the hydrogen clusters to ensure efficient production, given the local restrictions for the hydrogen/methane mixture (a problem we especially see come up in summertime). The latest hydrogen strategy involves plans for a 12,000 km grid until 2032, in which most of the capacity is probably taken from the current natural gas infrastructure. Next is the question of how to replace the gas, which is also a question of whether we are able to install the necessary H_2 capacity to utilize such a grid (given the amount of gas needed). We see in our hydrogen analysis that hydrogen is expensive, even if the power is excess (renewable) power that would be lost otherwise if we consider full-cycle costs. This also implies the need for a certain amount of water desalination capacity. Even if we choose this way, the required adjustments and installation of required capacity will take years. This is also true for CCS. Currently, no transportation infrastructure is installed in all relevant countries. In the case of Germany, both a regulation for any CCS activity (currently this is prohibited) and an export framework are missing. Most CCS projects will be exploration projects. Thus, we have timelines of seven to eight years for large infrastructures such as CO_2 pipelines and therefore all the timing and financing risk, which requires acceptance, despite all risks and problems, and a stable/adaptive investment environment. This is of course true for every part of the solution and especially one problem, namely, the current gap between energy prices and the necessary hydrogen price. This corresponds to the widely discussed risk of a dark lull. Our analysis shows that this can be handled only if we accept a certain fossil production and a backup capacity close to the current level. If we want to shut down all coal-fired power plants, this means investments in gas and hydrogen-based plants. Given the energy demand and efficiency loss, the latter option is minor. An open question is if it is necessary to step out of coal completely if we manage to speed up the implementation of CCS and use it in the decarbonization of power generation. Note that, due to the high CO_2 coefficient here, CCS would be an absolutely necessary condition. From the point of view of risk diversification—especially regarding politics and logistics—coal should remain an option. This is an old idea that has already been described [28].

As mentioned above, an all-renewable solution would mean an extensive overcapacity. It has to be doubted whether the necessary financial resources are available and whether the resulting bow wave in terms of resource demand of CO_2, human resources, and required production capacities is manageable. Considering our model assumptions for 2030, we see a considerable demand for investment and a lot of required measures not taken to that extent so far (and not even planned to be taken). In particular, the German and EU hydrogen strategy has to be revised; the electrolyzer capacity to store surplus power in the gas grid should be prioritized. This should be combined with the virtualization of the hydrogen market by using hydrogen certificates, which would also be a good approach to boosting hydrogen itself and the right approach to establishing a liquid tradable hydrogen market in the near future.

Finally, one has to consider that without subsidies and additional regulations, the price of European CO_2 certificates is unlikely to increase and remain reliable at a level of 100 to 150 EUR/t, which is required to attract investments into CCS and hydrogen. Such a price scenario will slightly affect each of the scenarios discussed above but will not change the major finding, i.e., that a new inter-seasonal structuring solution must be implemented.

6. Conclusions

According to the current German government (Year 2024), up to 6 million heat pumps are to be installed by 2030. This article is concerned with how to supply the required (renewable) energy and at what costs—taking into account that the challenge is not only providing this additional energy but also delivering it on any day at any demand level. For natural gas, a well-designed storage infrastructure exists, which is to be substituted by heat pumps without an alternative for structuring. Various possibilities, such as batteries and hydrogen, are discussed. We show that the shift to heat pumps induces a need to redesign the energy production itself and use all available technologies, i.e., renewables, batteries, hydrogen, the natural gas system, and CCS, in an integrated manner to maximize success and reduce cost. Otherwise, we risk unnecessary investments by factors 4 to 10 or effectively increase the demand for fossil power, also on an annual level. Using synergy and portfolio effects among all these technologies and combining them is not only a key factor but from our point of view, the only way to success.

In the current debate, the technologies and corresponding political measures are discussed separately. Based on our analysis, integrated and portfolio-based planning has to be intensified. In particular, the hydrogen strategy should be adjusted, and the hydrogen/gas combination to store and distribute hydrogen via the gas grid has to be strengthened since we see this as a key factor and the most realistic measure within the timeline between 2030 and 2035. In addition, this seems to be the only post-gas solution. Alternatives such as a certain overcapacity and, especially, CCS combined with a high fossil share are less expensive. Other inter-seasonal battery storage still shows a low technology readiness level and is simply too expensive at current price levels. Moreover, their resource demand is considerable and not consistent with the target of reducing the long-term dependency on foreign fossils. CCS is cheaper than the alternatives regarding pure numbers but critical with respect to the time frame and the fact that we talk about an extensive and expensive new infrastructure. Thus, we remain doubtful in that direction since we do not see enough measures so far to believe that CCS can become a realistic option. The market lacks scalable seasonal battery technology with adequate market readiness and at reasonable costs so far [5]. Using conventional batteries would be highly inefficient.

Finally, please note that this article is not an exact forecast but is based on multiple assumptions. At this point, many effects cannot be quantified exactly, and technologies such as CCS have to be applied first in order to actually show costs and performance. The adequate combination of different technologies needs to be analyzed for different regions like southern Europe or the US individually. An important question is also the timing, since a quick and cheap 50% solution, plus postponing the perfect solution for some years, may work better from an intertemporal point of view than waiting for the expensive 100% solution that might come up in some years. The winning strategy is avoiding as much CO_2 as possible for a given amount of money and spending as little money as possible for future solutions.

Author Contributions: Conceptualization, A.v.D. and S.S.; methodology, A.v.D. and S.S.; validation, A.v.D. and S.S.; formal analysis, A.v.D. and S.S.; investigation, A.v.D. and S.S.; resources, S.S.; data curation, A.v.D. and S.S.; writing—original draft preparation, A.v.D. and S.S.; writing—review and editing, A.v.D. and S.S.; visualization, A.v.D.; supervision, A.v.D. and S.S.; project administration, S.S. All authors have read and agreed to the published version of the manuscript.

Funding: This research received no external funding.

Data Availability Statement: The data presented in this study are available on request from the corresponding author. The data are not publicly available due to legal reasons.

Conflicts of Interest: This article solely represents the personal views of the authors, especially of Andreas von Döllen, and not necessarily the position of the Wintershall Dea AG, the employer of Andreas von Döllen. The company has not been involved in the research at all.

References

1. BMBF (2023a) Federal Ministry for Economic Affairs and Climate Action. The National Hydrogen Strategy. Available online: https://www.bmbf.de/bmbf/shareddocs/downloads/files/bmwi_nationale-wasserstoffstrategie_eng_s01.pdf?__blob=publicationFile&v=2 (accessed on 12 June 2024).
2. von Döllen, A.; Hwang, Y.S.; Schlüter, S. The Future is Colorful—An Analysis of the CO_2 Bow Wave and Why Green Hydrogen Cannot Do It Alone. *Energies* **2021**, *14*, 5720. [CrossRef]

3. Ganzer, C.; Pratama, Y.W.; Dowell, N.M. The role and value of inter-seasonal grid-scale energy storage in net zero electricity systems. *Int. J. Greenh. Gas Control* **2022**, *120*, 103740. [CrossRef]
4. Albertus, P.; Manser, J.S.; Litzelman, S. Long-Duration Electricity Storage Applications, Economics, and Technologies. *Joule* **2020**, *4*, 21–32. [CrossRef]
5. Armstrong, R.; Chiang, Y.-M.; Grünspecht, H.; Brushett, F.; Deutch, J.; Engelkemier, S.; Gençer, E.; Jaffe, R.; Joskow, P.; Mallapragada, D.; et al. *The Future of Energy Storage*; MIT Energy Initiative, Massachusetts Institute of Technology: Cambridge, MA, USA, 2022.
6. Papadakis, C.N.; Fafalakis, M.; Katsaprakakis, D. A Review of Pumped Hydro Storage Systems. *Energies* **2023**, *16*, 4516. [CrossRef]
7. Ruhnau, O.; Hirth, L.; Praktiknjo, A. Time Series of Heat Demand and Heat Pump Efficiency for Energy System Modeling. *Nat. Sci. Data* **2019**, *6*, 189. [CrossRef] [PubMed]
8. Reichel, C.; Müller, A.; Friedrich, L.; Herceg, S.; Mittag, M.; Neuhaus, D.H. CO_2 Emissions Of Silicon Photovoltaic Modules—Impact of Module Design and Production Location. In Proceedings of the WCPEC-8, 8th World Conference on Photovoltaic Energy Conversion, Milan, Italy, 26–30 September 2022.
9. Umweltbundesamt. (2023) Photovoltaik. Available online: https://www.umweltbundesamt.de/themen/klima-energie/erneuerbare-energien/photovoltaik#photovoltaik (accessed on 12 June 2024).
10. BMBF (2023b) Federal Ministry for Economic Affairs and Climate Action (2023a). Neuer Schwung für Erneuerbare Energien. Available online: https://www.bmwk.de/Redaktion/DE/Infografiken/Schlaglichter/2022/10/05-neuer-schwung-fuer-erneuerbare-energien-download.pdf?__blob=publicationFile&v=1 (accessed on 12 June 2024).
11. SMARD. (2023) Energy Market Data Provided by the Bundesnetzagentur. Available online: https://www.smard.de/home/downloadcenter/download-marktdaten/?downloadAttributes=%7B%22selectedCategory%22:1,%22selectedSubCategory%22:3,%22selectedRegion%22:false,%22selectedFileType%22:%22XLSX%22,%22from%22:1627596000000,%22to%22:1627682399999%7D (accessed on 12 June 2024).
12. OPSD (2023) Open Power System Data—A Free and Open Platform for Power System Modelling. Available online: https://open-power-system-data.org/ (accessed on 12 June 2024).
13. Watson, S.D.; Crawley, J.; Lomas, K.J.; Buswell, R.A. Predicting Future GB Heat Pump Electricity Demand. *Energy Build.* **2023**, *286*, 112917. [CrossRef]
14. Kost, C.; Shammugam, S.; Fluri, V.; Peper, D.; Memar, A.D.; Schlegl, T. Levelized Cost of Electricity Renewable Energy Technologies. In *Fraunhofer Institute for Solar Energy Systems ISE Working Paper*; Fraunhofer Institute for Solar Energy Systems ISE: Freiburg, Germany, 2021.
15. Jean, J.; Brown, P.R.; Jaffe, R.L.; Buonassisi, T.; Bulović, V. Pathways for solar photovoltaics. *Energy Environ. Sci.* **2015**, *8*, 1200–1219. [CrossRef]
16. Destatis Statistisches Bundesamt. *Umweltökonomische Gesamtrechnung, Berichtszeitraum 2000–2020*; Deutsches Statistisches Bundesamt: Wiesbaden, Germany, 2022. Available online: https://www.destatis.de/DE/Themen/Gesellschaft-Umwelt/Umwelt/UGR/private-haushalte/Publikationen/Downloads/haushalte-umwelt-pdf-5851319.pdf?__blob=publicationFile (accessed on 12 June 2024).
17. Lavoie, M. *Post-Keynesian Economics: New Foundations*; Edward Elgar Publishing: Gloucestershire, UK, 2022.
18. Destatis Statistisches Bundesamt. *Press Release No. N 012 of 1 March 2023*; Deutsches Statistisches Bundesamt: Wiesbaden, Germany, 2023. Available online: https://www.destatis.de/DE/Presse/Pressemitteilungen/2023/03/PD23_N012_43.html (accessed on 12 June 2024).
19. Carrara, S.; Alves Dias, P.; Plazzotta, B.; Pavel, C. *Raw Materials Demand for Wind and Solar PV Technologies in the Transition towards a Decarbonised Energy System*; Publication Office of the European Union: Luxembourg, 2020; EUR 30095 EN.
20. Elshkaki, A.; Graedel, T.E.; Ciacci, L.; Reck, B.K. Resource demand scenarios for the major metals. *Environ. Sci. Technol.* **2018**, *52*, 2491–2497. [CrossRef] [PubMed]
21. Gondal, I.A. Hydrogen integration in power-to-gas networks. *Int. J. Hydrogen Energy* **2019**, *44*, 1803–1815. [CrossRef]
22. Bard, J.; Gerhardt, N.; Selzam, P.; Beil, M.; Wiemer, M.; Buddensiek, M. *The Limitations of Hydrogen Blending in the European Gas Grid*; Fraunhofer Institute for Energy Economics and Energy System Technology: Kassel, Germany, 2022. Available online: https://www.iee.fraunhofer.de/content/dam/iee/energiesystemtechnik/en/documents/Studies-Reports/FINAL_FraunhoferIEE_ShortStudy_H2_Blending_EU_ECF_Jan22.pdf (accessed on 12 June 2024).
23. Mould, K.; Silva, F.; Knott, S.; O'Regan, B. A comparative analysis of biogas and hydrogen, and the impact of the certificates and blockchain new paradigms. *Int. J. Hydrogen Energy* **2022**, *47*, 39303–39318. [CrossRef]
24. White, L.; Fazeli, R.; Cheng, W.; Aisbett, E.; Beck, F.J.; Baldwin, K.G.H.; Howarth, P.; O'Neill, L. Towards emissions certification systems for international trade in hydrogen: The policy challenge of defining boundaries for emissions accounting. *Energy* **2021**, *215*, 119139. [CrossRef]
25. Bundesnetzagentur (2023) Rückblick: Gasversorgung im Jahr 2022. Available online: https://www.bundesnetzagentur.de/DE/Gasversorgung/a_Gasversorgung_2022/start.html (accessed on 12 June 2024).
26. Holst, M.; Aschbrenner, S.; Smolinka, T.; Voglstätter, C.; Grimm, G. *Cost Forecast for Low Temperature Electrolysis-Technology Driven Bottom-up Prognosis for PEM and Alkaline Water Electrolysis Systems*; Fraunhofer Institute for Solar Energy Systems ISE: Freiburg, Germany, 2021.

27. Chan, C.Y.; Rosner, F.; Samuelsen, S. Techno-Economic Analysis of Solid Oxide Fuel Cell-Gas Turbine Hybrid Systems for Stationary Power Applications Using Renewable Hydrogen. *Energies* **2023**, *16*, 4955. [CrossRef]
28. Ansolabehere, S.; Beer, J.; Deutch, J.; Ellerman, D.; Friedmann, S.J.; Herzog, H.; Jacoby, H.; Joskow, P.; Mcrae, G.; Lester, R.; et al. *Future of Coal*; MIT Energy Initiative, Massachusetts Institute of Technology: Cambridge, MA, USA, 2007.
29. Zoback, M.; Smit, D. Meeting the challenges of large-scale carbon storage and hydrogen production. *Proc. Natl. Acad. Sci. USA* **2023**, *120*, e2202397120. [CrossRef] [PubMed]
30. Schmelz, W.J.; Hochman, G.; Miller, K.G. Total cost of carbon capture and storage implemented at a regional scale: Northeastern and midwestern United States. *Interface Focus* **2020**, *10*, 20190065. [CrossRef] [PubMed]
31. Stern, A.; Berghout, N. *Is Carbon Capture to Expensive*; International Energy Agency: Paris, France, 2021. Available online: https://www.iea.org/commentaries/is-carbon-capture-too-expensive (accessed on 12 June 2024).
32. Lockwood, T. *Mapping the Cost of Carbon Capture and Storage in Europe*; Clean Air task Force: Boston, MA, USA, 2023. Available online: https://www.catf.us/2023/02/mapping-cost-carbon-capture-storage-europe/ (accessed on 12 June 2024).

energies

MDPI

Article

Effects of Removing Energy Subsidies and Implementing Carbon Taxes on Urban, Rural and Gender Welfare: Evidence from Mexico

Jorge Alberto Rosas Flores *, David Morillón Gálvez and Rodolfo Silva

Instituto de Ingeniería UNAM, Circuito Escolar, Ciudad Universitaria, Ciudad de México 04510, Mexico; damg@iingen.unam.mx (D.M.G.); rsilvac@iingen.unam.mx (R.S.)
* Correspondence: jorge.rosas.flores@gmail.com

Abstract: The demand for different energy goods and services is a fundamental component in a country's economic structure for development. Understanding it is vital in designing economic policies, such as taxes, that can improve the welfare of the population. A comprehension of the distributional effects of elasticities and the application of them to simulate household responses to price changes, as well as a calculation of the welfare impacts on poor and rich households in Mexico, should inform policy design. This paper uses the Household Income and Expenditure Survey (ENIGH) from 1996 to 2018 to estimate the demand of Mexican households for fuels, specifically electricity, liquefied petroleum gas, and gasoline. A Quasi Ideal Quadratic Demand System (QUAIDS) is employed to analyse the effects of removing energy subsidies and introducing a carbon tax. The results indicate that welfare losses would be regressive concerning electricity price increases, while changes in gasoline prices would be progressive. Redistributing the tax revenues accrued by removing energy subsidies and imposing the carbon tax would have more progressive effects on the economy of Mexican households, with welfare gains of up to 350% for the poorest households in the case of electricity consumption taxes.

Keywords: Mexico; QUAIDS; subsidies; carbon tax; welfare

1. Introduction

In 2018, the National Council for the Evaluation of Social Development Policy (CONEVAL) stated that 42% of the population of Mexico lives below the official poverty line, and 7.4% are in a situation of extreme poverty (https://www.coneval.org.mx/Medicion/MP/Documents/Pobreza_18/Pobreza_2018_CONEVAL, accessed on 20 December 2023). In Mexico, as in other developing countries with high CO_2 emissions, the socioeconomic welfare losses suffered by much of the population due to high energy prices are particularly critical. It is therefore important to focus on the development of optimal climate policy, poverty reduction, and further economic development. In recent years, the potential negative environmental and global warming effects of energy subsidies have been the subject of international attention. In Mexico, these subsidies, mainly on electricity and gasoline consumption, are among the highest in the world, and their costs go beyond the environmental effects that directly affect the population. In 2009, as part of the G20 and Asia-Pacific Economic Cooperation (APEC) agreements, Mexico committed to "phase out and rationalize over the medium-term inefficient fossil fuel subsidies, which encourage wasteful consumption... while providing targeted support for the poorest" (G20 (2009) Declaración de los Líderes del G-20 en la Cumbre de Pittsburgh y APEC (2009) APEC Summit Leaders' Declaration: Sustaining growth, connecting the region. https://www.apec.org/meeting-papers/leaders-declarations/2009/2009_aelm.aspx, accessed on 17 October 2023). In 2017, Mexico was one of twelve members of the World Trade Organization (WTO) to present a statement at the 11th Ministerial Conference to promote the debate on fossil fuel subsidies at the WTO (OMC (2017) Fossil Fuel Subsidies Reform Ministerial Statement https://www.wto.org/english/tratop_e/envir_e/fossil_fuel_e.htm, accessed on 1 March 2023).

In a World Bank study on global energy subsidies [1], it was stated that electricity tariffs in emerging markets and developing countries do not cover the suppliers' generation, transmission, and distribution costs. It was also noted that these costs are not calculated based on the provision of the service, but only on the price of the resources required. In the case of Mexico, electricity subsidies cost the Federal Government an average of MXN 98 billion (USD ~5 billion) every year, 0.9% of the gross domestic product (GDP).

In developing countries, such as Mexico, energy subsidies tend to benefit two groups of end users: low-income and high-income users. This creates two major problems: the effects are regressive (disproportionately benefiting those who consume most, rather than the poorest), and it reduces the opportunity for subsidies to be directed toward other development goals.

The purpose of this article is to analyse the residential demand for the fuels most used in Mexican homes, showing the price and income elasticities of demand. The analysis is carried out using a QUAIDS model; the results obtained help provide information relevant to the decision-making process for the design of more effective energy policies, in the area of energy-saving policies and the reduction in greenhouse emissions.

On the other hand, this paper aims to show the impacts on urban and rural households in Mexico caused by the rise in energy prices when fuel and electricity subsidies were eliminated, and/or when there was an increase in the carbon tax. Several simulations were carried out, showing that the present policy regarding subsidies benefits the poorest far less than is expected.

2. Overview of Energy Demand in Mexico

Analysing energy consumption in Mexico from 1996 to 2018, (Figure 1) in Mexico, it is seen that in 2018, transportation accounted for 46.5% of the total energy consumption, 31.8% was used in the industrial sector, and 14.3% was used in the residential sector. Over this period, the energy consumed by the residential sector increased 5.9%, although its share of the total energy used decreased from 20% to 14.3%, a decrease that may be, in part, due to the growth in the energy demand from the transportation sector.

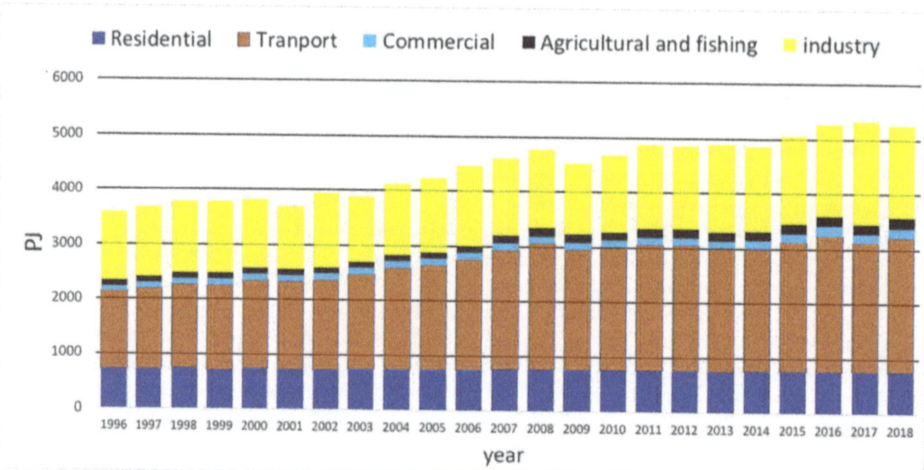

Figure 1. Energy consumption in Mexico, 1996–2018, by sector. Source: data from the Energy Ministry (http://sie.energia.gob.mx/bdiController.do?action=temas, accessed on 2 January 2023).

In the residential sector, the energy used includes that for cooking, heating, lighting, cooling, and entertainment. To meet these demands, various sources of energy are used, such as electricity, liquefied petroleum gas (LPG), natural gas, kerosene, and wood. Figure 2 shows the energy consumed in the residential sector in the same period, by fuel type. The amount of electricity used in this period grew by 3.2% per year on average, the highest growth rate of the fuels considered, its share in the overall consumption rising from 14% in 1996 to 23% in 2018. On the other hand, the share of LPG fell from 43% in 1996 to 38.5% in 2018, while firewood showed a more significant drop (from 39% to 30%).

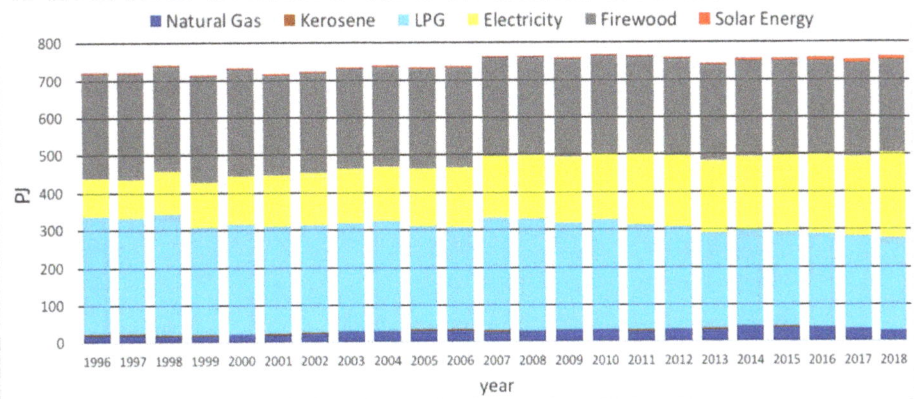

Figure 2. Energy consumption in the residential sector in Mexico, 1996–2018, categorized by fuel. Source: Energy Ministry data (http://sie.energia.gob.mx/bdiController.do?action=temas, access on 17 March 2023).

Official statistics (https://www.coneval.org.mx/Medicion/MP/Paginas/Pobreza-2018.aspx, accessed on 17 March 2023) show that almost 60 million people in Mexico were living below the poverty line in 2018. In urban and rural areas, the proportion of the poorest households was 14.3% and 31.9%, respectively. Within this group, 57.5% of urban households and 37.8% of rural households had to spend more than 10% of their income on energy (excluding gasoline) (https://www.inegi.org.mx/programas/enigh/nc/2018/, accessed on 16 March 2023).

2.1. Fuel Subsidies in Mexico

In Mexico, subsidies are granted by the government to the state-owned electricity generation company, the Federal Electricity Commission (CFE), to cover electricity generation costs. This subsidy increases as generation prices rise, benefitting 40 million households; between 2010 and 2020 the electricity subsidy accounted for 0.22% of the national GDP [2].

On the other hand, the price of gasoline has a Special tax on products and services (IEPS) that the government uses as a subsidy. By lowering this tax, the government controls the price paid by the consumer. In 2012 this subsidy was 1.28% of the GDP (DOF 2015) [3].

In the case of LPG, the price was regulated by the Federal Government through the Energy Regulatory Commission until 2017, when this commission accepted the market price of LPG with international references. This process was carried out gradually from 2014 to 2017, in line with the Energy Transition Law of 2015 (https://www.gob.mx/cms/uploads/attachment/file/409014/1._Liberaci_n_de_precios_finales_al_p_blico_de_gas_LP.pdf, accessed on 1 March 2023) [1].

In 2015 the Ministry of Finance and Public Credit (SHCP) updated the methodology to determine the fiscal stimulus on automotive fuels in terms of the special tax on production and services (IEPS). This regulates the price paid by the final consumer by reducing the IEPS paid by the consumer.

Energy subsidies in Mexico are questionable as a public policy, since they have collateral effects on both the population and the environment.

1. Energy subsidies are not an efficient means of providing support to the poor, and reinforce income inequalities, as most of the benefits of these subsidies go to those who can afford to consume more energy (rich households). This inequality is more acute in rural areas.
2. Energy subsidies consume resources that could be used to guarantee services that are human rights, such as health and education, or that could be used in Mexico's economic development, e.g., for infrastructure. Figure 3 shows the percentage of the GPD used in subsidies from 2010 to 2020.
3. The low energy prices, as a result of the subsidies, discourage investment from the private sector and reduce the stimulus for energy-efficient strategies to be used the consumer.
4. Inefficient energy consumption in turn produces higher CO_2 emissions, and thus indirectly increases pollution in cities.

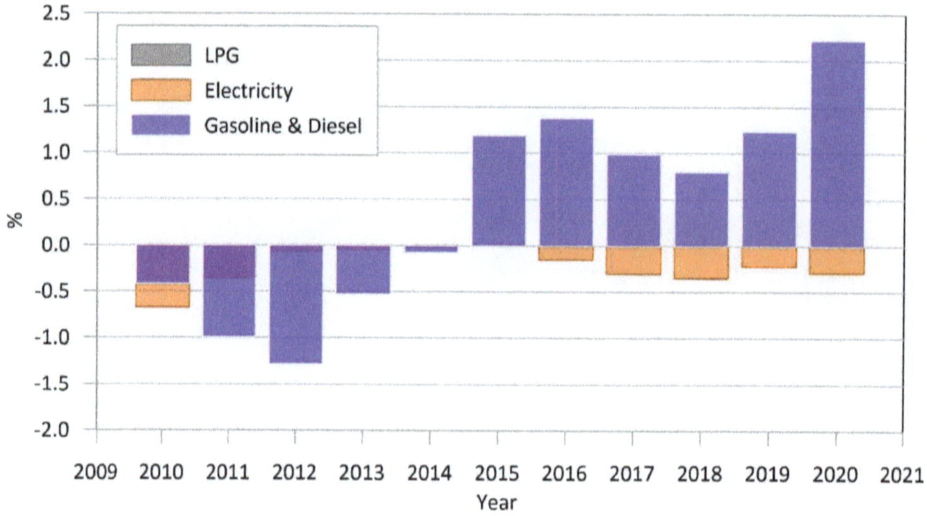

Figure 3. Fuel subsidies in Mexico as a percentage of GDP (negative value = subsidy). Data from [1,2].

2.2. Studies on Fuel Subsidies

There are several studies on developing countries showing that fuel subsidies result in a regressive distribution, meaning that high-income households receive higher amounts of subsidies than low-income households do. For example, Ref. [4] described carbon tax distribution in India, Ref. [5] analysed carbon tax distribution in Brazil, while carbon taxes and poverty in Thailand were discussed by [6].

A study on developing countries published by [7] concludes that the direct effect is either distributionally neutral, approximately (Bolivia and Mali), or regressive (Ghana, Jordan and Sri Lanka). There are very few studies on the distributional effects of fuel subsidies in Mexico, where it is underlined that the electricity and LPG consumed in Mexican households are normal goods, not luxuries, while gasoline is a luxury good. An initial study on the distributional effects of the elimination of subsidies and carbon taxes [8] found that the reduction in gasoline subsidies is progressive.

Other econometric studies in Mexico that use the general equilibrium model [9] show how distributional effects depend to a large extent on income recycling. While this does not explicitly cover the distributional effects of recycling, it does show various effects on the population [10]. These results have helped in the development of Mexico's climate policy.

Other studies note the new energy market prices following the Energy Transaction Law of 2015, while an estimate of the distributional effects due to the introduction of carbon tax from 2002 to 2014 was offered by [11–13]. Our study describes the distribution of welfare losses in Mexico due to the increase in fuel prices from 1996 to 2018, caused by the elimination of the fuel subsidy, as well as by the increase in the carbon tax. This study also highlights the gap between urban and rural households.

3. Data Description

Data from the National Households Income and Expenditure Survey (ENIGH, from the Spanish abbreviation Encuesta Nacional de Ingresos y Gastos de los Hogares) for 1996–2018 were used for the analysis carried out in this paper. The ENIGH survey is generally conducted every two years to collect data from a large sample of households across Mexico, in rural and urban areas, using the interview method (rural areas include households living in municipalities with fewer than 2500 inhabitants, while municipalities with over 2500 inhabitants are considered urban areas). A detailed description of the sampling methodology is described on the web page of the National Institute of Statistics and Geography (INEGI (household survey): http://www.inegi.org.mx/est/contenidos/proyectos/encuestas/hogares/default.aspx) accessed on 2 February 2023 (INEGI from the Spanish name) (INEGI 1997 to 2019).

The sampling design is based on stratified random sampling where the households are selected with equal probability, with a random start. Household consumption in the survey includes the consumption of goods and services through purchases, receipts in exchange of goods and services, and others. Household consumption–expenditure refers to the amount and value of household consumption for the first three months of the year during which the survey was conducted. The survey also comprises detailed information on demographic characteristics and the attributes of dwellings.

To make the data suitable for analysis, a series of transformations must be performed. First, all homes with no energy consumption are eliminated, because this is not real information. The operations were carried out in different ways. Some of the variables were first converted into logarithms and/or percentages before conducting the regression analysis. These variables include the price of different fuels, budget participation, and fuel expenditure. This includes the price of the various types of transport, budget shares, and public transport or expenditure on gasoline. All households that report null expenditure for bus, subway, gasoline, and taxi, as well as those with expenditure on each type of transport, were eliminated to avoid outliers. After transforming and filtering the data, 73,508 observations were kept for the estimation and the QUAIDS model could be created.

3.1. Energy Prices in Mexico

The cost of energy in Mexico has many variables, including those of the market, as well as regulations, collection methodologies, and particularities in the way each client consumes energy. This section presents the evolution of energy prices.

3.1.1. Electricity

The Mexican government has subsidized electricity costs for end consumers through tariffs determined by the Ministry of Finance and Public Credit (SHCP) for many years, using information on generation costs from the Federal Electricity Commission (CFE). These subsidies have a high economic and social cost for the country (see Figure 3).

In 2002, an energy reform made the tariff structure less regressive, introducing a domestic high consumption tariff (DAC), for households consuming over 250 kWh per month. This is a fixed monthly fee that varies by region, and the average monthly cost in 2018 was around USD 5.35 (MXN 100). In November 2017, the Energy Regulatory Commission (CRE) implemented a new means of devising electricity tariffs that is less regressive and will help recover generation and distribution costs.

3.1.2. Gasoline

Although there are other private companies, PEMEX is a federal government-owned public company that owns 85% of the national market. Gasoline prices vary depending on the type of gasoline and the point of sale. On average, the price of PEMEX Premium gasoline is 11–23% higher than that of PEMEX Magna gasoline. For reference, PEMEX Magna was MXN 21.71 per litre in 31 December 2018, equivalent to USD 1.10 per litre (USD 1 was worth MXN 19.65 on 31 December 2018). In 2012, the subsidy for fuels such as gasoline and diesel in Mexico reached a historical maximum of 1.2% of the country's GDP (see Figure 3).

3.1.3. Liquefied Petroleum Gas

Like other fuels in Mexico, LPG had a fixed price before 2017, which was lower than international market prices, in effect a subsidy for all end consumers. From 2010, the Mexican government gradually increased prices, bringing them closer to market prices, and by 11 August 2014, with the new hydrocarbons law (http://www.dof.gob.mx/nota_detalle.php?codigo=5355989&fecha=11/08/2014, accessed on 12 March 2023), market liberalisation came into force in 2017.

3.2. *Taxes*

In Mexico, there are two Pigouvian taxes that aim to internalize the negative externalities caused by an economic agent that affect the economic activities of other economic agents damaging the environment. The IEPS (Special Tax on Production and Services) is a federal tax applied to automotive gasoline, automotive diesel, and its biofuel equivalents. The revenues from these taxes go to all Mexican states and municipalities to mitigate environmental damage caused by climate change. The 2013 carbon tax has been applied since January 2014 (https://www.dof.gob.mx/nota_detalle.php?codigo=5547405&fecha=28/12/2018, accessed on 3 May 2023). It is internalized in the Special Tax Law on Products and Services.

3.3. *The QUAIDS Model: Background*

The empirical analysis uses a quadratic AIDS (almost ideal demand system) model of energy demand in Mexican households. Ref. [14] combined the Translog and Rotterdam models in the almost ideal demand system (AIDS), which together improve the properties, including the arbitrary approximation for any demand system to the first order, where the information from consumers is added and the homogeneity and symmetry restrictions are tested [15].

Thus, the AIDS has been used to model clustered asset systems, but with this model, it is difficult to obtain the effects of the nonlinear theory of Engel curves.

With these restrictions, a quadratic term is added in the logarithm of income to the AIDS model and thus the quadratic model is obtained, giving greater flexibility to the demand system.

This model was presented as a proposal to support the characteristics of congruence with microeconomic theory in terms of homogeneity, maximisation, and Slutsky symmetry, but through the inclusion of nonlinearity, it achieves a better approximation of the Engel curves of certain product groups. By adding quadratic and logarithmic terms in the model, goods are allowed to behave as luxury products for certain income levels and as necessities, or basic products, at other levels.

The representation is thus achieved reasonably well by incorporating the quadratic term. Ref. [16] showed that for demand models, the rank two generalized linear form is a necessary and sufficient measure. The quadratic almost ideal demand system (QUAIDS) is ranked third and can improve nonlinear Engel curves in an empirical analysis.

There has been work conducted for Latin American countries where, based on the estimation of a demand system, estimates of price and income elasticity are made. These works include those of [17–19]. Based on the 1987–1988 and 1995–1996 expenditure surveys in Brazil, Ref. [17] employed an AIDS model using the two-stage budget allocation technique to calculate elasticities for different population groups (the richest 50% and the poorest 50%) and the total elasticities for the basic food groups and subgroups. Ref. [18] estimated values for basic basket products, based on the AIDS model with time series data from the 1967–2007 period from the National Accounts for Colombia. Ref. [19] used the income survey and Colombia's 2006–2007 expenditures, and estimated various linear demand systems through cross-sectional models.

As seen in the international literature, various parametric and non-parametric specifications have been developed for the estimation of Engel curves. For this, different estimation methods have been considered, linearity or non-linearity, in expenditure or income, which shows the importance of making estimates using income strata since elasticities vary greatly between different groups.

Likewise, Refs. [20,21] estimated a complete food demand system for Italy and Indonesia, respectively, from the QUAIDS models, also incorporating demographic and regional variables.

4. The QUAIDS Model of Demand

The AIDS model, originally introduced by [14] has been widely adopted in estimating demand elasticities and energy demand systems [22–25]. On the other hand, Ref. [26] presents a generalisation of the QUAIDS model. The quasi-ideal quadratic demand system (QUAIDS) allows for a more flexible relationship between the share of spending on goods and total household spending. Studies carried out in various countries have used this method, for example Spain, Germany, and the United States, as shown by [25,27,28].

The QUAIDS model is a system of equations that represents the proportion or share of residential spending on a good as a function of prices in addition to the total spending on all the goods that make up the system. According to [15], the model is as follows:

$$w_i = \propto_i + \sum_{j=1}^{n} \text{In } p_j + \beta_i \text{ In} \left(\frac{x}{a(p)}\right) + \frac{\lambda_i}{b(p)} \left\{\text{In}\left[\frac{x}{a(p)}\right]\right\}^2 \tag{1}$$

where the subscripts i and j correspond to the different goods in the size budget, n, while identifying each household. The term w_i represents the proportion of expenditure on the good, i, that each household makes within the basket. The variables p_j and x refer to the price of good j and the total expenditure on all goods in the system in each household, respectively. The nonlinearity of the QUAIDS model is due to the indices of prices a and b, defined as follows:

$$\text{In } a(p) = \propto_0 + \sum_{i=1}^{n} \propto_i \text{In } p_i + \frac{1}{2}\sum_{i=1}^{n}\sum_{j=1}^{n} \gamma_{ij} \text{In } p_i \text{ In } p_j \tag{2}$$

where ln $a(p)$ is the transcendental log function

$$b(p) = \prod_{i=1}^{n} {p_i}^{\beta_i} \tag{3}$$

and $b(p)$ is the Cobb–Douglas price aggregator.

From Equation (1), it can be seen that the significance of parameter λ_i indicates that residential spending should be analysed using a QUAIDS model; otherwise, the AIDS model of [14] is the most suitable. As mentioned, one of the advantages of the QUAIDS model is that its functional form considers the rational behaviour of a consumer. In order for this to be satisfied, the model parameters are subject to the following restrictions:

$$\sum_{i=1}^{n} \propto_i = 1; \quad \sum_{i=1}^{n} \gamma_{ij} = 0; \quad \sum_{i=1}^{n} \beta_i = 0; \quad \sum_{i=1}^{n} \lambda_i = 0 \tag{4}$$

Homogeneity

$$\sum_{i=1}^{n} \gamma_{ij} = 0 \tag{5}$$

Symmetry

$$\gamma_{ij} = \gamma_{ji} \quad (6)$$

The adding up restriction can be satisfied by dropping one equation from the estimation. The homogeneity and symmetry restrictions can be imposed on the parameters to be estimated and assessed using likelihood ratio tests.

4.1. Demand Elasticities

To evaluate consumer elasticities and distribution quantities in relation to changes in spending and prices, spending elasticities, Marshallian price elasticities, and Hicks price elasticities were calculated using the indicated parameters of QUAIDS models during the study period. The budget elasticities could be calculated from the following equations.

It should be remembered that the coefficients that go with prices and total expenditure in the QUAIDS model do not correspond to the price and income elasticities of demand. These sensitivity measures are obtained through the expressions (7)–(9). This analysis was conducted using a QUAIDS model adapted to consider the censoring that arises because not all households consume all fuels at the same time.

The significance of the residuals of the auxiliary regression in all the equations of the system confirms that residential spending on the goods analysed is an endogenous variable. Furthermore, the significance in two equations of the inverse of the Mills ratio in the system of equations corroborates the importance of treating censoring in residential energy expenditure.

$$e_i = \frac{\mu_i}{w_i} + 1 \quad (7)$$

With

$$\mu_i = \frac{\delta w_i}{\delta \text{ In}_x} = \beta_i + \frac{2\lambda_i}{b(p)}\left\{\text{In}\left[\frac{x}{a(p)}\right]\right\} \quad (8)$$

The uncompensated price elasticity is given by the following:

$$e_{ij}^{u} = \frac{\mu_{ij}}{w_i} - \delta_{ij} \quad (9)$$

According to Roy's identity, the budget shares are given by the following:

$$\mu_{ij} = \frac{\delta w_i}{\delta \text{ In } p_i} = \gamma_{ij} = \mu_i\left(\alpha_j + \sum_{k}^{n} \gamma_{jk}\text{In}p_k\right) - \frac{\lambda_i \beta_j}{b(p)}\left\{\text{In}\left[\frac{x}{a(p)}\right]\right\}^2 \quad (10)$$

4.2. Welfare Effects (Simulation)

To determine the impacts of price changes on the well-being of the population, a simulation was carried out that made changes to the prices of energy systems [29]. The simulation illustrated the impacts on welfare resulting from the changes in energy prices caused by the implementation of carbon taxes, in three different scenarios, for several households (this work use stata files from the paper of [29], with data on Mexico).

Our measure of the effect on welfare of a price change is the compensating change: the amount of income that must be given to a household for it to be indifferent to the new price vector. This calculation is carried out using the following expression:

$$\text{In } V = \left\{\left[\frac{\text{In } x - \text{In } a(p)}{b(p)}\right]^{-1} + \lambda(p)\right\}^{-1} \quad (11)$$

On the other hand, for each household in the sample, to evaluate the effects on welfare of a specific set of increases in fuel prices, the following equation was used:

$$\frac{\Delta p_i}{p_i^0} = \frac{p_i^1 - p_i^0}{p_i^0} \quad (12)$$

where Δp_i is the price change of good i, p_i^0 is the price before the implementation of the price or rate increase, and p_i^1 is the new price resulting from a new price or tariff.

$$p_i^1 = \left(1 + \frac{\Delta p_i}{p_i^0}\right) p_i^0 \tag{13}$$

The literature on the welfare impacts of increases in energy prices and subsidy reforms generally focuses on first-order effects, as in [30]. These first-order effects, based on the earlier work of [31,32], only require observed demand, and no additional information on the substitution behaviour resulting from price changes. First-order welfare losses relative to income (total expenditures are used as a proxy) are calculated as follows:

$$FO = \sum_{i=1}^{n} w_i \left(\frac{\Delta p_i}{p_i^0}\right) \tag{14}$$

With the estimated coefficients, we calculated a second-order approximation of the compensatory variation (CV), which is the amount of money that the household needs to be compensated to reach utility level u0 (before the price changes) over total household expenses:

$$CV = \sum_{i=1}^{n} w_i \left(\frac{\Delta p_i}{p_i^0}\right) + \frac{1}{2} \sum_{i=1}^{n} \sum_{j=1}^{n} w_i e_{ij} \left(\frac{\Delta p_i}{p_i^0}\right) \left(\frac{\Delta p_j}{p_j^0}\right) \tag{15}$$

The CV is compared with the first-order effect to clarify the need to estimate a demand system in our context.

4.3. The Effects of Reducing Subsidies and Increasing Carbon Taxes

Microsimulations of price and tax increases have been used to evaluate the welfare and distributional effects of carbon tax in several countries, for example by [33] for the United States, [34] for Denmark, [28] for Italy, [35] for Ireland, [29] for South Africa, and [8,12,13], for Mexico.

The additional contribution offered here is the use of the QUAIDS model to evaluate the welfare loss from subsidy elimination and the increase in the carbon tax for fuels.

The first simulation considers the total elimination of energy subsidies for electricity, LPG, and gasoline, respectively. The second considers the implementation of a carbon tax (USD 25/ton) on unsubsidized electricity, LPG, and gasoline prices. The third simulates the effect of the first and second simulations. The ENIGH dataset, which details expenditure in the first quarter of 2018 for 14,265 households, was used for all the simulations.

Table 1 shows the percentage price changes for electricity, LPG, and gasoline for the three scenarios.

Table 1. Price changes in the three simulations.

Year: 2018	Electricity	LPG	Gasoline
Simulation 1 Price change (%)	150	0	14.67
Simulation 2 Price change (%)	2.3	5.7	5.5
Simulation 3 Price change (%)	152.3	5.7	20.17

5. Results

Table 2 shows the elasticities obtained using the income, prices, and expenditures for the dataset used; Table 3 shows the parameters obtained in the QUAIDS model. Budget elasticities for electricity, LPG, and public transport are less than one, because for the average household in Mexico, these items are necessities, which is consistent with the literature consulted. However, gasoline is considered a luxury good; its budget elasticity is greater than one. The results show that income is a key determinant of fuel demand when the price changes. Both uncompensated and compensated price elasticities show the expected negative signs; households in Mexico have an inelastic response to the price of electricity, gasoline, and public transport, while for LPG the response is elastic. Uncompensated price elasticities show that electricity and public transport are inelastic, and that LPG and gasoline are elastic. Cross-price elasticities show that electricity and LPG are substitutes and that the relationship is symmetric, probably due to the increased use of electricity for cooking. On the other hand, gasoline–electricity and gasoline–transportation are complementary.

Table 2. Demand elasticities for the average Mexican household.

		Price			
		Electricity	LPG	Gasoline	Public Transport
Budget elasticities		0.609277 (0.003815)	0.961675 (0.003279)	1.566975 (0.003515)	0.908607 (0.002601)
Compensated elasticity demand	Electricity	−0.27533 (0.013145)	0.269996 (0.010906)	−0.09399 (0.009485)	0.099318 (0.019876)
	LPG	0.404153 (0.016466)	−1.63677 (0.027055)	1.139219 (0.017102)	0.093396 (0.032158)
	Gasoline	−0.08182 (0.010259)	0.812188 (0.012208)	−0.64248 (0.012929)	−0.08789 (0.016985)
	Public Transport	0.071691 (0.016942)	0.05255 (0.018133)	−0.05599 (0.013388)	−0.06825 (0.033992)
Uncompensated elasticity demand	Electricity	−0.43764 (0.013393)	0.162819 (0.010817)	−0.24372 (0.00952)	−0.09074 (0.019862)
	LPG	0.147962 (0.016558)	−1.80593 (0.027013)	0.902879 (0.01717)	−0.20658 (0.032171)
	Gasoline	−0.49927 (0.010414)	0.536544 (0.012171)	−1.02757 (0.012916)	−0.57668 (0.017048)
	Public Transport	−0.17036 (0.017028)	−0.10728 (0.01809)	−0.27929 (0.01343)	−0.35167 (0.033992)

Table 3. Estimates of the QUAIDS model for demand.

		Electricity	**LPG**	**Gasoline**	**Public Transport**
constant αi	αi	0.325 *** (−0.0137)	0.1841 *** (−0.019)	−0.0259 *** (−0.0038)	0.0164 ** (−0.0088)
Expenditures	βi	−0.1284 *** (−0.0062)	0.1789 *** (−0.0065)	−0.0168 *** (−0.0011)	−0.0336 *** (−0.0035)
Price bus	γi1	−0.0325 * (−0.0181)	−0.1683 *** 0.0064541	−0.0037 ** 0.0044952	0.2046 *** 0.0131876
Price gasoline	γi2		0.2299 *** (−0.0060)	0.0003 * (−0.0016)	−0.0619 *** (−0.00415)
Price metro	γi3			−0.0078 *** (−0.00158)	
Price taxi	γi4			0.0112 * (−0.00324)	−0.1539 *** (−0.01052)
quadratic expenditures	λi	0.0029 *** (−0.00047)	−0.0038 *** (−0.00045)	−0.0003 (−0.0001)	0.0013 *** (−0.00035)

Table 3. *Cont.*

		Electricity	LPG	Gasoline	Public Transport
Age of head	ηa	0.00006 ***	−0.0000683	2.81×10^{-6}	4.85×10^{-6} ***
		(−0.00001)	(−0.00001)	(-2.11×10^{-6})	(-7.86×10^{-6})
Province	ηp	0.0071 ***	−0.0136 ***	0.0023 ***	0.0040 ***
		(−0.00029)	(−0.00046)	(−0.00010)	(−0.00017)
Settlement type	ηs	0.00095 ***	−0.002 ***	0.0003 ***	0.0013 ***
		(−0.00014)	(−0.00017)	(−0.00003)	(−0.00010)
Household size	ηh	0.0103 ***	−0.0117 ***	−0.0000381	0.0014 ***
		(−0.0004)	(−0.00045)	(−0.00003)	(−0.00010)

*** Significant at 1%, ** at 5%, and * at 10%.

5.1. Welfare Effects

As the distribution of the results for simulations 1 and 2 are similar, but with different magnitudes, only the results of simulation 3 are given in this section. The results of simulations 1 and 2 are presented in Appendix A. The changes experienced by households as a result of the removal of subsidies and the implementation of a carbon tax of USD 25/ton CO_2 equivalent to the energy price, shown in Table 1, are presented in this section. The first- and second-order effects on the welfare of the fuels studied are shown in Figure 4. There is no significant difference between the first- and second-order results at a 95% confidence interval for a change in electricity, LPG, and gasoline prices.

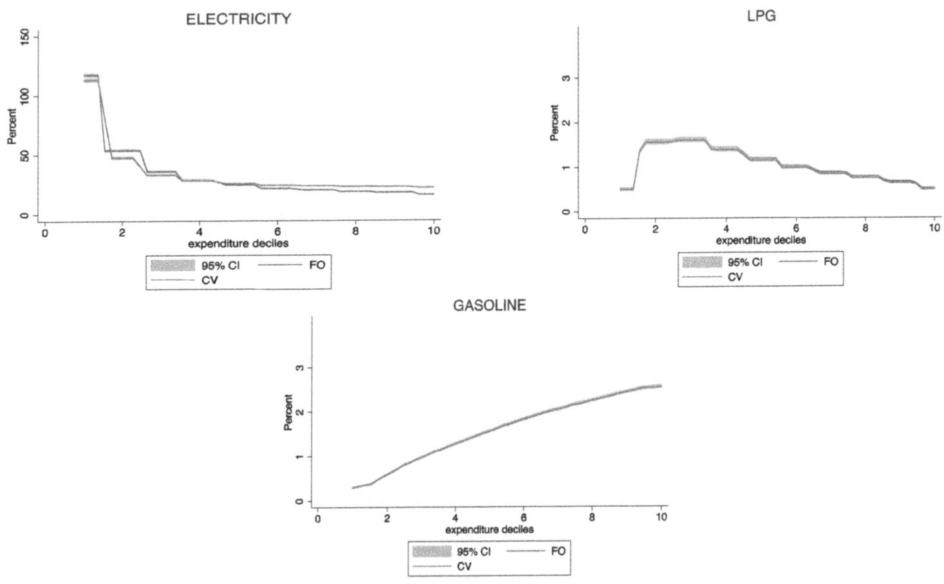

Figure 4. Welfare losses from the removal of subsidies, USD 25/ton CO_2.

The results also show that the increase in electricity prices is regressive, with the lowest income deciles losing 118%, while the highest income deciles lose only 22%. For LPG, the effect is also regressive, where decile 2 loses 1.6%, while the effect on decile 1 is very low (0.5%) and can be attributed to the use of firewood. In the case of decile 10, it loses 0.4%. On the other hand, the change in gasoline prices is progressive, and there is a significant difference between the losses in the first and last decile, 0.4 and 2.6%, respectively. It is important to mention that the calculation of the first welfare effect overestimates the welfare loss per income decile.

The welfare losses faced by households also depend on whether they are urban or rural. On average, urban households experience higher welfare losses from a rise in electricity prices than rural households do. This may be due to the household equipment they have, such as gas stoves, while rural households use firewood for cooking. In contrast, rural households experience greater welfare losses from the use of gasoline, as shown in Figure 5.

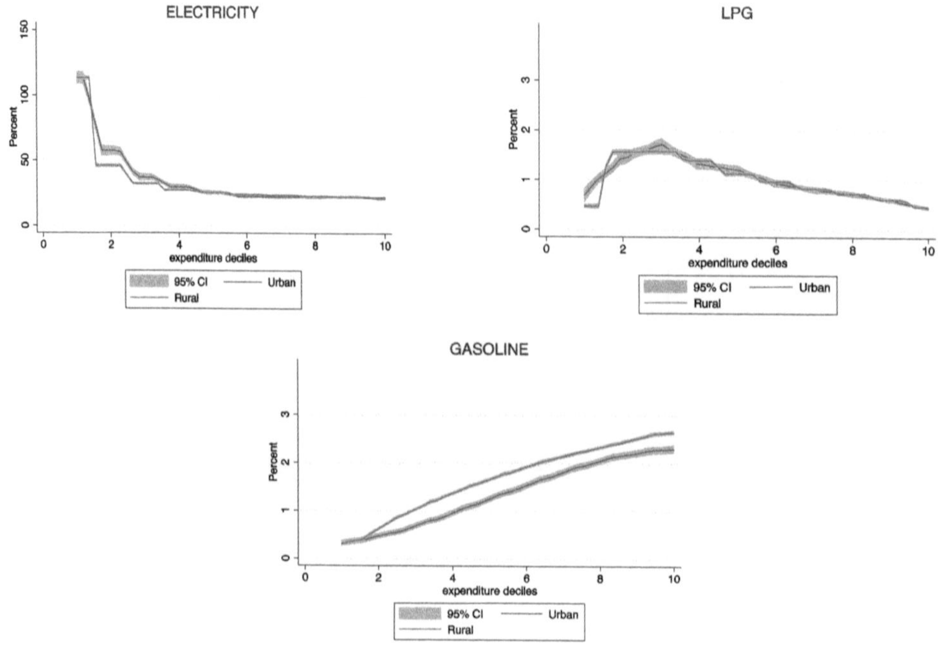

Figure 5. Welfare losses by type of settlement.

Figure 6 shows Welfare effect with a lump-sum transfer; revenue recycling is an important window of opportunity in the policies implemented by the Mexican government, as cutting fuel subsidies and implementing carbon taxes leads to more progressive welfare effects with welfare gains of up to 350% for the poorest households in the case of electricity taxes. Undoubtedly, the Mexican government should think of tax recycling as a policy to provide a global transfer of tax revenues to all households in accordance with their consumption of goods and services.

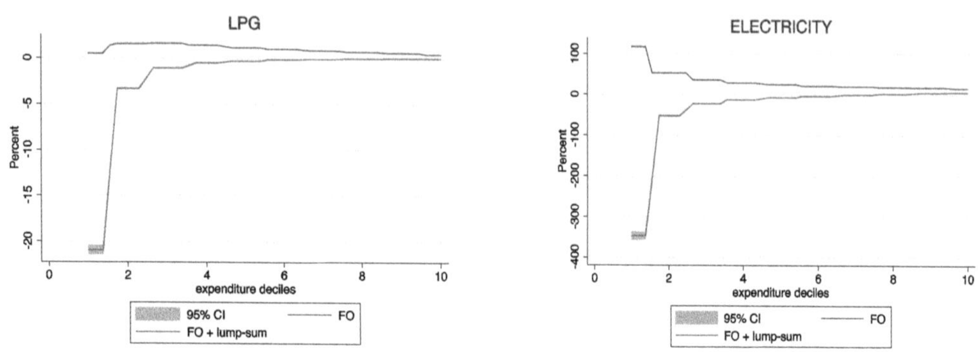

Figure 6. *Cont.*

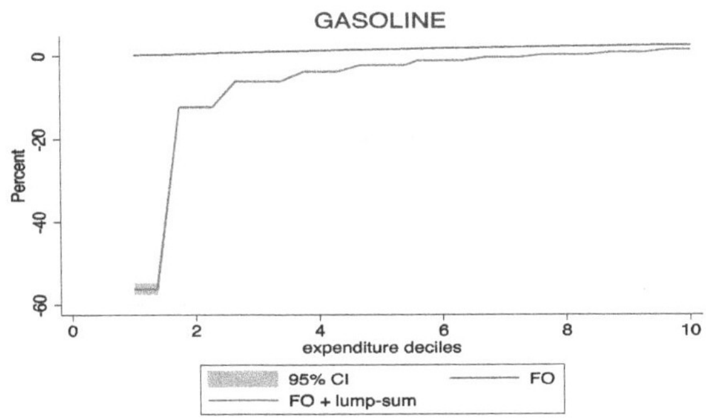

Figure 6. Welfare effect with a lump-sum transfer.

5.2. The Effect of Gender on the Welfare Losses of Households

When the gender of the head of the household is taken into account, the degree of the loss of welfare is not constant between male and female income deciles. On average, households headed by women experience greater welfare losses due to price changes in electricity, gas, and public transport compared with the price increase felt in households headed by men. The percentages of losses are presented in Figure 7. The results show that the welfare losses as a result of the increase in electricity and gas prices are regressive, significantly affecting female-headed households, while changes in gasoline prices are progressive and mainly affect male-headed households.

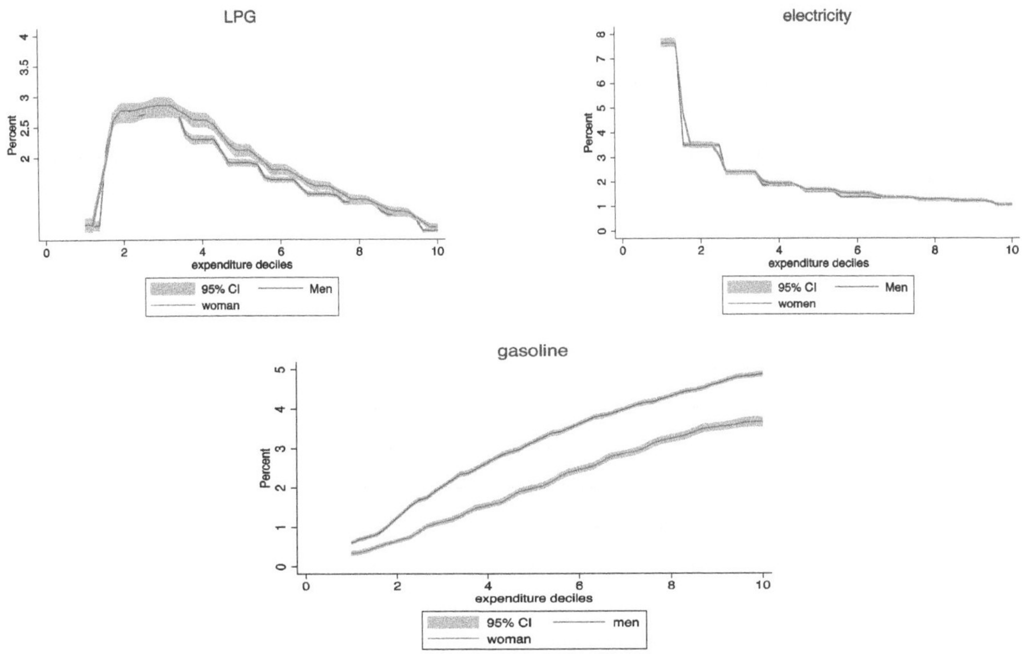

Figure 7. Welfare losses from an increase in prices.

6. Conclusions

This paper examines the potential effects on household welfare in Mexico if energy subsidies and carbon taxes were to be eliminated, as proposed. The empirical analysis estimates the residential energy demand in Mexico from 1996 to 2018 using econometric techniques. Data from the Mexican Household Income and Expenditure Survey are used, and a QUAIDS model is employed for estimation.

The results reveal that welfare losses due to an electricity price increase are regressive, while changes in gasoline prices lead to progressive losses. Furthermore, the analysis suggests that the initial estimation overstates welfare losses, and rural and urban households are impacted differently, with rural areas experiencing greater welfare losses, primarily due to higher gasoline prices. Despite 99% of the population in Mexico having access to electricity, the poorest families allocate a higher proportion of their budget to it compared with wealthier households, resulting in more significant welfare loss. Therefore, robust policies are necessary to support poorer households in the event of such policy changes.

Income is seen to be a crucial determinant of fuel demand responsiveness to price changes. Both uncompensated and compensated price elasticities exhibit the expected negative signs; households in Mexico display an inelastic response to electricity and gasoline prices, while the liquefied petroleum gas (LPG) demand is elastic. Cross-price elasticities indicate substitutability between electricity and LPG, likely due to increased electricity usage for cooking. Conversely, gasoline exhibits complementary relationships with electricity and public transportation.

Regarding gender, the results indicate that electricity and gas price increases disproportionately affect female-headed households, while gasoline price changes predominantly impact male-headed households. Low-income households would bear higher tax burdens from subsidy elimination, further reducing their already meagre incomes. Complementary policies such as social programs could mitigate welfare losses.

Electricity prices in Mexico are government-administered and not market-determined, resulting in subsidies. These subsidies distort the economy and negatively impact the environment by misguiding consumers on electricity usage. This study contributes to eliminating subsidies, avoiding the distortion of public finances, and aligning with Mexico's objectives for development and well-being with reduced greenhouse gas (GHG) emissions.

Redistributing tax revenues could lead to more progressive welfare effects, with substantial gains for the poorest households, particularly in electricity taxes. These resources could be reinvested in renewable energy projects and energy-saving initiatives to reduce GHG emissions and promote public transport usage.

These empirical results are applicable to countries with similar production and consumption profiles, particularly developing nations. The distributional consequences of energy consumption and taxes on air pollution (GHG) are contingent on income structures, even with different income recycling methods.

The Mexican Energy Ministry set goals for reducing electricity prices, developing strategies to have a supply of fuels with prices that encourage and increasing clean energy production by 25% for 2018, 30% by 2021, and 35% for 2024. The government has also pledged to reduce polluting emissions, through improvements in energy efficiency, and fuel substitution for individual transport that uses hydrocarbons. However, these objectives have only been partially achieved. Studies like this could aid Mexico and similar countries in redirecting subsidy policies to alleviate energy poverty, mitigate environmental impacts from fossil fuels, and foster more balanced economic development.

Author Contributions: The authors J.A.R.F., D.M.G. and R.S. have written, review and edited the paper. All authors have read and agreed to the published version of the manuscript.

Funding: Funding: This work was supported by CONACYT-SENER-Sustentabilidad Energética, under grant FSE-2014-06-249795 "Centro Mexicano de Innovación en Energía del Océano (CEMIE- Océano)".

Data Availability Statement: The data is available if requested from the authors by email.

Acknowledgments: This research was funded by the Mexican Centre for Ocean Energy Innovation (CEMIE-Océano, Centro Mexicano de Innovación en Energía del Océano) CONACYT-SENER Sustentabilidad Energética, project FSE-2014-06-249795, and Grupos Interdisciplinarios de Investigación, Instituto de Ingeniería-UNAM, project "Sostenibilidad del Caribe mexicano: Cambiando debilidades en fortalezas". Thanks go to Rebeca Mendez and Jorge Andres Rosas for editing and reviewing this paper and to Midnight Oil for their lyrics and music.

Conflicts of Interest: The authors have no conflicts of interest to declare.

Appendix A. Redistribution Policy

Welfare effect with a lump-sum transfer

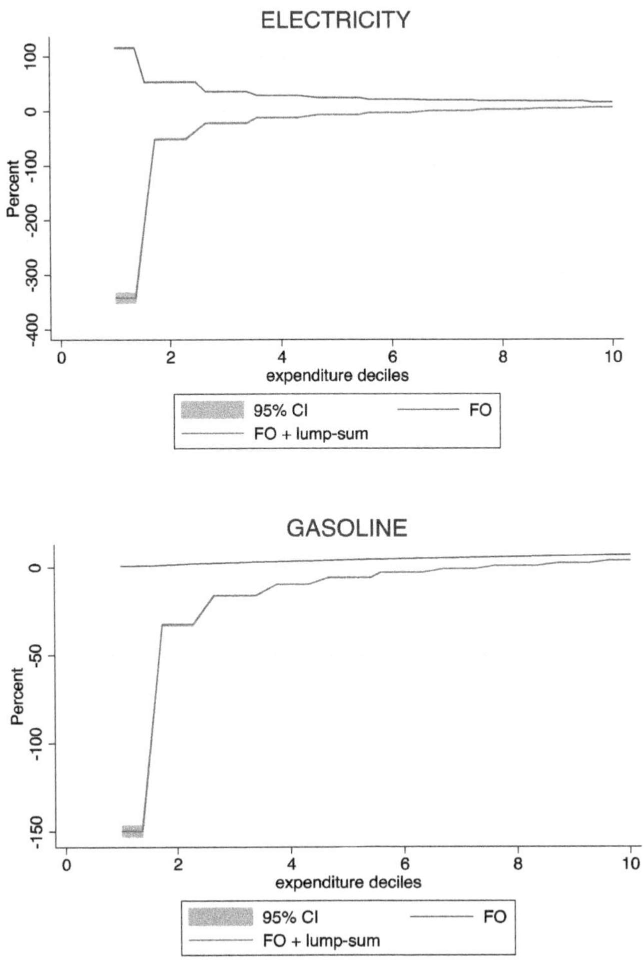

Figure A1. Simulation 1.

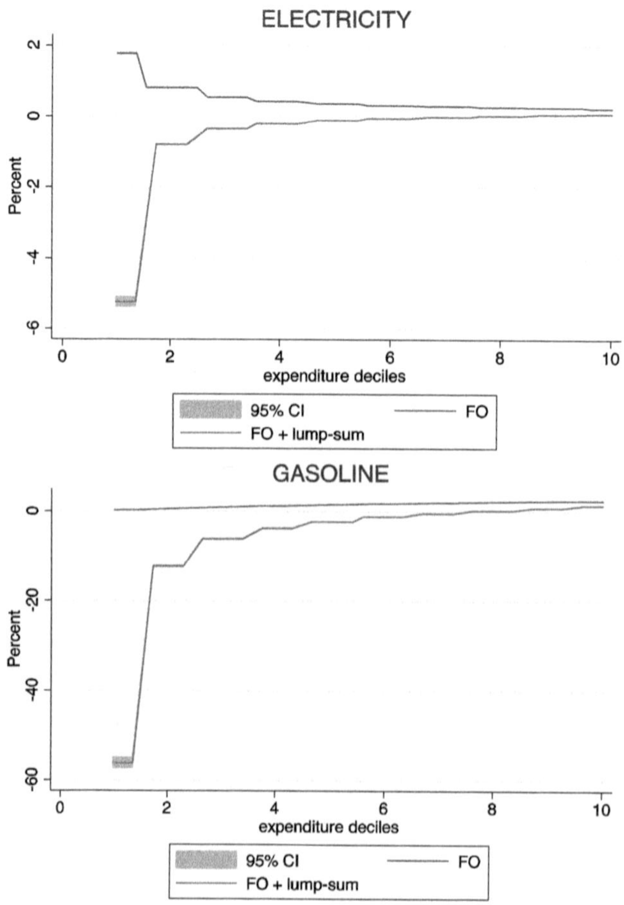

Figure A2. Simulation 2.

References

1. Kojima, M. *Identifying and Quantifying Energy Subsidies: Energy Subsidy Reform Assessment Framework (ESRAF) Good Practice Note 1*; ESMAP Paper; World Bank Group: Washington, DC, USA, 2017; Available online: https://openknowledge.worldbank.org/entities/publication/cb55bf07-e9d9-5195-abfe-2cdc7507b388 (accessed on 1 April 2024).
2. Sánchez, L.D.E.; Wooders, P.; Kjell, K.; Lean, T.; Beaton, C.; Shama, S.; Oharenko, Y. Mejorando y Refocalizando los Subsidios a la Electricidad. Alianza Energética. International Institute for Sustainable Development (IISD). 2018. Available online: https://www.energypartnership.mx/fileadmin/user_upload/mexico/media_elements/reports/SubsidiosElectricos-MEX.pdf (accessed on 2 March 2023).
3. Elgouacem, A.; Franzen, K.; Jensen Díaz, W.; Muth, J.; Paciorek, A. Eliminando Progresivamente Subsidios Ineficientes a los Combustibles Fósiles. 2017. Available online: https://www.oecd.org/fossil-fuels/Reporte%20IFFS%20Versi%C3%B3n%20Final.pdf (accessed on 5 March 2023).
4. Datta, A. The incidence of fuel taxation in India. *Energy Econ.* **2010**, *32*, S26–S33. [CrossRef]
5. Moz-Christofoletti, M.A.; Pereda, P.C. Winners and Losers: The Distributional Impacts of a Carbon Tax in Brazil. 2021. Available online: http://www.repec.eae.fea.usp.br/documentos/Moz-Christofoletti_Pereda_WP08.pdf (accessed on 15 March 2024).
6. Saelim, S. Carbon tax incidence on household demand: Effects on welfare, income inequality and poverty incidence in Thailand. *J. Clean. Prod.* **2019**, *234*, 521–533. [CrossRef]
7. Coady, D.; El-Said, M.; Robert Gillingham, K.; Kpodar, N.; Medas, P.; Newhouse, D. The Magnitude and Distribution of Fuel Subsidies: Evidence from Bolivia, Ghana, Jordan, Mali, and Sri Lanka. 2006. Available online: https://www.imf.org/external/pubs/ft/wp/2006/wp06247.pdf (accessed on 5 February 2023).
8. Rosas Flores, J.A.; Mohcine, B.; Rosas Flores, D.; Luis, F.Z.J. Distributional effects of subsidy removal and implementation of carbon taxes in Mexican households. *Energy Econ.* **2017**, *61*, 21–28. [CrossRef]

9. Gonzalez, F. Distributional effects of carbon taxes: The case of Mexico. *Energy Econ.* **2012**, *34*, 2102–2115. [CrossRef]
10. Orazio, A.; Vincenzo, D.M.; Valérie, L.; David, P. Welfare consequences of food prices increases: Evidence from rural Mexico. *J. Dev. Econ.* **2013**, *104*, 136–151.
11. Ramírez, J.C.; Ortiz-Arango, F.; Rosellón, J. Impact of Mexico's energy reform on consumer welfare. *Util. Policy* **2021**, *70*, 101191. [CrossRef]
12. Moshiri, M.; Santillan, M.A.M. The welfare effects of energy price changes due to energy market reform in Mexico. *Energy Policy* **2018**, *113*, 663–672. [CrossRef]
13. Renner, S. Poverty and distributional effects of a carbon tax in Mexico. *Energy Policy* **2018**, *112*, 98–110. [CrossRef]
14. Deaton, A.S.; Muellbauer, J. An almost ideal demand system. *Am. Econ. Rev.* **1980**, *70*, 312–326.
15. Banks, J.; Blundell, R.; Lewbel, A. Quadratic engel curves and consumer demand. *Rev. Econ. Stat.* **1997**, *79*, 527–539. [CrossRef]
16. Lewbel, A. Characterizing Some Gorman Engel Curves. *Econometrica* **1987**, *55*, 1454–1459. [CrossRef]
17. Menezes, T.; Azzoni, C.; Silveira, F. Demand elasticities for food products in Brazil: A two-stage budgeting system. *Appl. Econ.* **2008**, *40*, 2557–2572. [CrossRef]
18. Londoño, D.; Londoño, E.; Ramírez, A.H. Un sistema casi ideal de demanda para el gasto en Colombia: Una estimación utilizando el método generalizado de los momentos en el período 1968–2007. *Ecos De Econ.* **2011**, *15*, 39–58.
19. Cortés, D.; Pérez, J.E. El consumo de los hogares colombianos, 2006–2007: Estimación de sistemas de demanda. *Rev. Desarro. Soc.* **2010**, *66*, 7–44. [CrossRef]
20. Moro, D.; Sckokai, P. Heterogeneous preferences in household food consumption in Italy. *Eur. Rev. Agric. Econ.* **2000**, *27*, 305–323. [CrossRef]
21. Pangaribowo, E.H.; Tsegai, D.W. Food demand analysis of Indonesian households with particular attention to the poorest (Ausut 1, 2011). In ZEF-Discussion Papers on Development Policy, No. 151. Available online: https://www.econstor.eu/bitstream/10419/98222/1/770495125.pdf (accessed on 15 March 2024).
22. Ngui, D.; Mutua, J.; Osiolo, H.; Aligula, E. Household energy demand in Kenya: An application of the linear approximate almost ideal demand system (LA-AIDS). *Energy Policy* **2011**, *39*, 7084–7094. [CrossRef]
23. Mariana, L.; Helder, P.J.; Francisco, E. Automotive fuel consumption in Brazil: Applying static and dynamic systems of demand equations. *Energy Policy* **2009**, *37*, 5326–5333.
24. Gundimeda, H.; Kohlin, G. Fuel demand elasticities for energy and environmental policies: Indian sample survey evidence. *Energy Econ.* **2008**, *30*, 517–546. [CrossRef]
25. Labandeira, X.; Labeaga, J.M.; Rodríguez, M. A residential energy demand system for Spain. *Energy J.* **2006**, *27*, 87–111. [CrossRef]
26. Poi, B.P. Easy demand-system estimation with quaids. *Stata J.* **2012**, *12*, 433–446. [CrossRef]
27. Nikodinoska, D.; Schröder, C. On the emissions–inequality and emissions–welfare trade-offs in energy taxation: Evidence on the German car fuels tax. *Resour. Energy Econ.* **2016**, *44*, 206–233. [CrossRef]
28. Tiezzi, S.; Verde, S.F. Differential demand response to gasoline taxes and gasoline prices in the U.S. *Resour. Energy Econ.* **2016**, *44*, 71–91. [CrossRef]
29. Uju Okonkwo, J. Welfare effects of carbon taxation on South African households. *Energy Econ.* **2021**, *96*, 104903. [CrossRef]
30. Sterner, T. *Fuel Taxes and the Poor: The Distributional Effects of Gasoline Taxation and Their Implications for Climate Policy*; RFF Press: Washington, DC, USA, 2011.
31. Feldstein, M.S. Distributional equity and the optimal structure of public prices. *Am. Econ. Rev.* **1972**, *62*, 32–36. Available online: http://www.jstor.org/stable/1821470 (accessed on 12 February 2023).
32. Stern, N. The theory of optimal commodity and income taxation. In *The Theory of Taxation for Developing Countries*; Oxford University Press: Oxford, UK, 1987.
33. West, S.E.; Williams, R.C. Estimates from a Consumer Demand System: Implications for the Incidence of Environmental Taxes. *J. Environ. Econ. Manag.* **2004**, *47*, 535–558. [CrossRef]
34. Wier, M.; Birr-Pedersen, K.; Jacobsen, H.K.; Klok, J. Are CO_2 Taxes Regressive? Evidence from the Danish Experience. *Ecol. Econ.* **2005**, *52*, 239–251. [CrossRef]
35. Callan, T.; Lyons, S.; Scott, S.; Tol, R.S.; Verde, S. The distributional implications of a carbon tax in Ireland. *Energy Policy* **2009**, *37*, 407–412. [CrossRef]

energies

MDPI

Article

Toward a "Smart-Green" Future in Cities: System Dynamics Study of Megacities in China

Rui Hu [1,*] and Xinliang Han [2]

[1] School of Urban Economy and Public Administration, Capital University of Economics and Business, No. 21 Zhangjialukou, Fengtai District, Beijing 100070, China
[2] Asset Management Office, University of Science and Technology Beijing, No. 30, Xueyuan Road, Haidian District, Beijing 100083, China; redchina5@163.com
* Correspondence: hurui870929@hotmail.com

Abstract: This study investigates the development trend of smart-green cities, focusing on seven megacities in China. It addresses three issues that are common in urban green development, including the relationship between "smart" and "green", the scenario analysis of green development, and the uniqueness of megacities in green development. System dynamics modeling is applied. The simulation results reveal an "S"-shaped development curve for both aspects, indicating a gradual and accelerating growth pattern. Notably, the curve representing energy consumption lags behind the curve for smart city development by approximately three years. After 2030, when the smart city construction is expected to be completed, the proportion of the tertiary industry and investment in science and technology will play a significant role in limiting energy consumption. This study concludes by providing policy suggestions, including the need for long-term plans with phased targets, considering the specificity of megacities, and addressing external influences.

Keywords: smart-green city; system dynamics modeling; trend forecasting; megacities

1. Introduction

Assuming its role as a responsible global leader, China unveiled the "Dual Carbon" objective during the United Nations Climate Conference in September 2020. This objective signifies a commitment to reach a peak in carbon emissions by 2030 and achieve carbon neutrality by 2060. This initiative not only represents a solemn pledge to the international community but also holds a fundamental place within our economic and social development framework.

Currently, the global urbanization rate stands at 56% (data source: UN-Habitat, World Cities Report 2022: Envisaging the Future of Cities), with urban areas responsible for approximately 75% of total CO_2 emissions. In China, this percentage is even higher, reaching a staggering 80%. That is to say, cities are playing a critical role in achieving the "dual carbon" objective in China. The primary contributor to carbon emissions in cities is the consumption of fossil fuels in industry, transportation, and citizens' daily lives. To control the carbon emissions in urban areas, the concept of green cities has emerged.

The notion of green city development centers around striking a balance between ecological preservation, nature conservation, and human well-being and cultural enrichment [1]. China has the highest urbanization rate in the world. So, it is important to integrate the green development concept into the process. To achieve the goal of green development in Chinese urban areas, there are several key approaches to consider, like adopting a forward-thinking urban planning approach, optimizing the structure of energy supply and consumption, and promoting the green transformation of urban construction.

With the development of technology, the potential of informative innovation has drawn more and more attention to driving the green development of cities [2–4]. The convergence of technological advancements and urban environments has given rise to the concept of Smart Cities [5]. Smart cities represent the forefront of digital economy applications within urban spaces, offering an effective avenue for vertical and horizontal connectivity and serving as a catalyst for high-quality urban development. They have emerged as a powerful tool for emission control and the promotion of green development.

This study places particular emphasis on the impact of smart cities on urban green development. The concept of the smart city is dissected into four key dimensions: smart governance, smart living, smart economics, and smart labor. This study aims to address these four issues. Firstly, it seeks to ascertain the developmental trend of urban green development. Secondly, it aims to establish the correlation between a "smart city" and a "green city." Thirdly, it aims to explore how the trend of urban green development varies across diverse scenarios. Lastly, it tries to figure out the uniqueness of megacities' green development. To find the answers, a system dynamics model is constructed.

This paper's structure is organized as follows: Chapter One serves as the introduction, providing an overview of the study's objectives and context. Chapter Two offers a comprehensive review of the relevant literature, encompassing key studies and theories in the field. Chapter Three delves into the data sources and methodology employed in the research, outlining the approach taken for data collection and analysis. Chapter Four presents a detailed discussion of the simulation results, analyzing the findings and drawing insights from different scenarios. Finally, Chapter Five concludes the study, summarizing the key findings and presenting policy recommendations for enhancing urban green development.

2. Literature Review

2.1. Bibliometric Analysis

Both domestically and internationally, the promotion of urban development through the concept of "smart and green coordination" has gained significant attention and has become a hot topic. This approach emphasizes the integration of smart technologies and sustainable practices to achieve more efficient, environmentally friendly, and livable urban environments. The combination of smart and green strategies has the potential to address various urban challenges, including resource management, environmental protection, transportation efficiency, energy conservation, and quality of life for residents. As cities strive for sustainable development and aim to enhance their competitiveness on the global stage, the concept of "smart and green coordination" has emerged as a key focus area.

Over the past decade, research in the field of smart and green coordination has primarily concentrated on micro-level aspects. Several prominent research areas have emerged, including the exploration of smart-green communities [6–8], the impact of smart city initiatives on climate change [9–11], green infrastructure development [12,13], and the analysis of future challenges [14,15]. The prevalent research approach has predominantly involved modeling techniques to examine and forecast various scenarios and outcomes [16]. Among these research areas, the smart grid has experienced the longest period of sustained interest and burst intensity [17–20].

As we can see in Figure 1, the hot spots have dispersed and iterated rapidly. Most research focusing on detailed areas and macroscopic research is rare and emerging. The relationship between "smart" and "green" is one of the newly emerging topics.

The rapid development of the digital economy in China has led to a significant contribution of research outcomes by Chinese authors, as depicted in Figure 2. Over the past decade, research in this field conducted in China has been highly concentrated, with a focus on qualitative studies closely aligned with policy trends [21,22]. Given the frequent introduction of new policies, the research hotspots in this field are subject to frequent updates and changes.

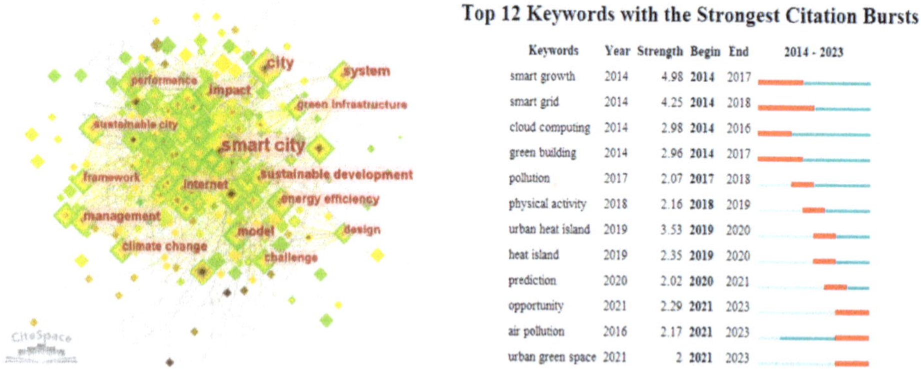

Figure 1. Keywords in Web of Science database search on smart-green areas.

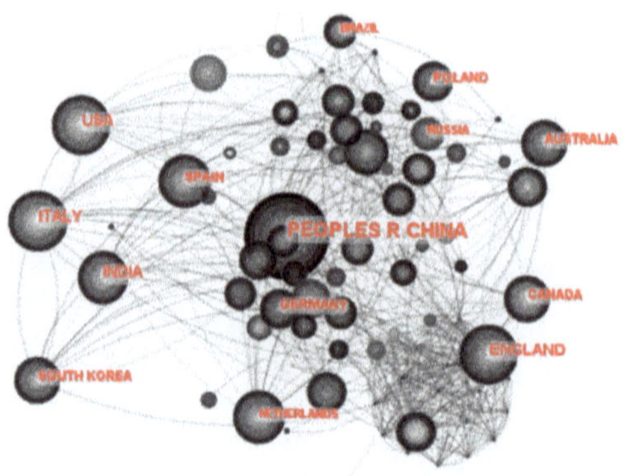

Figure 2. Authors' nationalities in smart-green areas.

Based on the analysis above, two prominent areas of focus are the interrelation between "smart" and "green", as well as the context of China. In this research, we have chosen to delve into the former topic as our main subject, while also considering the latter as an objective.

2.2. Study on the Effect Smart Cities Exert on Green Development

With the continuous deterioration of urban environments, research focus has gradually shifted towards exploring whether smart cities can promote sustainable development and enhance urban ecological benefits. Firstly, at the conceptual level, the incorporation of sustainability into smart city methods and models has garnered significant attention, leading to the emergence of compound concepts such as green smart cities and sustainable smart cities [23–25]. Secondly, there has been a substantial amount of research investigating the relationships between smart cities, environmental benefits, and sustainable development. Ahvenniemi et al. [26] argue that a truly "smart" city, in terms of its overall objectives, must be one that embraces sustainable development. Through the use of a difference-in-differences (DID) approach, other scholars [27] suggest that incorporating policies and planning centered around an understanding of the overarching vision of environmental quality and resilience can better achieve the integration of smart and green city concepts, thereby facilitating sustainable development.

Although some research [28] has theoretically affirmed the positive effects of smart cities on sustainable development and ecological efficiency, certain scholars still hold reservations about smart cities' environmental benefits [29]. There is also ongoing debate regarding whether smart cities are an effective approach to achieving sustainable development.

To explore the relationship between smart and green development, scholars often employ the difference-in-differences (DID) method. They typically select prefecture-level cities as their research subjects and examine urban energy consumption and emissions as the dependent variables. They have examined the impact of smart city pilot policies on ecological efficiency, revealing that smart city construction notably enhances urban ecological efficiency and that its promotive effect strengthens over time. Overall, research in China regarding the relationship between smart and green development is characterized by a close connection to policy trends, frequent updates on research hotspots, and the use of the DID method to assess the impact of smart city pilot policies on energy consumption and emissions at the prefecture-level city level.

2.3. Research Gaps

The existing research in the field of smart and green development, as shown in Table 1, has certain limitations that need to be addressed. Firstly, there is a lack of variable granularity, with "smart city" often being used as a general explanatory variable without further deconstruction of its underlying components. Secondly, the reliance on econometric research methods, while common, may have limitations in capturing the complexity of urban systems. Thirdly, there are few papers discussing the negative impact of smart city construction on green development [30]. Lastly, there is insufficient guidance for future development, as most studies are based on historical panel data with limited consideration of future trends.

Table 1. Main points of view on how smart cities affect green development.

	Main Idea	Source
Mechanism test	The increased proportion of science and technology expenditure in financial expenditure promotes the low-carbon development of cities.	[31–33]
	The development of the information industry promotes the upgrading of industrial structure and thus promotes the low-carbon development of cities.	[34–36]
	Promotes urban low-carbon development through technological innovation.	[31,37,38]
	Promotes urban low-carbon development by optimizing urban resource allocation.	[34,39,40]
	Reduces the scale of government to offset its impact on green development.	[41]
Heterogeneity	Smart city construction has a more significant effect on carbon reduction in cities with low financial development levels, high human capital levels, and low export-oriented economy levels.	[42]
	The effect is more significant in western cities, southern cities, non-environmentally friendly cities, resource-based cities, and non-provincial capital cities.	[34,43]
	The effect of carbon reduction in underdeveloped cities, megacities, and highly administrative cities is more significant.	[33,35,41]
	The effect of carbon reduction in cities with obvious economic agglomeration advantages and strict environmental regulations is more significant.	[37,44]

To address these limitations, this study proposes three solutions. Firstly, it suggests refining the explanatory variable "smart city construction" into the following four distinct dimensions based on the innovative quadruple theory: smart governance, smart living, smart economics, and smart labor. This approach allows for a more detailed understanding of the different aspects of smart city development and their impact on urban low-carbon development.

Secondly, this study proposes combining qualitative research methods with system simulation techniques. By integrating quantitative regression analysis and system dynamics modeling, this research can better account for the nonlinear, systematic, and dynamic nature of urban systems. This combined approach provides a more comprehensive understanding of the complex interactions and feedback loops within the urban environment.

Lastly, this study aims to incorporate trend prediction and dynamic scenario simulation. By focusing on overall trend changes and predicting urban green development trajectories under different scenarios, this research can elucidate the whole picture—both the positive and negative effects—and provide more targeted and forward-looking insights to support urban decision-making processes. This approach allows for a deeper understanding of potential future outcomes and will help inform policy and planning strategies accordingly.

3. Data and Methodology

3.1. Research Objects

This study focuses on seven megacities in China, namely, Shanghai, Beijing, Shenzhen, Chongqing, Guangzhou, Chengdu, and Tianjin. These cities represent complex and extensive urban systems characterized by various internal subjects, relationships, and external environments, all of which significantly influence the functioning of the overall system.

In terms of internal subjects and relationships, megacities demonstrate distinct "triple characteristics". Firstly, they experience scale expansion, with their populations, infrastructure, and economic activities rapidly growing and expanding. Secondly, diverse demands emerge as these cities accommodate the wide range of social, economic, and environmental needs of their large and diverse populations. Lastly, these cities exhibit feature emergence, where unique characteristics and challenges arise due to the presence of multiple subjects within their urban fabric.

In terms of complex relationships, megacities face significant management tasks, increased management difficulty, and intensified management risks. The sheer size and complexity of these cities create challenges in effectively managing various aspects such as urban planning, transportation, public services, and environmental sustainability. The interplay and interactions among different stakeholders and sectors further contribute to the complexity of managing these megacities.

Moreover, megacities are influenced by the dynamic external environment. Factors such as informationization impact, the pressure to achieve the "dual carbon" target (carbon peak and carbon neutrality), and policy constraints shape the context in which these cities operate. The rapid development of information and communication technologies, along with the increasing focus on sustainable development and carbon reduction, create both opportunities and challenges for these megacities in their pursuit of smart and green development.

By acknowledging and analyzing these triple characteristics, complex relationships, and dynamic external environments, this study aims to provide a comprehensive understanding of the challenges and opportunities faced by the seven megacities in China in their pursuit of smart and green development.

3.2. Research Framework

This study consists of four steps:

Firstly, the concept of a "smart city" is explained through four aspects based on the quadruple theory. These aspects include smart governance, smart living, smart economics, and smart human resources, which collectively form the components of a smart city system.

Secondly, an index system is designed. This study takes into full consideration the "Smart City Evaluation Model and Basic Evaluation Index System" issued by the National Standards Committee and designs indicators based on the availability of comprehensive data. Smart governance is assessed using indicators such as the urban auto cleaning rate and the online service adoption rate for civic administrative affairs (referred to as "E-governance" in the model). Smart living is assessed through indicators such as the internet adoption rate and the penetration rate of new energy vehicles. Smart economy is assessed using indicators such as the income of IT employees and the number of IT employees. Smart human resources are assessed through indicators such as the number of R & D staff in colleges and the number of universities in the city. Additionally, scenario settings are provided.

Thirdly, a simulation model, referred to as the "smart-green" model, is constructed. The results are analyzed from both a timeline perspective and in terms of scenario changes.

Lastly, based on the aforementioned analysis, policy recommendations are provided on how to develop a "smart-green" megacity.

The framework of this research is shown in Figure 3 below.

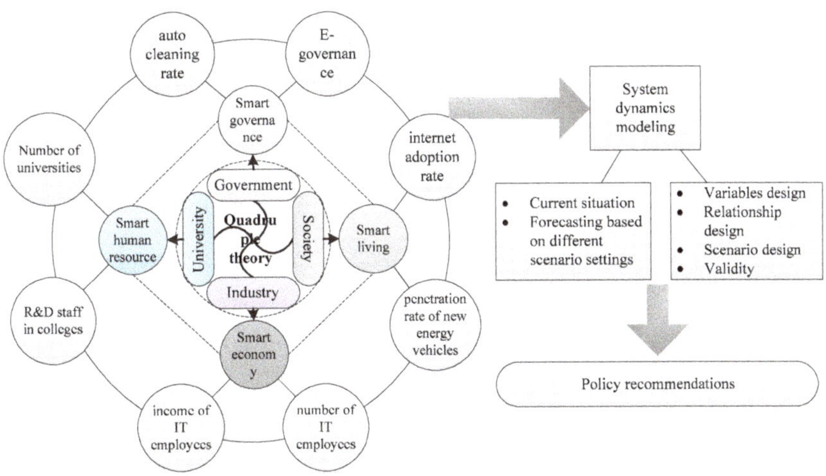

Figure 3. Framework of this study.

3.3. Data and Indicator System

There are 29 variables in "smart-green" system model. The dependent variable that we want to observe is green development level, which includes "Energy consumption" and "CO_2 emission". Their values are obtained directly from the energy section of the statistical yearbook.

As for the independent variables, there are three level variables, three flow variables, twenty auxiliary variables, and one shadow variable.

The level variables are external factors that influence the system model. They are the economic level, represented by GDP; industrial structure, represented by the proportion of the tertiary industry; and the social development level, represented by population. The flow variables are the level variables' growth rate.

There are two types of auxiliary variables: the smart city system and scenario variables. The smart city system is built based on the innovative quadruple theory in Figure 3, and includes four aspects: smart governance, smart life, smart economy, and smart human capital. The scenario variables are composed of environmental control investment and technology and science investment. Environmental control investment is represented by the proportion of environmental protection investment to GDP, and technological financial investment is represented by the proportion of R & D expenditure to GDP.

The values of some critical variables are listed in Table 2 below. The data were collected from the "*China City Statistical Yearbook*", city statistical bulletins and official news reporting. Some missing data were imputed using the method of nearby-year mean interpolation.

Table 2. Data used to build the simulation model.

		Smart Governance		Smart Living		Smart Economics		Smart Human Resource		Green Development Level
		Auto-Cleaning Ratio	E-Governance	Internet Coverage	E-Car Coverage	Income of IT Employees	Number of IT Employees	R & D Staff in Colleges	Number of Universities	Energy Consumption
Unit						CNY	10,000 People	People	Universities	Wtce
Beijing	2013	0.9	0.97	0.752	0.11	136,599	58.24	18,600	89	6723.9
	2015	0.9	0.97	0.765	0.11	159,486	68.007	19,540	90	6852.6
	2017	0.9	0.97	0.8	0.19	183,183	77.44	20,000	92	7132.8353
	2019	0.9	0.97	0.85	0.19	234,121	85.9131	22,000	93	7360.32
	2021	0.9	0.97	0.9	0.08	290,038	86.5	23,000	92	7103.617
Tianjin	2013	0.9	0.95	0.613	0.1	102,922	3.57	7440	55	7881.83
	2015	0.92	0.95	0.63	0.12	134,331	4.3541	7816	55	8319.38
	2017	0.92	1	0.7212	0.25	151,778	5.3464	8000	57	7831.72
	2019	0.92	1	0.7567	0.3105	144,510	6.5728	8800	56	8240.7
	2021	0.92	1	0.7633	0.3105	157,725	7	9200	56	8205.69
Guangzhou	2013	1	1	0.9203	0.1	120,000	9.39	17,000	80	5333.57
	2015	1	1	0.4975	0.1	160,000	10.0529	19,000	81	5688.89
	2017	1	1	0.5813	0.12	200,000	16.5897	21,000	82	5961.97
	2019	1	1	0.5933	0.205	220,000	22.6695	23,000	83	6294.2
	2021	1	1	0.6	0.3433	240,000	24	25,001	83	6575.64
Shenzhen	2013	0.974	0.9362	0.6888	0.1288	100,000	12.5	1600	10	3594.42
	2015	0.974	0.9362	0.4765	0.1288	110,000	13.4808	2100	12	3909.91
	2017	1	0.9362	0.5834	0.25	135,000	19.2466	2600	12	4272.64
	2019	1	0.95	0.6232	0.4	160,000	30.3677	3000	13	4534.14
	2021	1	0.95	0.6029	0.4809	150,000	32	3241	15	4756.67
Chengdu	2013	0.8406	0.9	0.4929	0.0528	98,000	14.54	18,600	53	17,774.57904
	2015	0.8406	0.9	0.5625	0.1034	110,000	16.9835	19,540	56	16,680.10628
	2017	0.8406	0.9	0.7067	0.25	130,000	31.4432	20,000	56	15,448.7
	2019	0.8406	0.93	0.7663	0.3225	150,000	26.1621	22,000	59	16,382.2
	2021	0.9	0.93	0.7354	0.3225	170,000	30	23,000	66	16,356
Chongqing	2013	0.9	0.997	0.439	0.1	73,598	13.95	18,100	63	6225.92
	2015	0.9	0.997	0.483	0.1	92,958	16.04	20,300	64	6924.77
	2017	0.9	1	0.5516	0.1524	112,043	4.7844	22,500	65	7251.59
	2019	0.93	1	0.8608	0.2033	131,356	4.736	24,700	65	7687.25
	2021	0.93	1	0.8542	0.2951	155,067	5	26,900	68	8046.31
Shanghai	2013	1	0.64	0.707	0.1066	153,989	49.43	18,765	68	10,890.39
	2015	1	0.64	0.731	0.2677	183,365	27.8195	23,453	67	10,930.53
	2017	1	0.73	0.8284	0.3	212,063	30.7312	25,478	64	11,381.85
	2019	1	0.73	0.7174	0.4784	237,405	41.768	36,000	64	11,696.46
	2021	1	0.73	0.715	0.4784	303,573	45	40,000	64	11,683.02

3.4. System Dynamic Model

Based on the analysis above, the following "smart-green" model was built and is illustrated in Figure 4.

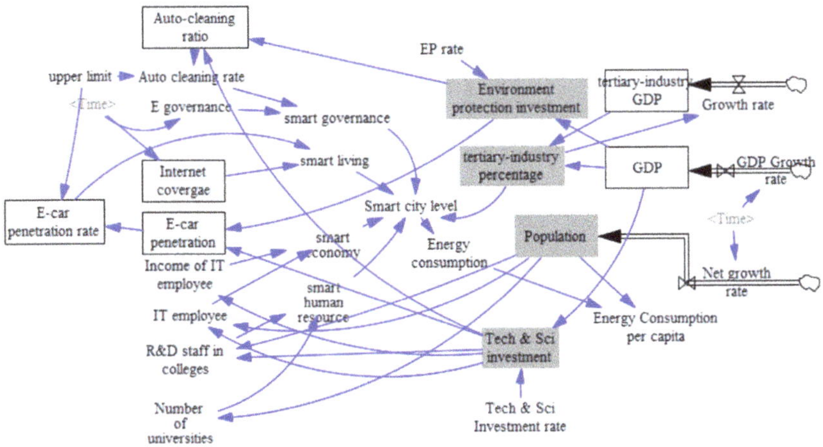

Figure 4. "Smart-green" system model.

The relationships between variables are expressed by equations. The variables listed in Table 3 are mostly assigned through a table function, which means their values come from real data, for example:

Internet coverage = WITH LOOKUP (Time, Look up ([(0.5,0) − (3000,10)], (2012,0.75), (2019,0.78), (2025,0.82), (2030,1), (2035,1)))

Population Net growth rate = WITH LOOKUP (Time, Look up ([(0,0) − (2035,50)],(2012,0),(2013,0.023),(2014,0.018),(2015,0.017),(2016,0.025),(2018,0.012), (2019,0.012),(2020,0.01),(2021,0.004),(2035,0.01)))

E-governance = WITH LOOKUP (Time, Look up ([(2012,−10) − (2035,10)],(2012,0.92),(2015,0.93),(2018,0.935),(2020,0.94),(2035,1)))

Others are assigned through equations based on regression analysis, for example:

$$\text{IT employee} = 0.0018 \times \text{Population} + 0.022 \times \text{"Tech \& Sci investment"} + 56$$

$$\text{Auto-cleaning rate} = (2 \times 10^{-5} \times \text{"Tech \& Sci investment"} + 6 \times 10^{-6} \times \text{Environmental protection investment} + 0.795) \times 0.25, \text{initial value } 0.78$$

$$\text{IT employee} = 0.0018 \times \text{Population} + 0.022 \times \text{"Tech \& Sci investment"} + 56$$

$$\text{"R \& D staff in colleges"} = -2.9 \times \text{Population} + 13.4 \times \text{"Tech \& Sci investment"} + 19261.7$$

As for the level variables, their values are controlled using flow variables, for example:

$$\text{Population} = \text{INTEG (Population} \times \text{Net growth rate), Initial Value} = 17464.25$$

$$\text{GDP} = \text{INTEG}(2 \times \text{GDP} \times \text{GDP Growth rate}), \text{Initial Value} = 13658$$

$$\text{tertiary-industry GDP} = \text{INTEG}(2 \times \text{"tertiary-industry GDP"} \times \text{Growth rate}), \text{Initial Value} = 2155.3$$

The scenario settings are shown in Table 3:

Table 3. Scenario settings.

	Tertiary-Industry GDP Growth Rate	Environmental Protection Investment Rate	Science and Technology Investment Rate
High	IF THEN ELSE ("tertiary-industry percentage" <0.7, 0.09, 0.02)	China's target value is 3%.	Highest rate is now in Israel, which is 5%.
Medium	IF THEN ELSE ("tertiary-industry percentage" <0.7, 0.08, 0.01)	Current value of megacities is around 2%.	Current rate in seven megacities 2.4%.
Low	IF THEN ELSE ("tertiary-industry percentage" <0.7, 0.07, 0.01)	Current national value is around 1.5%.	Average value for OECD countries in 2019 was 2%.

4. Results and Discussion

The simulation, covering the period from 2012 to 2035, was validated by comparing the simulation results with real data from 2012 to 2021. The average error was found to be 3.1%, with the largest error occurring in 2020 at 6.5%. It is worth noting that the COVID-19 pandemic started in 2020, resulting in significant reductions in economic and social activities in most cities, leading to a sudden decrease in energy consumption. Considering that the model is considered reliable when the error is below 5%, the "smart-green" model can be deemed acceptable.

The traditional DID method involves only one single factor, the policy, and it is based on historical facts. The system dynamics model is a "black box" model. All relevant factors can be included in this system. Additionally, it is a model that represents the future. This is the main advantage of the SD method.

4.1. Development Level of Smart Cities

In seven megacities, the trend of smart city construction continues to rise until the end of the forecasting period. However, after 2029, the growth rate becomes very low, indicating that after approximately 17 years of construction, the smart city reaches a certain maturity level, leading to a deceleration in its progress.

In terms of the four dimensions of a smart city, namely, smart governance and smart human resources, we observe consistent progress. Specifically, the smart human resources dimension maintains a relatively stable trajectory. This is mainly due to the Education Bureau of China promoting vocational education and new universities choosing to establish themselves in other cities due to high land rents and other costs in these seven megacities. Additionally, these cities have already attracted the most talented technology professionals in the country, resulting in limited potential for further development. On the other hand, smart governance continues to grow until the early 2030s. Citizens can fulfill most of their needs through online apps or by calling the 12345 public service hotline. However, administrative approvals still require visits to the service hall, and the administrative reform involved in this process is a time-consuming task.

Smart living and smart economy, on the other hand, are the main driving forces behind the fluctuations in the smart level curve. In terms of smart living, new energy vehicles are transforming people's lives in these cities. Currently, the average penetration rate of new energy vehicles in the seven cities has reached 30%, but the overall retention rate is only 4%. There is still a significant gap that must be filled to achieve the target of 30% to 40%. The Ministry of Industry and Information Technology (MIIT) has released the "*Development Plan for the New Energy Vehicle Industry (2021–2035)*", which indicates that by 2025, the national penetration rate of new energy vehicles will reach 20%. As megacities, the penetration rate in these cities will be even higher. Therefore, there will be a noticeable increase in the smart living curve in 2025, and the trend will become smoother in 2030 when over 50% of new vehicles will be powered by new energy.

4.2. Trend of Green Development

As discussed earlier, the overall trend of smart cities can be described as "fast-growing, slow-growing, and then stabilizing", and correspondingly, the energy consumption in these cities has followed a pattern of "fast growth, slow growth, stabilization, and finally a decrease". Based on the current scenario, prior to 2030, energy consumption continues to increase steadily. Between 2030 and 2033, it reaches its peak level. Afterward, it begins a gradual decline. As shown in Figure 5 below. And correspondingly, the green development level experiences a "steady-decrease-bottom-increase" process.

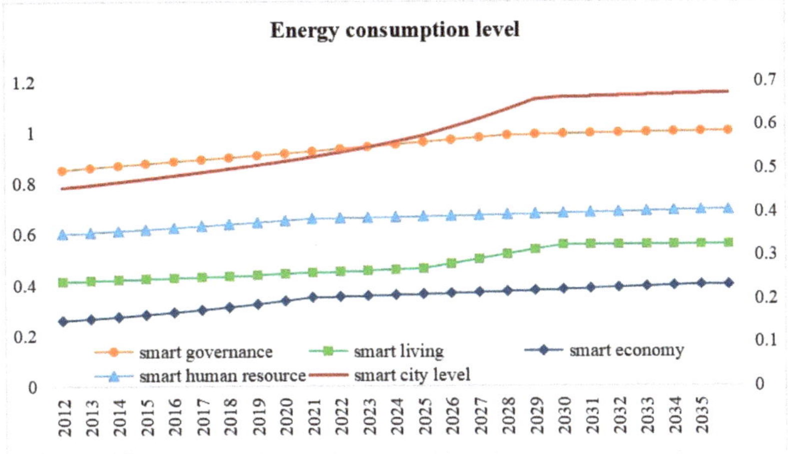

Figure 5. Development level of smart city.

4.2.1. Industrial Structure Scenarios

When the third industry proportion is higher, the initial energy consumption increases until 2028, and starts to decrease after that. Compared with the current curve, the peak is reached earlier (about 4 years earlier), and the maximum value is lower (about 3.1% lower). Additionally, in a low scenario, energy consumption would be lower for the first 18 years, but exceed the baseline after 2019 and continue rising towards the end of the forecast period, as shown in Figure 6 below.

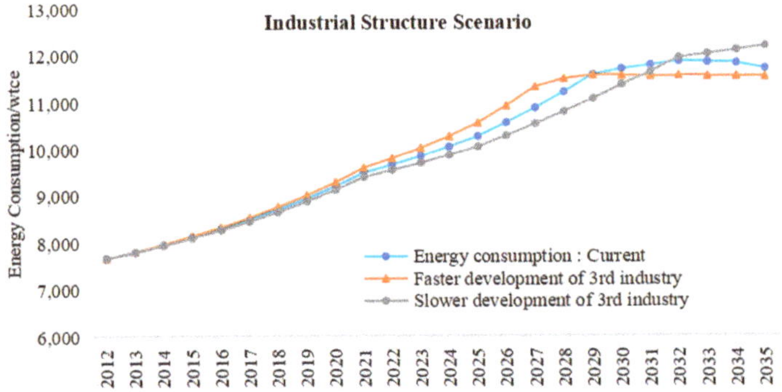

Figure 6. Energy consumption trend under industrial structure scenario.

While the growth of the tertiary sector can improve economic efficiency and competitiveness to some extent, it can also lead to increased energy consumption at the beginning. There are several reasons for this:

(1) Increased demand for services: As the tertiary sector develops, there is an increased demand for services such as transportation, logistics, communication, tourism, and entertainment. These services often require energy, such as fuel and electricity, for transportation and energy consumption by information technology equipment. For megacities, the increased demand for services may exacerbate the "urban diseases", which may have an amplifying effect on energy consumption.
(2) Energy consumption in commercial and office buildings: With the growth of the service industry, the number of commercial and office buildings also increases. These buildings require lighting, heating, ventilation, and air conditioning infrastructure, which often consume significant amounts of energy. The developers tend to build skyscrapers that rely on air conditioners and fresh air ventilation systems in megacities. This kind of building costs much more energy than traditional ones.
(3) Energy consumption in digitization and information technology: With the digitization and widespread use of information technology in the tertiary sector, there is a corresponding increase in energy consumption from various devices, servers, and data centers. Technologies such as cloud computing, big data analytics, and artificial intelligence all require substantial energy support.

But after the updating of industrial structure, the energy consumption will decrease. So, the optimization of industrial structure has demonstrated a lag effect on energy consumption.

4.2.2. Science and Technology Scenarios

A higher investment rate in science and technology initially results in increased overall energy consumption. However, after the year 2026, the curve starts to shift below the original curve and reaches its peak around 2033, followed by a gradual decrease. By the end of the forecasting period, there is a reduction of approximately 3.5% in energy consumption compared to the original curve. Conversely, a lower investment rate leads to lower energy consumption for approximately 16 years, but it continues to grow, surpassing the original curve. At the end of the forecasting period, energy consumption is approximately 4% higher, as shown in Figure 7 below.

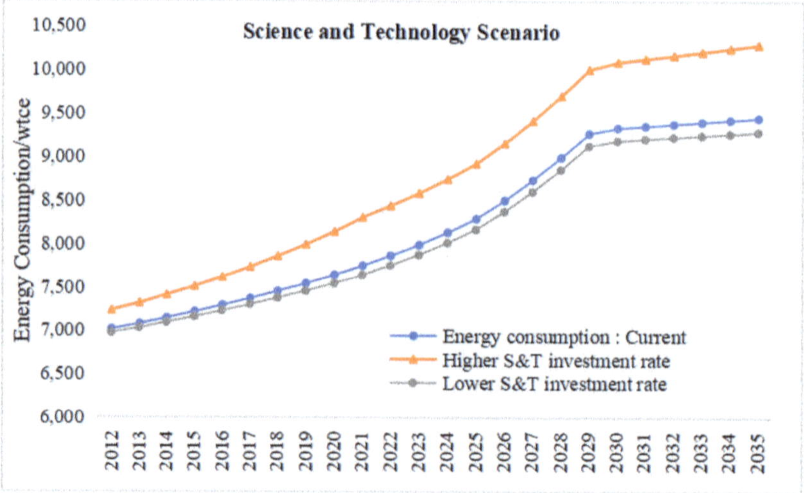

Figure 7. Energy consumption trend under science and technology investment scenario.

Increasing investment in technology can lead to an increase in energy consumption in the initial period, which might be due to the following reasons:

(1) Increased investment in technology often leads to advancements in production and consumption activities, resulting in higher energy demands. For example, technological progress can lead to the emergence of new energy-intensive industries and products in megacities, such as data centers, cloud computing, and smart devices, all of which require significant energy supply.
(2) Increased quantity of equipment and facilities: An increase in technology investment is often accompanied by an expansion in the quantity of equipment and facilities, such as factories, laboratories, and offices. The operation of this equipment and these facilities requires energy supply, thereby increasing energy consumption.
(3) Manufacturing and disposal of electronic devices: Increased technology investment means more electronic devices and computer systems, which consume energy during their manufacturing, usage, and disposal processes. The manufacturing of electronic devices involves the extraction, processing, and transportation of raw materials, while the usage and disposal of devices require electricity and other resources.

Environmental investment rate has little effect on consumption.

Similar to the industrial structure scenario, an energy consumption effects is shown after a relatively long period of preparation. A lag effect also exists here.

4.2.3. Environmental Investment Rate

Surprisingly, the direct change in environmental investment has little to do with the energy consumption. There will be only a 5% difference at the end of the forecasting period. There are some reasons for this:

Increasing environmental protection investment may not have an immediate significant impact on energy consumption due to several reasons:

(1) Small difference in scenario settings: China is already one of the top countries that invest in environmental protection, especially its megacities, which represent the frontier of China's development. So, the difference between the current and high scenarios is very small.
(2) Technology and equipment upgrade cycles: Environmental protection investment often involves upgrading and replacing existing technologies and equipment to improve energy efficiency and reduce pollutant emissions. However, equipment replacement and technology upgrades often require time and money, especially for large-scale production facilities. In such cases, the impact of environmental protection investment may require a longer cycle to be reflected in energy consumption.
(3) Environmental protection investment often entails high costs, such as purchasing energy-efficient equipment, implementing cleaner production processes, or adopting renewable energy sources. These costs may impose financial burdens on businesses, particularly during the initial investment phase, which could lead to significant decreases in energy consumption.

Although the impact of increased environmental protection investment on energy consumption may not be immediately evident, it remains an essential means to achieve sustainable development and reduce environmental impacts. As environmental technologies continue to develop and mature, coupled with increasing environmental awareness, the effects of environmental protection investment will gradually become more apparent, making more substantial contributions to reducing energy consumption. Therefore, sustained environmental protection investment and long-term commitment are key to achieving energy consumption reduction and environmental improvement, as shown in Figure 8 below.

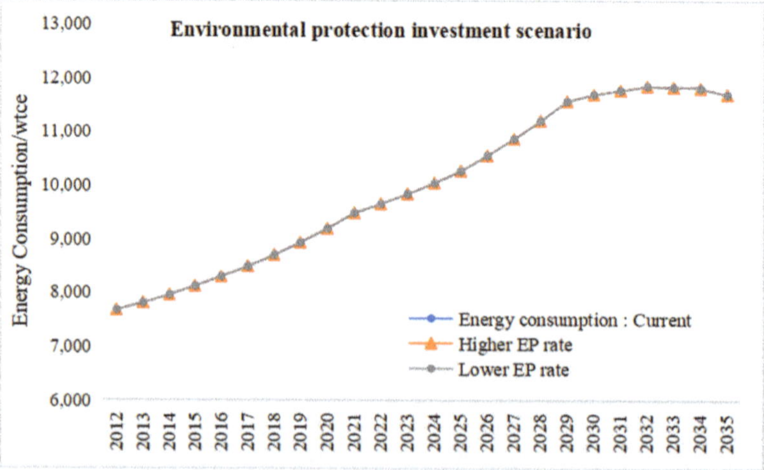

Figure 8. Energy consumption trend under environmental protection investment scenario.

4.2.4. Per Capita Scenario

When taking population into consideration, the decrease around 2030 becomes more significant. All megacities in China are now trying to control their populations. But, except for Beijing, the other cities' populations are still rising. So, the changing energy consumption trend becomes more significant, as shown in Figure 9 below.

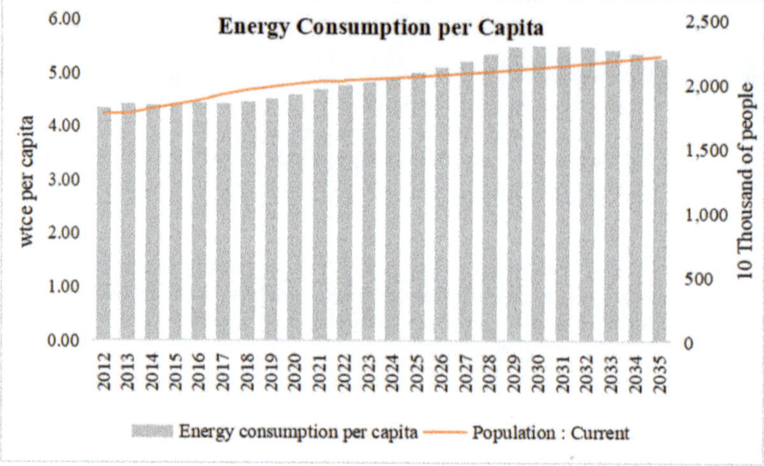

Figure 9. The trend of per capita energy consumption.

4.3. Relationship between "Smart" and "Green" Development

As we can see in Figures 3–8, the trends of smart cities and energy consumption are in synchronization. The energy curve lags behind the smart curve by approximately three years. They all meet their peak at around the early 2030s. The smart city curve horizontally shifts at that point, and energy consumption slightly drops, as shown in Figure 10 below.

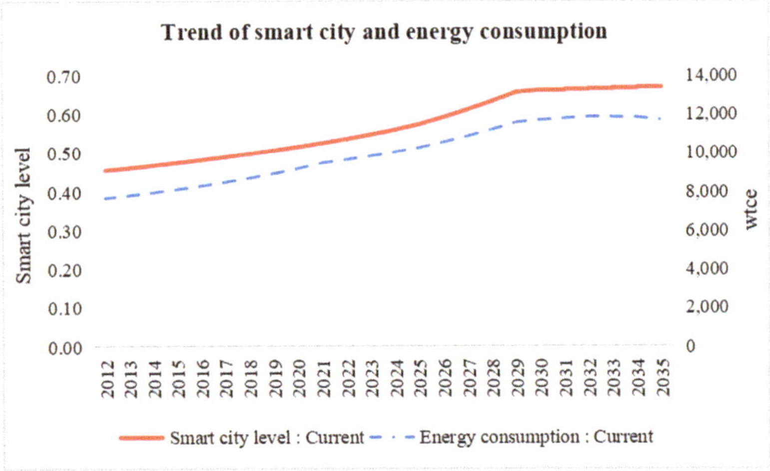

Figure 10. Trend of smart cities and energy consumption.

That is to say, when the construction period of a smart city finishes, unless there are other intervening factors or circumstances that cause a deviation from the expected pattern, the smart city level will maintain stability for about 5 years. Not until then will its positive effects on energy consumption starts to occur, and until the end of our forecasting period, the influence does not reach its limits. Among the four aspects of smart cities, smart governance and smart living are the main driving forces.

5. Conclusions and Policy Implications

5.1. Conclusions

From the analysis above, the following conclusions can be drawn:

1. Both smart development and energy consumption will experience a growth pattern characterized by an S-shaped curve in megacities. Correspondingly, the urban green development level will experience a "steady-decrease-bottom-increase" process. There are mainly two stages.

Stage one: smart city construction and energy increase. The smart city construction process costs more energy in megacities. Governors used to believe that a smart city can reduce its energy consumption in a short time. But our simulation results show that this is not true. During their construction, smart cities cost more energy, especially in megacities. The reasons for this are, on one hand, smart city requires high-performance infrastructure support, sensors, surveillance devices, data centers, etc., which all cost large amount of power. On a large scale, this is a much bigger problem in megacities than in smaller ones. On the other hand, smart cities rely on the collection, transmission, and processing of big data. Large amounts of data need to be transmitted through networks and processed and analyzed in data centers. These data transmission and processing processes require a significant amount of electrical power, especially when dealing with large-scale cities and massive data volumes. This leads to further increases in energy consumption.

Stage two: smart city construction has been completed and the energy consumption starts to drop. As we can see in the simulation results, when the construction has finished and all the smart city facilities are put into operation, the energy consumption will decline significantly. Seeking quick results is not a good idea here.

2. External factors influence the system in different ways. Economic structure, science and technology investment, and environmental protect investment were considered as external factors in this model. As discussed in Section 4.2, the first two factors both show influences on megacity energy consumption. A higher proportion of tertiary industry

in an economic system means an earlier and lower energy peak; a lower proportion of tertiary industry in an economic system leads to a later and higher energy peak. So, during the phase of industrial restructuring, there may be a temporary increase in energy consumption levels, which is considered a necessary cost and a transitional phase for long-term sustainable development. S & T investment rate has similar effects to the industrial structure factor. Environmental protection investment has little influence on energy consumption in megacities.

3. The energy consumption levels in megacities are higher than the national average level. In 2022, the energy consumption was 3.83 tce per capita. The average number in these seven cities was 4.43 tce per capita. And the peak will be 5.51 tce per capita in 2030 and 2031 according to the forecast. The reasons for this might be as follows:

First, high population density: Big cities usually have larger populations concentrated in relatively small areas, which leads to higher population density. The increase in population density will increase the demand for energy, including energy consumption in residential heating, household electricity, transportation, etc. Second, the high degree of urbanization: Big cities are more urbanized and have more businesses, offices, and public facilities, which require a lot of energy to keep them running. Lighting, heating, cooling and other forms of energy consumption in commercial areas and office areas are relatively large. At the same time, large cities also have more transport networks, road lighting, and public transport systems, which also increase energy demand. Third, the diversification of economic activities: Larger cities usually have more economic activities and industrial sectors. These activities involve many fields such as manufacturing, the service industry, the financial industry, the technology industry, etc. These industries usually have high demand for energy. For example, industrial facilities such as factories, office buildings, and large commercial centers consume large amounts of energy. Fourth, the high standard of living: Big cities usually have a higher standard of living, and people's spending habits may be more luxurious. People may have more electronic devices, household appliances, private cars, and other energy-consuming products, and at the same time, use these devices and services more frequently.

5.2. Policy Implications

Three aspects of policy implications are given.

1. A long-term smart city development plan with phased targets needs to be made, and it is not rational to pursue immediate results. In China, the plan is the guidance for all the tasks. As discussed above, the smart city plays a role in urban green development that starts with restraint and ends with promotion. So, it takes time to meet the "dual carbon" target for megacities.

When a long-term plan is developed, it is necessary to conduct it in stages. It has been ten years since the smart city pilot policy was implemented in the cities studied herein. The policy during that time was to support the smart-city-related industry, construct the infrastructure, and make an early plan. For the next 8 to 10 years, until the early 2030s, the key focus should be integrating the fragmented components, developing up-level applications, and building self-learning organic cities. So, the policy should also put more efforts into encouraging the development of these areas.

After that, the construction process will transition to a maintenance and optimization phase. And correspondingly, the financial investment in smart infrastructure should be tightened and funding for efficiency evaluation should be increased.

With a well-designed long-term plan, smart cities will have better performance in promoting urban green development.

2. To be smart and green, the specificity of megacities should be fully considered. As discussed in Chapter 4, this specificity manifests in the following ways. First, scale and complexity. Megacities have large scales, large populations, vast territories, complex urban systems and infrastructure networks, and large-scale management and operational challenges. Second, data processing and analysis. The huge and diverse amounts of

data generated by megacities, including data from sensors, devices, and citizens, require powerful data processing and analysis capabilities to extract useful information and insights. Third, infrastructure requirements. The intelligent construction of megacities requires strong infrastructure support, such as high-speed networks, communication base stations, data centers, etc., to meet large-scale data transmission and processing needs.

So, policies need to encourage multi-agent involvement and regional collaboration. The smart construction of megacities requires the participation and cooperation of the government, enterprises, academia, and social organizations. It is necessary to establish a good cooperation mechanism and sharing platform to achieve resource sharing, information exchange, and collaborative innovation.

Also, it is difficult for megacities to achieve this alone, so regional collaboration is needed. Cities in the same region should share resources, including sharing experiences, technologies, data, and best practices. Different cities can learn from each other, jointly solve problems in smart construction and green development, and avoid redundant construction and wasting resources. And cities can share complementary advantages and industrial cooperation. Through regional synergy, cities can give full play to their respective advantages and characteristics and form complementary industrial layouts and cooperative relationships. For example, one city may have strengths in renewable energy, while another may have unique experience in smart transportation or environmental governance. Through cooperation and coordinated development, the optimal allocation of resources and the complementary development of industries can be realized. Additionally, geographically close cities are able to share infrastructure interconnection and transportation connectivity. Regional coordination promotes the interconnection of urban infrastructure and transportation connectivity. This includes building efficient transportation networks, shared infrastructure, and information platforms to achieve data sharing, service collaboration, and resource integration across cities. This can improve the overall benefits of urban smart construction and green development. Last but not least, cities can share strategic coordination and policy support. Regional synergy can promote strategic coordination and policy support among cities. Cities can jointly formulate and promote consistent policy frameworks and regulations, forming unified standards and guidance to promote the common goal of smart construction and green development. At the same time, through regional cooperation, cities can also obtain more policy support and resource input to accelerate the process of smart construction and green development.

3. Policies should be more inclined toward the field of scientific and technological support. Financial support in the field of environmental protection can temporarily maintain the status quo. Based on our simulation results, increasing science and technology investment shows more obvious control of urban energy consumption. China's current environmental protection investment is already world-leading, so there is not much potential in this area. Also, the tertiary industry aspect should be promoted.

5.3. Future Work

There are two directions that future research can focus on:

(1) The scope of the research can be expanded. This research is based on seven megacities in China, but future work could encompass broader objectives beyond China. This research framework holds applicability for all developing countries currently undergoing rapid urbanization processes.

(2) The scenarios can be further enriched. As indicated in Section 2.1, various factors can influence the level of green development. While this study includes industrial structure, environmental protection investment, and R & D investment, it would also be intriguing to explore additional variables such as population dynamics, policy impacts, and other relevant factors.

Author Contributions: Conceptualization, R.H.; Methodology, R.H.; Software, X.H.; Validation, X.H.; Formal analysis, R.H.; Data curation, X.H.; Writing—original draft, R.H.; Writing—review & editing, X.H.; Visualization, X.H.; Supervision, R.H. All authors have read and agreed to the published version of the manuscript.

Funding: Beijing Municipal Social Science Foundation (19JDGLB027) and Beijing Social Science Fund (20GLA005).

Data Availability Statement: All data used in this research are public. Readers can get access to them on "China City Statistical Yearbook", city statistical bulletins and official news reporting.

Conflicts of Interest: The authors declare no conflict of interest.

References

1. Corbusier, L. *The City of Tomorrow and Its Planning: The City of Tomorrow and Its Planning*; John Wiley & Sons: Malden, MA, USA, 1987.
2. Angelidis, M. Smart green just city actions versus urban planning in European union. *Sustain. Dev. Cult. Tradit. J.* **2021**, *1*, 60–80. [CrossRef]
3. Xie, Z.; Wu, R.; Wang, S. How technological progress affects the carbon emission efficiency? Evidence from national panel quantile regression. *J. Clean. Prod.* **2021**, *307*, 127133.
4. Pan, J. Scientific planning as the key to new-type of ecologically friendly urbanization. *Environ. Prot.* **2014**, *42*, 14–18.
5. Gibson, D.V.; Kozmetsky, G.; Smilor, R.W. (Eds.) *The Technopolis Phenomenon: Smart City, Fast Systems, Global Networks*; Rowman & Littlefield: Lanham, MD, USA, 1992.
6. Li, F.; Yang, H.; Gao, X.; Han, H. Towards IoT-based sustainable digital communities. *Intell. Converg. Netw.* **2022**, *3*, 14. [CrossRef]
7. De Aguiar, C.; Leshed, G.; Pinch, T.; Green, K. Evaluation of communIT, a large-scale, cyber-physical artifact supporting diverse subgroups building community. *J. Smart Cities Soc.* **2022**, *1*, 165–178. [CrossRef]
8. Cristiano, E.; Deidda, R.; Viola, F. The importance of green roofs in an urban Water-Energy-Food-Ecosystem nexus context. In *EGU General Assembly Conference Abstracts*; EGU: München, Germany, 2021. [CrossRef]
9. Otieno, J.P.; Duncan, M.; Oluoch, P. Climate Neutral Icpac (Smart City, Transport & Energy, Software & Data). 2021. Available online: https://www.researchgate.net/publication/354585953_CLIMATE_NEUTRAL_ICPACSMART_CITY_TRANSPORT_ENERGYSOFTWARE_DATA (accessed on 15 January 2022).
10. Sembiring, M. Smart City Development: Supporting Climate Resilience in Southeast Asia? Non-Traditional Security in the Asia-Pacific. In *Non-Traditional Security in the Asia-Pacific: A Decade of Perspectives*; World Scientific: Singapore, 2021.
11. Hadidi, L.A.; Rahman, S.M.; Maghrabi, A.T. Smart city—A sustainable solution for enhancing energy efficiency and climate change mitigation in Saudi Arabia. *Int. J. Glob. Warm.* **2021**, *24*, 91–107. [CrossRef]
12. Wang, Y.; Jiulin, L.I.; Kangkang, G.U.; Lin, L.U. Optimization of urban green infrastructure network layout based on MSPA-CIRCUIT: Case of Hefei. *Acta Ecol. Sin.* **2022**, *42*, 2022–2032.
13. Ou, X.; Lyu, Y.; Liu, Y.; Zheng, X.; Li, F. Integrated multi-hazard risk to social-ecological systems with green infrastructure prioritization: A case study of the Yangtze River Delta, China. *Ecol. Indic.* **2022**, *136*, 108639. [CrossRef]
14. Gulsecen, S.; Gezer, M.; Celik, S.; Koolu, F.N. Future needs from the Smart Mega City (SMC) plans smart green city the case of Istanbul. In *Sustainable Mega City Communities*; Butterworth-Heinemann: Oxford, UK, 2021.
15. Lee, S.H.; Lee, D.W. Smart Convergence—The Future of Green City. *J. Digit. Converg.* **2012**, *10*, 233–237.
16. Anthopoulos, L. Evaluating Green Smart City's Sustainability with an Integrated System Dynamics Model. In Proceedings of the CeDEM Asia 2014: Conference for E-Democracy an Open Government, Hong Kong, China, 4–5 December 2014; MV-Verlag: Greifswald, Germany, 2015.
17. Oad, A.; Ahmad, H.G.; Talpur, M.S.H.; Zhao, C.; Pervez, A. Green smart grid predictive analysis to integrate sustainable energy of emerging V2G in smart city technologies. *Optik* **2023**, *272*, 170146. [CrossRef]
18. Qureshi, K.N.; Hussain, R.; Jeon, G. A distributed software defined networking model to improve the scalability and quality of services for flexible green energy internet for smart grid systems. *Comput. Electr. Eng.* **2020**, *84*, 106634. [CrossRef]
19. Zhang, X.; Manogaran, G.; Muthu, B. IoT enabled integrated system for green energy into smart cities. *Sustain. Energy Technol. Assess.* **2021**, *46*, 101208. [CrossRef]
20. Mukherjee, P.; Patra, S.S.; Pradhan, C.; AL-Amodi, S.; Barik, R.K. Green Energy Powered by Smart Grids: A Business Model for Long Term Sustainability. In Proceedings of the Cyber Technologies and Emerging Sciences: ICCTES 2021, Uttarakhand, India, 17–18 December 2021; Springer: Singapore, 2022.
21. Cao, W.; Zhang, Y.; Qian, P. The effect of innovation-driven strategy on green economic development in China—An empirical study of smart cities. *Int. J. Environ. Res. Public Health* **2019**, *16*, 1520. [CrossRef]
22. Tian, H.; Qing, S. Can innovation-driven approaches promote the sustainable economic development of smart cities? *Sci. Technol. Prog. Policy* **2018**, *35*, 6–12.
23. Bibri, S.E.; Krogstie, J. Smart sustainable cities of the future: An extensive interdisciplinary literature review. *Sustain. Cities Soc.* **2017**, *31*, 183–212. [CrossRef]

24. Silva, B.N.; Khan, M.; Han, K. Towards sustainable smart cities: A review of trends, architectures, components, and open challenges in smart cities. *Sustain. Cities Soc.* **2018**, *38*, 697–713. [CrossRef]
25. Martin, C.; Evans, J.; Karvonen, A.; Paskaleva, K.; Yang, D.; Linjordet, T. Smart-sustainability: A new urban fix? *Sustain. Cities Soc.* **2019**, *45*, 640–648. [CrossRef]
26. Ahvenniemi, H.; Huovila, A.; Pinto-Sepp, I.; Airaksinen, M. What are the differences between sustainable and smart cities? *Cities* **2017**, *60*, 234–245. [CrossRef]
27. Gazzola, P.; Del Campo, A.G.; Onyango, V. Going green vs. going smart for sustainable development: Quo vadis? *J. Clean. Prod.* **2019**, *214*, 881–892.
28. Yao, T.; Huang, Z.; Zhao, W. Are smart cities more ecologically efficient? *Evidence from China. Sustain. Cities Soc.* **2020**, *60*, 102008. [CrossRef]
29. Viitanen, J.; Kingston, R. Smart cities and green growth: Outsourcing democratic and environmental resilience to the global technology sector. *Environ. Plan. A* **2014**, *46*, 803–819.
30. Shi, D. Evolution of industrial development trend under digital economy. *China's Ind. Econ.* **2022**, *11*, 26–42.
31. Jiang, X.; Wang, L. Industrial structure upgrading effect of smart city policy: An empirical study based on multi-period DID. *Forum Sci. Technol. China* **2021**, *12*, 31–40.
32. Lichao, W.; Li, W.; Chen, L.; Li, J. The impact of smart city construction on green technology innovation: An empirical study based on panel data of prefecture-level cities. *Technol. Econ.* **2022**, *41*, 1–16.
33. Cheng, Z.; Wang, L.; Zhang, Y. Does smart city policy promote urban green and low-carbon development? *J. Clean. Prod.* **2022**, *379*, 134780.
34. Huang, H.; Xie, Y.; Li, N. Does Smart City Construction Promote Low-carbon Development? *Urban Dev. Stud.* **2022**, *29*, 105–112.
35. Fan, H.; Mi, X. Research on the construction of smart cities and green economy transformation of cities. *Urban Probl.* **2021**, *11*, 96–103.
36. Liu, K.; Meng, C.; Tan, J.; Zhang, G. Do smart cities promote a green economy? Evidence from a quasi-experiment of 253 cities in China. *Environ. Impact Assess. Rev.* **2023**, *99*, 107009. [CrossRef]
37. Lin, X.; Wang, D. Can smart city construction promote green development—Evidence based on "quasi natural experiment". *Technol. Econ.* **2022**, *41*, 104–113.
38. Filiou, D.; Kesidou, E.; Wu, L. Are smart cities green? The role of environmental and digital policies for Eco-innovation in China. *World Dev.* **2023**, *165*, 106212.
39. Tian, H.; Song, Q. Can innovation-driven promote the green development of smart city economy—An empirical analysis based on the panel data of 47 cities in China. *Sci. Technol. Prog. Policy* **2018**, *35*, 6–12.
40. Liu, Y.; Li, Q.; Zhang, Z. Do Smart Cities Restrict the Carbon Emission Intensity of Enterprises? Evidence from a Quasi-Natural Experiment in China. *Energies* **2022**, *15*, 5527.
41. Du, J.; Wang, Y.; Zhao, A. The Effects and Machanism of Smart City Construction on Urban Green Development. *Soft Sci.* **2020**, *34*, 59–64.
42. Zhang, J.; Li, Q. Influence of Smart City Construction on Urban Technology Innovation. *Technol. Prog. Policy* **2020**, *37*, 38–44.
43. Wang, K.L.; Pang, S.Q.; Zhang, F.Q.; Miao, Z.; Sun, H.P. The impact assessment of smart city policy on urban green total-factor productivity: Evidence from China. *Environ. Impact Assess. Rev.* **2022**, *94*, 106756.
44. Zhang, E.; He, X.; Xiao, P. Does Smart City Construction Decrease Urban Carbon Emission Intensity? Evidence from a Difference-in-Difference Estimation in China. *Sustainability* **2022**, *14*, 16097.

energies

MDPI

Article

Robust Multiobjective Decision Making in the Acquisition of Energy Assets

Rafael Bambirra [1,2,*], Lais Schiavo [3], Marina Lima [3], Giovanna Miranda [3], Iolanda Reis [4], Michael Cassemiro [4], Antônio Andrade [4], Fernanda Laender [4], Rafael Silva [5], Douglas Vieira [3] and Petr Ekel [2,6]

1. Programa de pós Graduação em Engenharia Elétrica, Universidade Federal de Minas Gerais—UFMG, Belo Horizonte 31270-901, Brazil
2. ASOTECH, Belo Horizonte 30535-630, Brazil; petr.ekel2709@gmail.com
3. ENACOM, Belo Horizonte 31275-100, Brazil; lais.schiavo@enacom.com.br (L.S.); douglas.agv@gmail.com (D.V.)
4. Aliança Energia, Belo Horizonte 30170-050, Brazil; antonio.andrade@aliancaenergia.com.br (A.A.)
5. VALE, Nova Lima 34000-000, Brazil; rafael.ferreira.silva@vale.com
6. Programa de pós Graduação em Informática, Pontifícia Universidade Católica de Minas Gerais—PUC/MG, Belo Horizonte 30535-901, Brazil

* Correspondence: bambirrarafael@ufmg.br

Abstract: In asset management for energy portfolios, quantitative methodologies are typically employed. In Brazil, the NEWAVE computational model is universally used to generate scenarios of hydraulic production and future prices, which result in revenue distributions. These distributions are then used to estimate the portfolio's revenue and assess its risk. Although this is a well-established analysis, it has some shortcomings that are not always considered. The validity of the revenue series constructed by NEWAVE, especially in long-term analysis, is a real problem for agents concerning the acquisition of assets such as power plants. Another issue is the disregard for other objectives that are important for the operationality of the management task and are often ignored, such as operational risk. To address these limitations, this work combines the areas of multicriteria decision making under uncertainty and risk management and presents a methodology for evaluating the acquisition of long-term energy assets, as well as a practical application of the proposed method. Investment alternatives are evaluated in multiple developed scenarios, so it is possible to measure how robust a given option is. By analyzing several scenarios simultaneously, a larger region of uncertainties can be covered, and therefore, decision making becomes more secure. The proposed methodology includes six objectives, designed to address a wider range of stakeholder needs. This approach is applied to an illustrative portfolio, producing results that allow for a more comprehensive understanding of decision attributes. Therefore, this work not only addresses the current limitations in the field but also adds an original contribution by considering simultaneously several scenarios and integrating multiple objectives in a robust and secure decision-making framework.

Keywords: energy investments; energy trade; risk management; long-term planning; multicriteria decision making

1. Introduction

Electric power is a commodity with high volatility and uncertainty in its price. With the evolution of energy markets, the possibility was created for market agents to freely trade this commodity. This fact makes markets more dynamic and competitive [1,2]. With the changes in the electric sector and advances in opening the energy market, which led to an increase in the number of transactions, agents realized that efficient risk management is essential for the healthy performance of this activity [3].

Risk analysis is strategically important as it is necessary to identify factors that will influence the financial return of an investment [4]. The acquisition of new energy assets by a generating agent cannot be decided solely based on their costs, whether investment or generation, as it is an activity with long-term impacts. During decision making about a change in the portfolio, risks of different natures are to be considered [5–7].

To define a portfolio, market participants adhere to two objective functions: maximizing the expected net revenue of the portfolio (E[NPV]) and maximizing revenue in the worst scenarios (written through the conditional value at risk (CVaR)) [8,9]. The use of revenue at risk (RaR) as an operational limit is also common, in an attempt to control the size of revenue dispersion. However, other ways of measuring risk could be used, such as the lower partial moments used in [10] or the cash flow at risk approach presented in [6].

An important point in decision making under conditions of uncertainty is the absence of optimal solutions (regardless of whether the problem is monocriteria or multicriteria): the solution optimal for one scenario is not optimal for another scenario. So, what is a decision under conditions of uncertainty? The solution under conditions of uncertainty is a robust (or non-dominated) solution, which permits one to attend any scenario in the best way [11]. How is it possible to find a robust solution? The discipline of Operations Research cannot help her. It is based on the concept of existing optimal solutions. All approaches, methods and strategies of Operations Research are directed at obtaining optimal solutions, which do not exist. Considering this, a strategy applied in this work is directed at finding the worst solutions, which are dominated by other solutions. By cutting off the worst solutions and applying any type of preference information, it is possible to reduce decision uncertainty regions.

This paper proposes an evolution of a methodological inheritance for the selection of energy portfolios. The Brazilian energy market operates and is oriented towards a periodic autoregressive model used by the National System Operator for its control and operation [12]. In turn, other market participants also use this tool for their commercial and operational risk management. However, due to the characteristics of the sector, the price and generation forecasts returned by this system are sensitive to initial conditions and climate forecasts. This sensitivity can influence the accuracy of the model's result, and relying on a single forecast may not be sufficient for decision making.

Therefore, the objective of this work is to use the decision-making techniques proposed in [11,13] for the construction of portfolios and their analysis in multiple scenarios. To achieve this, the performance of an agent's portfolio in different constructed scenarios was evaluated. This performance is assessed by applying the $\langle X, F \rangle$ decision-making model [11,13] using the functions of expected net present value, CVaR, and revenue at risk, which are already used by the market. This article also proposes the use of new objective functions—insurance index (InS), alternative synergy with the current portfolio and operational risks—in addition to the classical objective functions for making decisions on energy portfolios. By stressing the cases and using decision-making methodologies, it is possible to construct a robust portfolio for the company's position.

Characterizing the originality and contribution of this work, it is necessary to stress that the majority of methods related to energy trade and portfolio management are based on the probabilistic approach [14]. However, it is not possible to talk about the future and construct the future based on past information and trends, at least for long-term planning (as evident from events such as the global financial crisis of 2008 and the COVID pandemic in 2020). Considering this, in the present work, we try to use the possibilistic approach, which consists in constructing and using representative combinations of initial data, states of nature, or scenarios. This work proposes the use of six objective functions, designed specifically to address the unique demands of agents engaged in asset acquisition during the process of energy portfolio management.

The paper is structured as follows. Section 2 presents a review of correlated works and theoretical references. Section 3 describes the proposed method and objective functions. Section 4 presents the characteristics of the problem, the problem formulation, the initial data, and the evaluated alternatives. The results of this paper are illustrated by considering a case study given in Section 5. Finally, in Section 6, the main conclusions of this study are presented.

2. Theoretical Reference

2.1. Private Investments in Energy

In the context of energy market operations, companies buy and sell energy to manage their energy balance over time and ensure their operational stability while seeking the lowest possible cost. To improve the way they perform this task, companies have started applying financial risk management models and began trading energy based on a multicriteria approach, considering revenues and market risks directly in the construction of their energy portfolios.

Considering that this work, from a practical application perspective, is dedicated to the risk management of energy portfolios, it is necessary to indicate the results of [2,3,5,6]. These works can be considered as the initial theoretical bases of the discussion of the problem of risk management in energy portfolios, which continues today. These works present the reason why energy prices are volatile parameters and discuss financial engineering approaches to measuring risks in the energy market.

The results presented in [2] addressed the modeling of the risk metric value at risk (VaR). The authors of [2] explain how the methodology should be used, presenting requirements such as return forecasts. The VaR metric is also addressed in [5], as well as CVaR and other metrics. It exposes some examples and comparisons of results from the used methods. In [6], different risk metrics are presented, and how market risks can be measured and managed using real options models and stochastic optimization techniques is explored. Moreover, in [3], there is a discussion about how energy prices can fluctuate and the associated risks when pricing and hedging electricity derivatives.

Regarding trading, in general, agents seek commercial strategies that contemplate seasonal productions of each source at each period of the year, in order to offer the best negotiation. The work of Camargo et al. [15] deals with the management of energy trading contracts through the formation of portfolios composed of renewable energy sources, from the point of view of the generator, consumer, and trader. From the generator's perspective, this work analyzes risk management policies defined based on the periodic accounting of CVaR and their influence on contracting strategies. Additionally, this work explores how Swap contracts are used to provide security for parties during times when the short-term market is vulnerable, given the conditions of purchase and sale prices, the hydraulic generator's contractual balance, the agent's risk aversion, and the projections of the settlement price for differences (*Preço de Liquidação das Diferenças* (PLD)) and generation scaling factor (GSF) in the planning horizon.

To define these projections in the Brazilian market, agents use a computational model called NEWAVE. However, since it is a tool undergoing constant changes, using it as the only reference in decision making can be inadequate [12]. The fact that the electrical system undergoes constant regulatory evolutions and the use of new technologies can also invalidate these projections in the long term [16].

In order to address the limitations of conventional risk assessment methods, the work of Ilbahar et al. [17] incorporates the subjective judgment of decision makers to map uncertainties in the risk factors of renewable energy projects. Overall, the paper presents a new and comprehensive approach to assessing risks in renewable energy investments that overcomes the limitations of traditional methods. However, the proposed approach may require a higher level of expertise and effort in data collection and analysis, which may limit its practical application in some contexts.

Another work dealing with decision making in energy investments is [18]. This work proposes an approach that combines a fuzzy analytical hierarchy process (FAHP) and a technique for order preference by similarity to the ideal solution (TOPSIS) to assess the suitability of new potential services in the renewable energy sector. The challenges and complexities associated with the process of developing new services in the renewable energy sector are also discussed, including technological uncertainty and the need for collaboration among stakeholders.

On the same path, the authors of [19] propose a stochastic decision support model for renewable energy investments in Brazil that considers expected returns and CVaR as objective functions and NEWAVE output data for scenario creation. The study reveals that the risk associated with intermittent sources can be managed through CVaR assessments, although the level of the decision maker's risk aversion significantly influences the company's market position. The authors found that a diversified firm's asset base, along with the complementary nature of generation sources, can significantly reduce the financial risks of the investor's portfolio. These results showed that the decision of a new investment must consider the current portfolio of the company.

Recent studies discussing multiobjective decision making in energy trading also include works [20,21]. The authors of [20] propose a hybrid trading mechanism that operates on multiple time scales, taking into account the transmission speeds and limits of various energy sources, while developing a strategy based on the Markov decision process. On the other hand, the authors of [21] explore energy trading strategies in a residential energy system. Although both studies present strategies that improve energy costs for the involved agents, these studies tend to overlook other needs of the agents, thereby compromising the construction of a robust portfolio. Additionally, they focus on short-term analysis, which may limit the comprehensiveness and real-world applicability of their strategies.

When making long-term decisions, it is essential to consider the uncertainty and variability of the future. This is where possibilistic information becomes crucial. Unlike probabilistic information, which relies on statistical analysis and provides the likelihood of certain outcomes occurring, possibilistic information considers the uncertainty of the future and focuses on the range of possible outcomes without assigning probabilities [11]. This is necessary since the future is constantly evolving, with regulatory, technological, and other changes constantly altering the range of possible outcomes, and is what makes this work different from those mentioned above. Therefore, decision makers should be aware that probabilities can quickly become outdated and should be prepared to adapt their plans accordingly. In this sense, a possibilistic view of decision making, focusing on the range of possible outcomes instead of assigning probabilities, may be more reliable in the long term [22].

According to [23], to account for the revenue of an energy portfolio, let W be the set of energy assets. It follows that $x_{i,t} \in W$ is the volume resulting from the purchase (or sale, if negative) of an energy asset i in period t. Therefore, the result (revenue or expense) of trading an energy asset can be described as follows:

$$\varrho_t^E(x_t) = h_t\left(\sum_{c=1}^{C} p_{c,t}v_{c,t} + \sum_{i=1}^{I} p_{i,t}x_{i,t}\right), \forall i \in W_c, \tag{1}$$

where h_t is the number of hours within period t, W_c is the set of assets that are energy contracts, C is the total number of contracts, I is the total number of assets of a certain type, v is the volume of energy bought/sold already existing in the portfolio, and p is the price of the contract/asset.

The operation and maintenance cost of the agent's assets can be described as:

$$\varrho_{t,s}^M(x_t) = h_t \sum_{i=1}^{I}(-\theta_i(g_{i,t,s} + x_{i,t})), \forall i \in W \wedge \neg W_c, \tag{2}$$

where θ_i is the operation and maintenance cost of the plant, modeled by a constant whose unit is expressed in $R\$/MWh$, and $g_{i,t,s}$ is the volume of energy existing in the portfolio.

The revenue, or expense, resulting from energy exposed to the spot market can be described as:

$$\varrho^{P}_{t,s}(x_t) = h_t PLD_{t,s}\xi_{t,s}(x_t), \tag{3}$$

where the energy exposure to the short-term market is calculated as follows:

$$\xi_{t,s}(x_t) = \sum_{i=1}^{I} \gamma_{t,s} \cdot x_{i,t} + \sum_{i=1}^{I} \tilde{g}i,t,s + \hat{V}t,s - \check{V}_{t,s}. \tag{4}$$

In (4), $\gamma_{t,s}$ equals 1 for all assets, except for hydraulics, participating in the MRE, since γ is the generation scaling factor (GSF). The $\tilde{g}_{i,t,s}$ is the adjusted physical guarantee of the agent, if the asset is hydraulic plants participating in the MRE, since in this case, the resource considered for the plant is its physical guarantee multiplied by the GSF factor. The physical guarantee value, in general, defines the maximum amount of energy that the project can trade [23].

The total volumes resulting from energy purchase or sale contracts are represented by the terms $\hat{V}_{t,s}$ and $\check{V}_{t,s}$, respectively. In other words, the portion that composes the exposure result consists of the balance value between the resource and the requirement.

Therefore, the total revenue of an agent with multiple assets of different types can be defined by the following equation:

$$\varrho_{t,s}(x_t) = \varrho^{E}_{t}(x_t) + \varrho^{M}_{t,s}(x_t) + \varrho^{P}_{t,s}(x_t). \tag{5}$$

Since this is a long-term investment evaluation approach, seasonality and energy modulation, practices used by the Brazilian market agents for portfolio operation in monthly and daily time frames, will not be considered.

2.2. Generalization of the Classical Approach to Dealing with Information Uncertainty

The classical approach [24–26] for dealing with information uncertainty is based on the assumption that the analysis is carried out for a given number K of solution alternatives $X_k, k = 1, \ldots, K$ and a given number J of representative combinations of initial data (the states of nature or scenarios) $Y_j, j = 1, \ldots, J$, which define the corresponding payoff matrix. The payoff matrix, presented in Table 1 (the first six columns), reflects the effects (or consequences) of an action X_k for the corresponding state of nature.

Table 1. Payoff matrix with characteristics estimates.

	Y_1	...	Y_j	...	Y_J	$F^{min}(X_k)$	$F^{max}(X_k)$	$\bar{F}(X_k)$	$R^{max}(X_k)$
X_1	$F(X_1, Y_1)$	...	$F(X_1, Y_j)$	...	$F(X_1, Y_J)$	$F^{min}(X_1)$	$F^{max}(X_1)$	$\bar{F}(X_1)$	$R^{max}(X_1)$
...	...	...	...	...	...	...	...	...	...
X_k	$F(X_k, Y_1)$	...	$F(X_k, Y_j)$	...	$F(X_k, Y_J)$	$F^{min}(X_k)$	$F^{max}(X_k)$	$\bar{F}(X_k)$	$R^{max}(X_k)$
...	...	...	...	...	...	...	...	...	...
X_K	$F(X_K, Y_1)$	...	$F(X_K, Y_j)$	...	$F(X_K, Y_J)$	$F^{min}(X_K)$	$F^{max}(X_K)$	$\bar{F}(X_K)$	$R^{max}(X_K)$

Source: [22].

The analysis of payoff matrices and the choice of rational solution alternatives are based on the use of choice criteria [24–26]. The application of the choice criteria of Wald, Laplace, Savage, and Hurwicz, which are of a general nature, is discussed next. There are other choice criteria, for example, Bayes, maximum probability, minimum dispersion, maximum measure of Bayesian sets, maximum integral power, etc. [27,28]. However, these criteria assume certain informational situations about the states of nature.

To better understand the use of the criteria of Wald, Laplace, Savage, and Hurwicz, Table 1 includes the following characteristics estimates for a solution alternative:

- The minimum level of objective function:

$$F^{min}(X_k) = \min_{1 \le j \le J} F(X_k, Y_j), \tag{6}$$

which is the most optimistic estimate if the objective function is to be minimized or the most pessimistic estimate if the objective function is to be maximized;
- The maximum level of objective function:

$$F^{max}(X_k) = \max_{1 \le j \le J} F(X_k, Y_j), \tag{7}$$

which is the most optimistic estimate for the maximized objective function or the most pessimistic estimate if the objective function is to be minimized;
- The average level of objective function:

$$\bar{F}(X_k) = \frac{1}{J} \sum_{j=1}^{J} F(X_k, Y_j); \tag{8}$$

- The maximum level of regret:

$$R^{max}(X_k) = \max_{1 \le j \le J} R(X_k, Y_j), \tag{9}$$

where $R(X_k, Y_j)$ is an excess of expenses that occur under the combination of the state of nature Y_j and the choice of the solution alternative X_k instead of the solution alternative that is locally optimal for the given Y_j.

To determine the regrets $R(X_k, Y_j)$, it is necessary to define the minimum value of the objective function for each combination of the state of nature:

$$F^{min}(Y_j) = \min_{1 \le k \le K} F(X_k, Y_j). \tag{10}$$

On the other hand, if the objective function is to be maximized, it is necessary to define its maximum value for each combination of the state of nature (for each column of the payoff matrix):

$$F^{max}(Y_j) = \max_{1 \le k \le K} F(X_k, Y_j). \tag{11}$$

The regret for any alternative solution X_k and any state of nature Y_j can be assessed as:

$$R(X_k, Y_j) = F(X_k, Y_j) - F^{min}(Y_j), \tag{12}$$

if the objective function is to be minimized, or

$$R(X_k, Y_j) = F^{max}(Y_j) - F(X_k, Y_j), \tag{13}$$

if the objective function is to be maximized.

The choice criteria, based on the use of characteristic estimates, are represented as (14)–(17) under the assumption that the objective function is to be minimized. The Wald choice criterion uses the estimate $F^{max}(X_k)$ and allows choosing the solution alternatives X^W, for which the estimate is minimum:

$$\min_{1 \le k \le K} F^{max}(X_k) = \min_{1 \le k \le K} \max_{1 \le j \le J} F(X_k, Y_j). \tag{14}$$

The use of this criterion generates solution alternatives, assuming the most unfavorable combination of initial data. It ensures that the level of the objective function is not greater than a certain value under any possible future conditions. On the other hand, the focus on the most unfavorable combination of initial data is extremely cautious (pessimistic or conservative) [26].

The Laplace choice criterion, $\bar{F}(X_k)$, uses the estimate (8) and is aimed at choosing the solution alternatives X^L, for which the estimate is minimum:

$$\min_{1 \le k \le K} \bar{F}(X_k) = \min_{1 \le k \le K} \frac{1}{J} \sum_{j=1}^{J} F(X_k, Y_j). \tag{15}$$

This criterion corresponds to the principle of "insufficient reason" [26], that is, the assumption that we have no basis for distinguishing a particular combination of initial data. Therefore, it is necessary to act as if they are equally likely, which is a disadvantage. However, the average score is sufficiently important.

The Savage choice criterion is associated with the use of the estimate $R^{max}(X_k)$ and allows choosing the solution alternatives X^S, for which the estimate is minimum:

$$\min_{1 \le k \le K} R^{max}(X_k) = \min_{1 \le k \le K} \max_{1 \le j \le J} R(X_k, Y_j). \tag{16}$$

As in the case of the Wald choice criterion, the use of Equation (16) is based on the *minmax* principle. Therefore, the Savage choice criterion can also be considered conservative. However, the experience of [26] shows that the recommendations based on the application of Equation (16) can be inconsistent with the decisions obtained with the use of Equation (14). Operating with values of $R^{max}(X_k)$, we obtain a slightly different assessment of the situation, which could lead to more "daring" (less conservative) recommendations.

Finally, the Hurwicz choice criterion uses a convex combination of $F^{max}(X_k)$ and $F^{min}(X_k)$ and allows choosing the solution alternatives X^H that produce the minimum for:

$$\min_{1 \le k \le K} \left(\beta F^{max}(X_k) + (1-\beta) F^{min}(X_k) \right) = \min_{1 \le k \le K} \left(\beta \max_{1 \le j \le J} F(X_k, Y_j) + (1-\beta) \min_{1 \le j \le J} F(X_k, Y_j) \right), \tag{17}$$

where $\beta \in [0, 1]$ is the "pessimism–optimism" index whose magnitude is defined by the decision maker. If $\beta = 1$, the Hurwicz choice criterion is transformed into the Wald choice criterion, and if $\beta = 0$, Equation (17) is transformed into the criterion of "extreme optimism" (*minmin*) for which the combination of initial data is most favorable. The author of [26] recommends choosing a range of $0.5 < \beta < 1$.

The choice criteria discussed above have found widespread practical applications, as in [26,29] in both mono-objective and multiobjective problems.

The definition of the solution alternatives can be based on applying the modification [30] of the Bellman–Zadeh approach to decision making in a fuzzy environment [31]. This approach is used for solving multiobjective problems for each scenario, with the modification providing constructive lines for obtaining harmonious solutions to such problems [32]. These solution alternatives, also referred to as locally optimal solution alternatives, serve as the basis for constructing the payoff matrices. Nonetheless, in some instances, the decision makers themselves may also propose solutions for constructing these matrices, such as in this paper.

The generalization of the classical approach to dealing with information uncertainty [11,32,33] is associated with the analysis of the problems defined by Equations (14)–(17) for a given objective function in an environment with multiple states of nature $Y_j, j = 1, \ldots, J$. Therefore, considering the Wald, Laplace, Savage, and Hurwicz choice criteria, respectively, as objective functions, we consider:

$$F^W(X_k) = F^{min}(X_k) = \min_{1 \le j \le J} F(X_k, Y_j), \tag{18}$$

$$F^L(X_k) = \bar{F}(X_k) = \frac{1}{J} \sum_{j=1}^{J} F(X_k, Y_j), \tag{19}$$

$$F^S(X_k) = R^{max}(X_k) = \max_{1 \le j \le J} R(X_k, Y_j), \tag{20}$$

$$F^H(X_k) = \beta F^{max}(X_k) + (1-\beta)F^{min}(X_k) = \beta \max_{1 \le j \le J} F(X_k, Y_j) + (1-\beta) \min_{1 \le j \le J} F(X_k, Y_j). \tag{21}$$

This consideration of the choice criteria allows constructing q problems that generally include four or fewer objective functions (if not all choice criteria are used in the analysis) as follows:

$$F_{r,p}(X) \to \underset{X \in L}{\text{extr}}, r = 1, \ldots, t \le 4, p = 1, \ldots, q, \tag{22}$$

where L represents the feasible region for choosing solutions.

Thus, the analysis of solution alternatives and consequent choice of rational solution alternatives can be carried out within the $\langle X, F \rangle$ models [11,13,33].

The analysis, performed in this way, ensures the choice of rational solution alternatives according to the principle of Pareto optimality [32]. Considering this, the payoff matrix with characteristic estimates, presented in Table 1, is transformed into the matrix of choice criteria estimates in Table 2.

Table 2. Matrix with the choice criteria for the objective function of number p.

	$F_p^W(X_k)$	$F_p^L(X_k)$	$F_p^S(X_k)$	$F_p^H(X_k)$
X_1	$F_p^W(X_1)$	$F_p^L(X_1)$	$F_p^S(X_1)$	$F_p^H(X_1)$
...	...	...	...	...
X_k	$F_p^W(X_k)$	$F_p^L(X_k)$	$F_p^S(X_k)$	$F_p^H(X_k)$
...	...	...	...	...
X_K	$F_p^W(X_k)$	$F_p^L(X_k)$	$F_p^S(X_k)$	$F_p^H(X_k)$
	$\min_{1 \le k \le K} F_p^W(X_k)$	$\min_{1 \le k \le K} F_p^L(X_k)$	$\min_{1 \le k \le K} F_p^S(X_k)$	$\min_{1 \le k \le K} F_p^H(X_k)$
	$\max_{1 \le k \le K} F_p^W(X_k)$	$\max_{1 \le k \le K} F_p^L(X_k)$	$\max_{1 \le k \le K} F_p^S(X_k)$	$\max_{1 \le k \le K} F_p^H(X_k)$

Source: [33].

Therefore, using q matrices for the estimates of the choice criteria, we can construct q modified matrices of the choice criteria estimates, as shown in Table 3, by applying the relations:

$$\mu_{A_p}(X) = \left(\frac{\max_{X \in L} F_p(X) - F_p(X)}{\max_{X \in L} F_p(X) - \min_{X \in L} F_p(X)} \right)^{\lambda_p}, \tag{23}$$

for the objective functions that must be minimized and relations:

$$\mu_{A_p}(X) = \left(\frac{F_p(X) - \min_{X \in L} F_p(X)}{\max_{X \in L} F_p(X) - \min_{X \in L} F_p(X)} \right)^{\lambda_p}, \tag{24}$$

for the objective functions that must be maximized. In (23) and (24), μ_{A_p} is the membership function of the p-th objective function, and λ_p is the importance coefficient of the p-th objective function.

Table 3. Matrix with the modified choice criteria for the objective function of number *p*.

	$\mu_{A_p}^W(X_k)$	$\mu_{A_p}^L(X_k)$	$\mu_{A_p}^S(X_k)$	$\mu_{A_p}^H(X_k)$
X_1	$\mu_{A_p}^W(X_1)$	$\mu_{A_p}^L(X_1)$	$\mu_{A_p}^S(X_1)$	$\mu_{A_p}^H(X_1)$
…	…	…	…	…
X_k	$\mu_{A_p}^W(X_k)$	$\mu_{A_p}^L(X_k)$	$\mu_{A_p}^S(X_k)$	$\mu_{A_p}^H(X_k)$
…	…	…	…	…
X_K	$\mu_{A_p}^W(X_k)$	$\mu_{A_p}^L(X_k)$	$\mu_{A_p}^S(X_k)$	$\mu_{A_p}^H(X_k)$

Source: [33].

Finally, in the presence of q modified matrices of the choice criteria estimates, applying the results from [11,33], we can construct the aggregated matrix of the choice criteria estimates, as shown in Table 4. This matrix includes the estimates calculated based on [11] and can be used to select non-dominated or robust solution alternatives.

Table 4. Matrix with the aggregated levels of the fuzzy choice criteria.

	$\mu_D^W(X_k)$	$\mu_D^L(X_k)$	$\mu_D^S(X_k)$	$\mu_D^H(X_k)$
X_1	$\mu_D^W(X_1)$	$\mu_D^L(X_1)$	$\mu_D^S(X_1)$	$\mu_D^H(X_1)$
…	…	…	…	…
X_k	$\mu_D^W(X_k)$	$\mu_D^L(X_k)$	$\mu_D^S(X_k)$	$\mu_D^H(X_k)$
…	…	…	…	…
X_K	$\mu_D^W(X_k)$	$\mu_D^L(X_k)$	$\mu_D^S(X_k)$	$\mu_D^H(X_k)$
	$\max\limits_{1 \le k \le K} \mu_D^W(X_k)$	$\max\limits_{1 \le k \le K} \mu_D^L(X_k)$	$\max\limits_{1 \le k \le K} \mu_D^S(X_k)$	$\max\limits_{1 \le k \le K} \mu_D^H(X_k)$

Source: [33].

Where μ_D is the aggregated membership function, which is obtained using the minimum operator. Taking into account the results presented above, it is possible to suggest the general scheme of multicriteria decision making under conditions of information uncertainty that is associated with the following steps, in the general case:

- The first step involves constructing q payoff matrices, corresponding to the number of objective functions. These matrices account for all combinations of solution alternatives $X_k, k = 1, \ldots, K$ and the representative states of nature $Y_j, j = 1, \ldots, J$. To construct payoff matrices, it is necessary to solve q multicriteria problems formalized within the framework of $\langle X, F \rangle$ models. By solving them, it is possible to obtain the solution alternatives $X_k, k = 1, \ldots, K$ $(K \le J)$. After that, $X_k, k = 1, \ldots, K$ are substituted $F_p(X), p = 1, \ldots, q$ for $Y_j, j = 1, \ldots, J$. These substitutions generate q payoff matrices;
- The second step is related to the analysis of the obtained payoff matrices. The execution of this phase is based on the approach proposed in [11] discussed above. However, the insufficient resolving capacity of this phase may lead to non-unique or not well-distinguished solutions, and this circumstance demands the application of the third phase;
- The third step is associated with the construction and analysis of $\langle X, R \rangle$ models [11,22] for the subsequent contraction of decision uncertainty regions. The use of $\langle X, R \rangle$ models allows taking into account indices of quantitative character and qualitative character, based on the knowledge, experience, and intuition of the involved experts

2.3. $\langle X, R \rangle$ *Decision-Making Models*

Decision-making models based on qualitative information are a type of decision-making tool that relies on non-numerical information to evaluate, compare, select, order, and/or prioritize solution alternatives on the basis of the corresponding preferences of decision makers. These models are particularly useful when there is a lack of reliable data or when the decision encounters complex or ambiguous situations. One of the main features of decision-making models that use qualitative information is that they rely on subjective judgments and opinions, rather than objective data. These models permit one to include things such as personal experience, specialized knowledge, and intuition. Although this may seem less reliable than quantitative data, qualitative information can be invaluable in situations where there are no reliable data available or where the decision is based on human behavior, emotions, or beliefs [34].

Another feature of qualitative information-based decision-making models is that they are often used in situations where the decision maker faces uncertainty or ambiguity of information. One of the advantages of qualitative decision-making models is that they can be more flexible than quantitative models. Since they do not rely on specific numerical data, they can be adapted to suit different situations and circumstances. This can be particularly useful when making decisions in dynamic or changing environments, where data may be incomplete or unreliable [35].

Moreover, many situations requiring the application of the multicriteria approach are associated with the problems that can initially be solved based on a single criterion or multiple criteria. However, if the uncertainty of information does not allow defining a unique solution, it is possible to use additional criteria to distinguish the alternatives.

There exist diverse formats of preference representation, as discussed, for instance, in [22]. Taking this into account, it is necessary to indicate that the results of [13] permit one to transform different formats as well as information of qualitative character to non-reciprocal fuzzy preference relations, applying so-called transformation functions. It allows us to concentrate attention on procedures of decision making in a fuzzy environment.

Suppose there is a set X of alternatives coming from the decision uncertainty region and/or predetermined alternatives, which are to be evaluated on q criteria. The decision-making problem can be presented by the pair $\langle X, R \rangle$ where $R = \{R_1, R_2, \ldots, R_p, \ldots, R_q\}$ is a vector of fuzzy preference relations [11,22], which can be presented as:

$$R_p = \left[X \times X, \mu_{R_p}(x_k, x_l)\right], p = 1, 2, \ldots, q, x_k, x_l \in X, \tag{25}$$

where $\mu_{R_p}(x_k, x_l)$ is the membership function of the p-th fuzzy preference relation.

In (25), R_p is defined as a fuzzy set of all pairs from the Cartesian product $X \times X$, where the membership function $\mu_{R_p}(x_k, x_l)$ represents the degree to which x_k weakly dominates x_l, and consequently, the degree to which x_k is not worse than x_l for the p-th criterion. It should be noted that non-reciprocal fuzzy preference relations and fuzzy estimates are somewhat equivalent. In particular, if two alternatives $x_k \in X$ and $x_l \in X$ have fuzzy estimates with membership functions $\mu(x_k)$ and $\mu(x_l)$, then the quantity $R(x_k, x_l)$ is the preference degree $\mu(x_k) \succeq \mu(x_l)$, while the quantity $R(x_l, x_k)$ is the preference degree $\mu(x_l) \succeq \mu(x_k)$. According to [13,36], the quantities $R(x_k, x_l)$ and $R(x_l, x_k)$ can be evaluated as follows:

$$R(x_k, x_l) = \sup_{\substack{x_k, x_l \in X \\ x_k \leq x_l}} \min\{\mu(x_k), \mu(x_l)\}, \tag{26}$$

$$R(x_l, x_k) = \sup_{\substack{x_k, x_l \in X \\ x_l \leq x_k}} \min\{\mu(x_k), \mu(x_l)\}. \tag{27}$$

If the indicator has a maximization character, (26) and (27) should be written for $x_k \geq x_l$ and $x_l \geq x_k$, respectively. More information on the construction of R_p can be found in [11,13].

The fuzzy preference relation matrices can be processed to construct strict preference relation matrices according to the following equation:

$$R^S = R \setminus R^{-1}, \quad (28)$$

where R^{-1} is the inverse fuzzy preference relation.

The membership function corresponding to (28) can be described as follows:

$$\mu_R^S(X_k, X_l) = \max\{\mu_R(X_k, X_l) - \mu_R(X_l, X_k), 0\}. \quad (29)$$

The use of (29) allows constructing the set of non-dominated alternatives with the membership function that allows evaluating the non-dominance level of each alternative X_k according to the following equation:

$$\mu_R^{ND}(X_k) = \inf_{X_l \in X}\left[1 - \mu_R^S(X_k, X_l)\right] = 1 - \sup_{X_l \in X} \mu_R^S(X_k, X_l). \quad (30)$$

Considering that it is natural to choose alternatives that provide the highest level of non-dominance, one can choose alternatives X^{ND} according to the following equation:

$$X^{ND} = \left\{ X_k^{ND} | X_k^{ND} \in X, \mu_R^{ND}(X_k^{ND}) = \sup_{X_l \in X} \mu_R^{ND}(X_k) \right\}. \quad (31)$$

Equations (29)–(31) can be used for the solution of choice problems, as well as for evaluation, comparison, ranking, and/or prioritization of alternatives for some criterion. These equations can also be applied when R is a vector of fuzzy preference relations, under different approaches for multiattribute analysis. The application used in this work consists of the flexible approach with an optimism degree adjustment, although other approaches can be found in [11,32].

This application approach is performed using the ordered weighted average (OWA) operator, originally introduced in [37], as follows:

$$\mu^{ND}(X_k) = OWA\left(\mu_{R_1}^{ND}(X_k), \mu_{R_2}^{ND}(X_k), \ldots, \mu_{R_q}^{ND}(X_k),\right) = \sum_{i=1}^{q} w_i B_i(X_k), \quad (32)$$

where $B_i(X_k)$ is the largest value among $\mu_{R_1}^{ND}(X_k), \mu_{R_2}^{ND}(X_k), \ldots, \mu_{R_q}^{ND}(X_k)$. The weights in Equation (32) need to satisfy the following constraints: $w_i > 0, i = 1, 2, \ldots, q$ and also $\sum_{i=1}^{q} w_i = 1$. These weights can be indirectly defined by decision makers as described in [11,32].

3. Objectives in Energy Investments

The decision to invest in a new asset alongside an existing portfolio is often evaluated using net present value (NPV), one of the most common factors in investment appraisal. The NPV depicts the result of the portfolio's cash flow over a period, including the initial capital investment for the acquisition of the asset, the estimate of profit related to this investment, and the residual value of the investment. The adopted equation is expressed as follows:

$$NPV_s(x_t) = -CI(x) + \sum_{t=1}^{T} \frac{\varrho_{t,s}(x_t)}{(1+\tau)^t} + \frac{VR(x)}{(1+\tau)^T}, \quad (33)$$

where τ is the discount rate, the investment costs CI are discounted at $t = 1$, and the residual value VR is added at T. Therefore, the first considered objective function is the expected net present value of the portfolio [19,38], which can be calculated as follows:

$$E[NPV] = \frac{1}{S} \sum_{s=1}^{S} NPV_s(x_t). \quad (34)$$

The portfolio risk analysis in this study is based on the conditional value at risk (CVaR), a special case of the value at risk (VaR), proposed in [39]. The CVaR represents, among the scenarios studied, the expected revenue in the worst $\alpha\%$ cases, generating a conservative decision that focuses on the least profitable conditions.

The approach considers that for a given histogram of revenues, one should identify the worst $\alpha\%$ revenues (highlighted area in Figure 1). The value of VaR represents the revenue that delimits this area, and the value of $CVaR$ reflects the average of the worst revenue values. The revenue at risk (RaR) can be calculated from the difference between the expected NPV and the $CVaR$. Therefore, the second considered objective function is the $CVaR$, and the third one is the RaR. The choice of the $CVaR$ as the second objective function is explained by its advantages over the VaR [5,39].

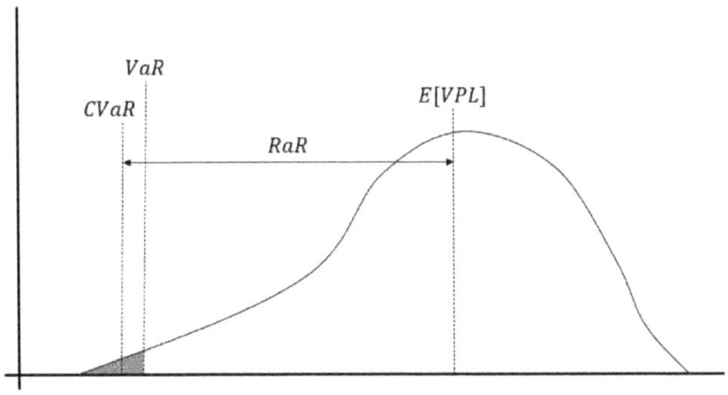

Figure 1. Revenue distribution estimates.

Moreover, it is rational to introduce the insurance index (InS) as the fourth objective function. This index quantifies the degree of improvement in the $CVaR$ compared to the $E[NPV]$ when transitioning from one position to another and is useful when evaluating alternatives of different volumes [11]. It is calculated for positive $CVaR$ variations as follows:

$$InS = \frac{\Delta CVaR}{\Delta E[NPV]}, \ \Delta CVaR > 0, \ \Delta E[NPV] \neq 0, \tag{35}$$

where $\Delta CVaR$ is the variation of $CVaR$ compared to the current position, and $\Delta E[NPV]$ is the variation of the expected revenue compared to the current position.

To complement the decision-making process, if there are doubts about the robustness of the alternatives, for example, the following additional objectives of the qualitative character are proposed:

- Prioritize alternatives that have the greatest synergy with the portfolio's resources;
- Prioritize alternatives with the lowest operational risk.

The first additional objective function is designed to encapsulate the issues raised in [23], whereas the second additional objective function responds to the operational concerns specified in [40]. However, unlike these works, the functions in this paper have been designed with a qualitative approach. Therefore, the evaluation of alternatives according to these criteria relies on expert opinions, which can be expressed in any preference format. Since any format can be translated into fuzzy preference relations [13], this evaluation takes place within the $\langle X, R\rangle$ decision-making models.

4. Application Example

An example illustrates the practical application of the results described above. In this example, two scenarios obtained through the NEWAVE model applied in the long term are evaluated. These scenarios are official results of the energy expansion studies of the *Empresa de Pesquisa Energética* [41,42]. The study horizon from 2023 to 2033 was considered.

Figure 2 shows the average behavior of the prices over time in each scenario, as well as the adopted over-the-counter (OTC) price definition. The OTC forward price will represent the price profile that agents are willing to negotiate in the free market over the horizon.

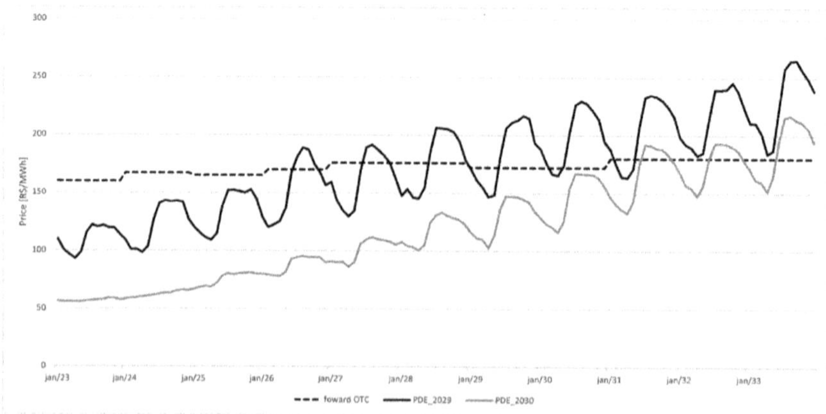

Figure 2. Average PLD of scenarios and forward over-the-counter price.

The existing mix of assets for the power producer examined in this study can be seen in Table 5. This table compiles the company's resources (where physical guarantee ≥ 0) and requirements (where physical guarantee ≤ 0), categorizing them by their source type.

Table 5. Portfolio composition.

Type	Total Physical Guarantee [MWm]	Average Cost of Operation and Maintenance [R$/MWh]	Concession Expiration
Hydraulic power plants—Group 1	38.3	0.54	31 May 2028
Hydraulic power plants—Group 2	36.0	0.54	31 July 2032
Hydraulic power plants—Group 3	128.6	0.54	31 December 2035
Hydraulic power plants—Group 4	122.8	0.54	25 August 2036
Wind power plants	67.5	0.21	31 December 2033
Sales contracts 1	−325.0	225.00	31 December 2033
Sales contracts 2	−23.0	230.00	31 December 2033

The resource considered for the plants depends on their associated generation profile, as shown in Figure 3. The hydroelectric plants in the portfolio depend on the evaluated scenario since their resource is characterized by the GSF projection for the future (shown in Figure 4), as they belong to the MRE.

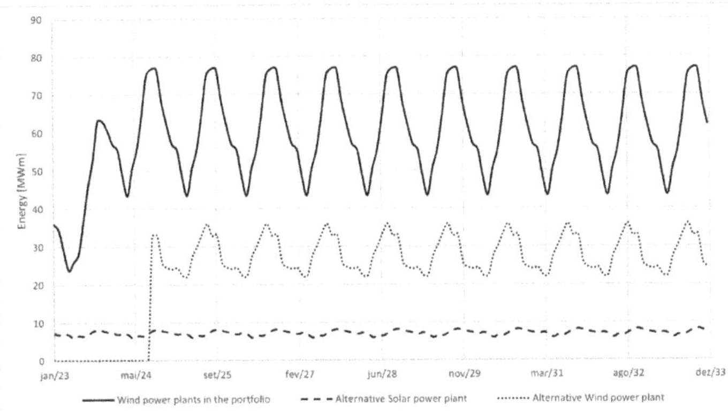

Figure 3. Predicted generation profile of the plants.

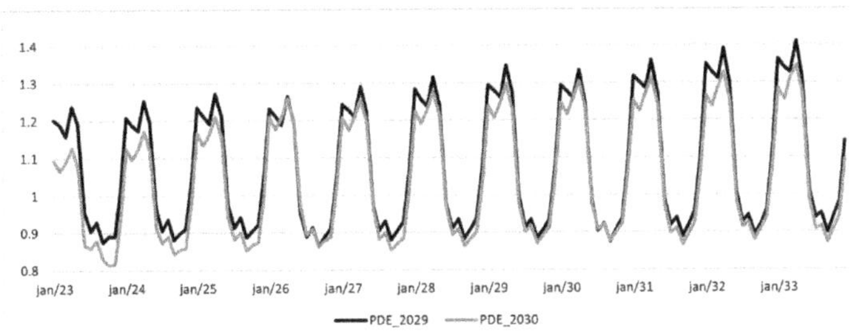

Figure 4. Average GSF in the scenarios.

The investment alternatives are described in Table 6, and the characteristics of the plants in question are given in Table 7. Therefore, these alternatives will be evaluated considering either the individual purchase of each plant combined with the total or partial energy sale, or the acquisition of the plants together for forming the portfolio composition. In all contracting cases, only firm contracts are considered, which are contracts with fixed volume and price over the contracting horizon, without any flexibility or premium.

Table 6. Description of alternatives.

Alternative Type	Physical Guarantee (MWm)	Operation Cost (R$/MWh)	Portfolio Entry	Concession Expiration
Wind power plant	28.0	0.00	1 September 2024	31 May 2057
Solar power plant	8.0	0.00	1 January 2022	31 December 2033
Wind power plant sales contract	−28.0	210.00	1 January 2022	31 December 2033
Solar power plant sales contract	−8.0	197.50	1 September 2024	31 December 2033

Table 7. Characteristics of the alternatives.

Name	Installed Capacity (MW)	Capacity Factor	Investment Cost (M R$)
Wind power plant	53.7	0.52	250.00
Solar power plant	47.0	0.17	172.00

The final alternatives for decision making are defined as follows:

- X_0: Maintain the current portfolio;
- X_1: Current portfolio with the addition of the solar plant with a contract sale of 100% of its physical guarantee;
- X_2: Current portfolio with the addition of the wind plant with a contract sale of 100% of its physical guarantee;
- X_3: Current portfolio with the addition of the wind plant with a contract sale of 80% of its physical guarantee;
- X_4: Current portfolio with the addition of the wind plant with a contract sale of 50% of its physical guarantee;
- X_5: Current portfolio with the addition of the wind and solar plants with a contract sale of 100% of their physical guarantees;
- X_6: Current portfolio with the addition of the solar plant with a contract sale of 75% of its physical guarantee;
- X_7: Current portfolio with the addition of the solar plant with a contract sale of 75% of its physical guarantee and the wind plant with a contract sale of 80% of its physical guarantee.

All alternatives are simulated to obtain their performance in each objective for each of the scenarios. The NPV is calculated considering an annual discount rate of 8%.

5. Results

The portfolio's behavior, as depicted in Figure 5, showcases the relationship between revenue variation and the CVaR for each scenario. Additionally, the risk associated with revenue is presented in Figure 6. The construction of the market line is carried out by manipulating the contract volume in the portfolio along a contracting range from −100 MWm to 100 MWm, where *PDE_2029* is transcribed as Y_1 and *PDE_2030*, as Y_2. This calculation is carried out considering that the trading will be carried out according to the over-the-counter price (Figure 2). The market line is not considered in the decision-making process; however, it can indicate a comparison basis that can be used to invalidate some choices.

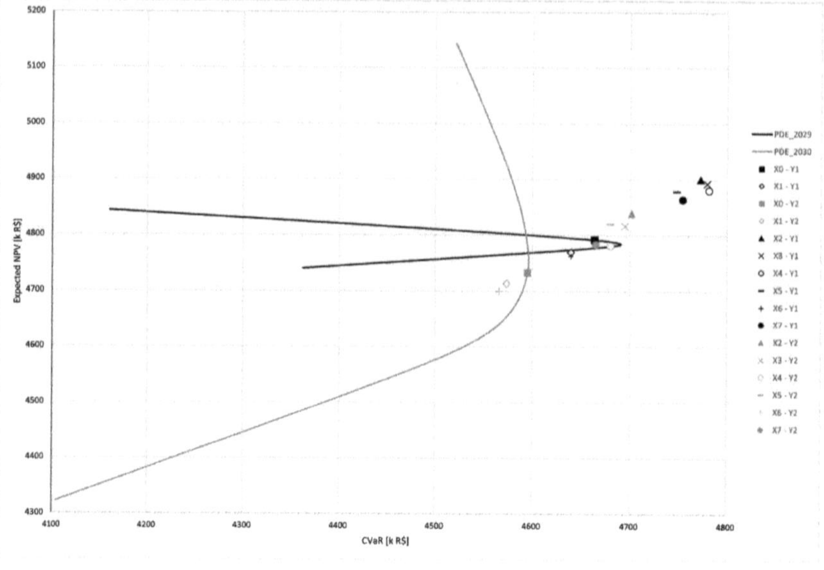

Figure 5. NPV vs. CVaR.

Evaluating the portfolio's behavior in the scenarios shown in Figure 5, we notice that the portfolio's $CVaR$ value is more responsive in the scenario with higher prices. In this scenario, options X_2, X_3, and X_4 outperform all other alternatives. These choices symbolize the purchase of the wind plant with different committed sales volumes in contracts. However, options that solely involve purchasing the solar plant, such as X_1 and X_6, are outperformed by the current position and should consequently be excluded from the consideration. Yet, if the solar plant is purchased along with the wind plant, as in X_5 or X_7, they outperform the current position.

The options that include the standalone purchase of solar plants continue to perform poorly in scenario Y_2. They are outperformed by the current position and fall within this scenario's market curve. The standout options in this scenario are those where a substantial portion of energy from the purchased plants is committed, such as X_2, X_3, and X_5. Among these, option X_2 outperforms all others in this scenario.

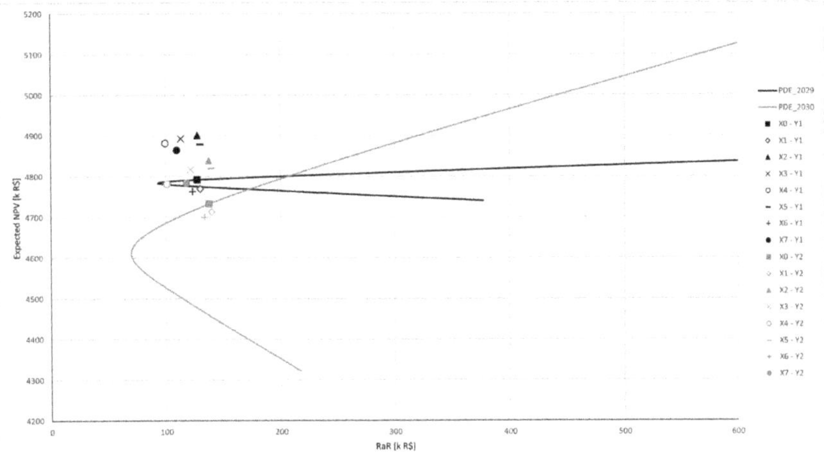

Figure 6. NPV vs. RaR.

Similarly, in both scenarios in Figure 6, the alternatives that stood out the most were those that consisted of the acquisition of the wind plant or simultaneously with the solar plant with the partial sale of their energy. For this figure, the total sale of the solar plant's energy is dominated by the current position in both scenarios, and X_5 is also dominated by other alternatives.

To evaluate the robustness of the alternatives in the considered scenarios, payoff matrices were constructed for the objectives in question. These matrices can be seen in Tables 8–11.

Table 8. Payoff matrix of expected NPV.

Alternative	PDE_2029	PDE_2030
X_0	4791.51	4732.74
X_1	4769.46	4713.26
X_2	4900.36	4839.31
X_3	4892.69	4816.45
X_4	4880.84	4781.12
X_5	4878.31	4819.83
X_6	4762.62	4699.49
X_7	4863.80	4783.20

Table 9. Payoff matrix of CVaR.

Alternative	PDE_2029	PDE_2030
X_0	4664.09	4594.92
X_1	4639.48	4573.23
X_2	4772.78	4701.65
X_3	4779.64	4694.74
X_4	4781.34	4680.04
X_5	4748.16	4679.91
X_6	4639.42	4565.59
X_7	4754.29	4665.14

Table 10. Matriz payoff of RaR.

Alternative	PDE_2029	PDE_2030
X_0	127.42	137.82
X_1	129.98	140.03
X_2	127.58	137.66
X_3	113.05	121.71
X_4	99.50	101.08
X_5	130.14	139.92
X_6	123.20	133.90
X_7	109.50	118.06

Table 11. Payoff matrix of InS.

Alternative	PDE_2029	PDE_2030
X_0	0.00	0.00
X_1	143.45	143.45
X_2	119.91	119.91
X_3	111.46	111.46
X_4	94.05	94.05
X_5	125.14	125.14
X_6	133.94	133.94
X_7	116.34	116.34

The results presented in Table 12 can be considered as the final response to the analysis of the $\langle X, F \rangle$ models. According to the results in this table, alternatives X_0, X_1, X_2, X_5, and X_6 are to be discarded from the decision-making process. On the other hand, alternative X_3 received an intermediate score in all the selection criteria. Meanwhile, it is not possible to define a significant relevance between alternatives X_4 and X_7, making it necessary to evaluate these two alternatives using additional criteria. Decision makers assessed these two alternatives according to the two additional criteria using fuzzy estimates presented in Figure 7, where the vertical axis represents $\mu(X_k)$ [13].

Table 12. $\langle X, F \rangle$ model result.

Alternative	Wald	Laplace	Savage	Hurwicz
X_0	0.00	0.00	0.00	0.00
X_1	0.00	0.00	0.00	0.00
X_2	0.06	0.07	0.06	0.06
X_3	0.47	0.51	0.47	0.49
X_4	0.58	0.66	0.58	0.65
X_5	0.00	0.00	0.00	0.00
X_6	0.00	0.00	0.00	0.00
X_7	0.56	0.61	0.56	0.59

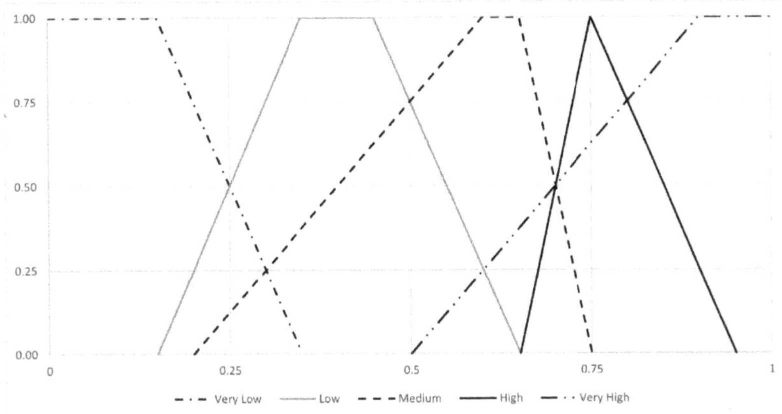

Figure 7. Qualitative scales based on fuzzy sets used for objectives based on qualitative information. Source: [13].

For the criterion of prioritizing alternatives that have greater synergy with the portfolio's resources, the following were indicated: X_4—high; X_7—very high. As for the criterion of prioritizing alternatives with the lowest operational risk, the following estimates were indicated: X_4—medium; X_7—high. Using (26) and (27) for these estimates it is possible to contruct the following non-reciprocal fuzzy preference relation as:

$$R_1 = \begin{bmatrix} 1 & 0.75 \\ 1 & 1 \end{bmatrix}, \quad R_2 = \begin{bmatrix} 1 & 0.5 \\ 1 & 1 \end{bmatrix}, \tag{36}$$

where R_1 is the preference relation matrix concerning prioritizing alternatives that have the greatest synergy with the portfolio's resources, and R_2 relates to prioritizing alternatives with the lowest operational risk.

Considering that the criteria have the same importance and since the evaluations of the alternatives resulted in a convergence of preference for X_7, the non-strict fuzzy preference relation matrix for the development of the $\langle X, F \rangle$ model is equal to R_2. Therefore, one can obtain the membership function of the strict fuzzy preference relation as follows:

$$P = \begin{bmatrix} 0 & 0 \\ 0.5 & 0 \end{bmatrix}, \tag{37}$$

and finally, the use of Equation (31) allows us to obtain the following membership function of the fuzzy set of non-dominated alternatives:

$$ND = \begin{bmatrix} 0.5 & 1 \end{bmatrix}, \tag{38}$$

that justifies the choice of X_7.

6. Conclusions

The present work reflects research results related to applying the techniques of multi-criteria decision making under conditions of uncertainty to long-term investment planning in the electricity sector. By analyzing the energy portfolio of the company based on the proposed approach, it is possible to quantify the expected revenue and the financial risks involved in acquiring energy assets. The measurement of portfolio revenue from investment options utilizes the expected NPV across different scenarios, with risk quantified using the $CVaR$. An evaluation of the portfolio's revenue at risk is also conducted. To improve the investment analysis, three further objectives have been introduced, two of which possess a qualitative nature.

In the given illustrative example, evaluations of investments in alternatives potentially impacting the portfolio's performance over time are conducted. The selection of an alternative impacts the energy balance of the agent, modifying exposure in the long-term scenarios under assessment. This, in turn, influences the values of the objective functions. It is observed that the outcomes display sensitivity to the PLD and GSF projections sourced from NEWAVE. A robust solution in the application example involves the incorporation of both solar and wind plants into the existing portfolio, without committing all energy from these plants to contract sales, thus retaining a percentage exposed to the spot price.

Employment of the proposed methodology facilitates the evaluation of alternatives beyond their associated costs, considering factors pertinent to the agent such as the insurance index (InS), synergy with the resources in the portfolio, and operational risks. Utilizing multiple scenarios enables the simultaneous consideration of different specific situations, constituting a significant step forward in addressing issues associated with using NEWAVE as input for the long-term decision model. The use of the objectives of the qualitative character allows for the distinction and selection of non-dominated alternatives previously evaluated within the $\langle X, F \rangle$ decision models. Although, in this work, the approach to constructing robust (non-dominated) solutions in multicriteria (multiobjective) decision making under conditions of uncertainty was applied to the problem of energy portfolio management, it is of a universal character and can be applied to solving wide classes of problems related to decision making in conditions of uncertainty. However, it is recognized that future work would benefit from considering additional scenarios derived from different information sources.

Author Contributions: Conceptualization, R.B., L.S., M.L., I.R., M.C., A.A., F.L., R.S. and D.V.; Data curation, L.S.; Formal analysis, R.B.; Funding acquisition, R.S., D.V. and P.E.; Methodology, R.B., D.V. and P.E.; Project administration, A.A.; Software, R.B., L.S. and M.L.; Supervision, L.S., I.R. and F.L.; Validation, M.L. and G.M.; Visualization, M.L.; Writing—original draft, R.B., L.S. and P.E.; Writing—review & editing, R.B., G.M., A.A., D.V. and P.E. All authors have read and agreed to the published version of the manuscript.

Funding: This study was financed in part by Aliança Energia; National Council for Scientific and Technological Development (CNPq)—Grants 311922/2021-0, 314800/2021-2, 424572/2021-4; Minas Gerais State Agency for Research and Development (FAPEMIG)—Grant APQ-04571-22; Vale S.A. (within the Research, Development, and Innovation Partnership Agreement).

Data Availability Statement: The scenario data is available at: https://github.com/bambirrarafael/data_PDE2029_PDE2030 (accessed on 14 August 2023).

Conflicts of Interest: The authors declare no conflict of interest. The funders had no role in the design of the study; in the collection, analyses, or interpretation of data; in the writing of the manuscript; or in the decision to publish the results

Abbreviations

The following abbreviations are used in this paper:

CVaR	conditional value at risk
GSF	generation scaling factor
InS	insurance index
MRE	energy reallocation mechanism
NPV	net present value
PLD	settlement price for differences
OTC	over the counter
RaR	revenue at risk
VaR	value at risk

References

1. Božić, Z.; Dobromirov, D.; Arsić, J.; Radišić, M.; Ślusarczyk, B. Power Exchange Prices: Comparison of Volatility in European Markets. *Energies* **2020**, *13*, 5620. [CrossRef]
2. Sadeghi, M.; Shavvalpour, S. Energy risk management and value at risk modeling. *Energy Policy* **2006**, *34*, 3367–3373. [CrossRef]
3. Weron, R. Energy price risk management. *Phys. A Stat. Mech. Its Appl.* **2000**, *285*, 127–134. [CrossRef]
4. Chawda, S.; Bhakar, R.; Mathuria, P. Uncertainty and risk management in electricity market: Challenges and opportunities. In Proceedings of the 2016 National Power Systems Conference (NPSC), Bhubaneswar, India, 19–21 December 2016; pp. 1–6. [CrossRef]
5. Dahlgren, R.; Liu, C.C.; Lawarree, J. Risk assessment in energy trading. *IEEE Trans. Power Syst.* **2003**, *18*, 503–511. [CrossRef]
6. Denton, M.; Palmer, A.; Masiello, R.; Skantze, P. Managing market risk in energy. *IEEE Trans. Power Syst.* **2003**, *18*, 494–502. [CrossRef]
7. Pérez Odeh, R.; Watts, D.; Negrete-Pincetic, M. Portfolio applications in electricity markets review: Private investor and manager perspective trends. *Renew. Sustain. Energy Rev.* **2018**, *81*, 192–204. [CrossRef]
8. Fu, Y.; Sun, Q.; Wennersten, R. Effectiveness of the CVaR method in risk management in an integrated energy system. *Energy Rep.* **2020**, *6*, 1010–1015. [CrossRef]
9. Peixoto, G.M.; Pereira, R.B.; da Silva Amorim, L.; Machado-Coelho, T.M.; Soares, G.L.; Parma, G.G. Analysis of the Participation of Small Hydroelectric Power Plants in the Energy Reallocation Mechanism (in portuguese). In Proceedings of the 14° Simpósio Brasileiro de Automação Inteligente, Ouro Preto, Brazil, 27–30 October 2019; pp. 1–7. [CrossRef]
10. Gökgöz, F.; Atmaca, M.E. Portfolio optimization under lower partial moments in emerging electricity markets: Evidence from Turkey. *Renew. Sustain. Energy Rev.* **2017**, *67*, 437–449. [CrossRef]
11. Ekel, P.; Pedrycz, W.; Pereira, J. *Multicriteria Decision-Making under Conditions of Uncertainty: A Fuzzy Set Perspective*; John Wiley & Sons: Hoboken, NJ, USA, 2020.
12. Maceiral, M.; Penna, D.; Diniz, A.; Pinto, R.; Melo, A.; Vasconcellos, C.; Cruz, C. Twenty Years of Application of Stochastic Dual Dynamic Programming in Official and Agent Studies in Brazil-Main Features and Improvements on the NEWAVE Model. In Proceedings of the 2018 Power Systems Computation Conference (PSCC), Dublin, Ireland, 11–15 June 2018; pp. 1–7. [CrossRef]
13. Ramalho, F.D.; Ekel, P.Y.; Pedrycz, W.; Pereira Júnior, J.G.; Soares, G.L. Multicriteria decision making under conditions of uncertainty in application to multiobjective allocation of resources. *Inf. Fusion* **2019**, *49*, 249–261. [CrossRef]
14. Abba, Z.; Balta-Ozkan, N.; Hart, P. A holistic risk management framework for renewable energy investments. *Renew. Sustain. Energy Rev.* **2022**, *160*, 112305. [CrossRef]
15. Camargo, L.A.S.; Leonel, L.D.; Rosa, P.S.; Ramos, D.S. Optimal Portfolio Selection of Wind Power Plants Using a Stochastic Risk-Averse Optimization Model, Considering the Wind Complementarity of the Sites and a Budget Constraint. *Energy Power Eng.* **2020**, *12*, 459–476. [CrossRef]
16. Bradshaw, A. Regulatory change and innovation in Latin America: The case of renewable energy in Brazil. *Util. Policy* **2017**, *49*, 156–164. [CrossRef]
17. Ilbahar, E.; Kahraman, C.; Cebi, S. Risk assessment of renewable energy investments: A modified failure mode and effect analysis based on prospect theory and intuitionistic fuzzy AHP. *Energy* **2022**, *239*, 1–11. [CrossRef]
18. Martínez, L.; Dinçer, H.; Yüksel, S. A hybrid decision making approach for new service development process of renewable energy investment. *Appl. Soft Comput.* **2023**, *133*, 1–17. [CrossRef]
19. de Freitas, R.A.; Vogel, E.P.; Korzenowski, A.L.; Oliveira Rocha, L.A. Stochastic model to aid decision making on investments in renewable energy generation: Portfolio diffusion and investor risk aversion. *Renew. Energy* **2020**, *162*, 1161–1176. [CrossRef]
20. Yang, L.; Sun, Q.; Zhang, N.; Li, Y. Indirect Multi-Energy Transactions of Energy Internet With Deep Reinforcement Learning Approach. *IEEE Trans. Power Syst.* **2022**, *37*, 4067–4077. [CrossRef]
21. Wang, L.; Zhou, Q.; Xiong, Z.; Zhu, Z.; Jiang, C.; Xu, R.; Li, Z. Security constrained decentralized peer-to-peer transactive energy trading in distribution systems. *CSEE J. Power Energy Syst.* **2022**, *8*, 188–197. [CrossRef]
22. Pedrycz, W.; Ekel, P.; Parreiras, R. *Fuzzy Multicriteria Decision-Making: Models, Methods and Applications*; John Wiley & Sons: Hoboken, NJ, USA, 2011.
23. Santos, F.F.G.D.; Vieira, D.A.G.; Saldanha, R.R.; Lisboa, A.C.; de Castro Lobato, M.V. Seasonal energy trading portfolio based on multiobjective optimisation. *Int. J. Logist. Syst. Manag.* **2014**, *17*, 180. [CrossRef]
24. Luce, R.; Raiffa, H. *Games and Decisions*; John Wiley & Sons: Hoboken, NJ, USA, 1957.
25. Raiffa, H. *Decision Analysis: Introductory Lectures on Choices under Uncertainty*; Addison-Wesley: Boston, MA, USA, 1968.
26. Belyaev, L.S. A Practical Approach to Choosing Alternative Solutions to Complex Optimization Problems under Uncertainty. *Int. Inst. Appl. Syst. Anal.* **1977**, *1*, 1–60.
27. Hodges, J.L.; Lehmann, E.L. The use of Previous Experience in Reaching Statistical Decisions. *Ann. Math. Stat.* **1952**, *23*, 396–407. [CrossRef]
28. Trukhaev, R. *Models of Decision Making in Conditions of Uncertainty*; Nauka: Moscow, Russia, 1981.
29. Figueiredo, L.R.; Frej, E.A.; Soares, G.L.; Ekel, P.Y. Group Decision-Based Construction of Scenarios for Multicriteria Analysis in Conditions of Uncertainty on the Basis of Quantitative and Qualitative Information. *Group Decis. Negot.* **2021**, *30*, 665–696. [CrossRef]
30. Ekel, P. Fuzzy sets and models of decision making. *Comput. Math. Appl.* **2002**, *44*, 863–875. [CrossRef]

31. Bellman, R.E.; Zadeh, L.A. Decision-Making in a Fuzzy Environment. *Manag. Sci.* **1970**, *17*, 141–164. [CrossRef]
32. Ekel, P.; Kokshenev, I.; Parreiras, R.; Pedrycz, W.; Pereira, J., Jr. Multiobjective and multiattribute decision making in a fuzzy environment and their power engineering applications. *Inf. Sci.* **2016**, *361-362*, 100–119. [CrossRef]
33. Pereira, J.; Ekel, P.; Palhares, R.; Parreiras, R. On multicriteria decision making under conditions of uncertainty. *Inf. Sci.* **2015**, *324*, 44–59. [CrossRef]
34. Ralston, B.; Wilson, I. *The Scenario Planning Handbook*; South-Western: Nashville, TN, USA, 2006.
35. Amer, M.; Daim, T.U.; Jetter, A. A review of scenario planning. *Futures* **2013**, *46*, 23–40. [CrossRef]
36. Ekel, P.; Martins, C.; Pereira, J.; Palhares, R.; Canha, L. Fuzzy set based multiobjective allocation of resources and its applications. *Comput. Math. Appl.* **2006**, *52*, 197–210. [CrossRef]
37. Yager, R. On ordered weighted averaging aggregation operators in multicriteria decisionmaking. *IEEE Trans. Syst. Man Cybern.* **1988**, *18*, 183–190. [CrossRef]
38. Pinheiro Neto, D.; Domingues, E.G.; Coimbra, A.P.; de Almeida, A.T.; Alves, A.J.; Calixto, W.P. Portfolio optimization of renewable energy assets: Hydro, wind, and photovoltaic energy in the regulated market in Brazil. *Energy Econ.* **2017**, *64*, 238–250. [CrossRef]
39. Rockafellar, R.T.; Uryasev, S. Optimization of conditional value-at-risk. *J. Risk* **2000**, *2*, 21–41. [CrossRef]
40. Carrano, E.G.; Tarôco, C.G.; Neto, O.M.; Takahashi, R.H. A multiobjective hybrid evolutionary algorithm for robust design of distribution networks. *Int. J. Electr. Power Energy Syst.* **2014**, *63*, 645–656. [CrossRef]
41. Empresa de Pesquisa Energética. *Plano Decenal de Expansão de Energia 2029*; Ministério de Minas e Energia: Lisboa, Portugal, 2020.
42. Empresa de Pesquisa Energética. *Plano Decenal de Expansão de Energia 2030*; Ministério de Minas e Energia: Lisboa, Portugal, 2021.

energies

MDPI

Perspective

The Q-NPT: Redefining Nuclear Energy Governance for Sustainability

Hassan Qudrat-Ullah

School of Administrative Studies, York University, Toronto, ON M3J 1P3, Canada; hassanq@yorku.ca or hassanqu@gmail.com

Abstract: Global peace, security, and sustainable energy development depend on effective nuclear energy governance. While the Nuclear Non-Proliferation Treaty (NPT) has served as a cornerstone in this domain, it faces challenges such as trust deficits, inequitable access to nuclear technologies, and regional instability. This paper proposes the Qudrat-Ullah Nuclear Peace and Trust (Q-NPT) framework, a dynamic implementation roadmap designed to address these issues. The framework focuses on fostering trust among stakeholders, ensuring equitable access to nuclear technologies, and promoting inclusivity in governance structures. A key theoretical contribution is the integration of trust-building measures with sustainable energy transitions, highlighting nuclear energy's role in decarbonization and global energy security. The paper outlines actionable pathways for implementing the Q-NPT framework, including enhanced oversight by the International Atomic Energy Agency (IAEA), capacity-building initiatives, and training programs to enable safe and sustainable nuclear cooperation, particularly in developing nations. By operationalizing nuclear programs through this approach, the Q-NPT framework aligns nuclear energy governance with global sustainable energy objectives.

Keywords: nuclear energy; nuclear energy governance; decarbonization; sustainable energy; trust-building; developing nations; IAEA

1. Introduction

The governance of nuclear energy stands at a pivotal juncture, reflecting tensions between technological advancement, security imperatives, and stark inequities in access and opportunity. As humanity grapples with the intertwined crises of climate change and energy insecurity, nuclear energy emerges as a paradoxical solution: a potential enabler of decarbonization and a focal point of geopolitical contention. Despite its transformative potential, nuclear energy remains underutilized in many developing nations, where structural, political, and economic barriers restrict access to this critical technology. This disparity raises pressing questions: how can the global community reconcile the need for equitable energy transitions with the imperatives of non-proliferation and global stability?

The urgency of these questions is underscored by the dual necessity for reliable, large-scale, low-carbon energy production and the mitigation of climate vulnerabilities. Nuclear energy offers a unique convergence of opportunity and necessity, yet its promise remains elusive for those most affected by climate change. As Hendricks and Ziegler (2019) [1] argue, nuclear diplomacy must evolve to bridge the gap between energy security and environmental stewardship. However, existing governance frameworks often fall short. These frameworks, as evidenced by research from Höffken and Ramana (2020) [2], not only fail to dismantle inequities but in some cases exacerbate them, perpetuating systemic barriers that disproportionately disadvantage developing nations. For example, the restricted dissemination of nuclear technologies often limits the ability of these nations to participate meaningfully in sustainable energy transitions, reflecting deeper patterns of environmental injustice.

The challenges extend beyond inequitable technological access. Governance mechanisms frequently lack sufficient emphasis on trust-building, a critical component for fostering international cooperation and ensuring compliance with non-proliferation norms. Duursma et al. (2021) [3] highlight trust as a cornerstone of effective global agreements, a principle often overlooked in existing governance models. Similarly, Bunn and Wier (2020) [4] emphasize the need for collaborative frameworks to address proliferation risks while enabling equitable energy transitions. These gaps underscore the necessity for a reimagined governance approach that integrates equity, trust, and inclusiveness.

This study introduces the Qudrat-Ullah Nuclear Peace and Trust (Q-NPT) framework as a bold response to these challenges. The Q-NPT framework emphasizes three core pillars: fostering trust among stakeholders, facilitating equitable technology transfer, and building capacity in developing nations. By integrating these elements, the framework seeks to transform nuclear energy governance into a mechanism for achieving decarbonization and enhancing global energy security. Moreover, the Q-NPT framework addresses a critical research gap by offering a dynamic implementation roadmap that operationalizes these principles, bridging the theoretical foundations of governance with actionable policy strategies.

This paper advances the discourse on nuclear energy governance by proposing a paradigm shift toward equity-oriented and trust-based cooperation. It outlines the Q-NPT framework as a transformative solution to the interconnected crises of climate change, energy inequality, and global instability. By challenging the limitations of existing governance structures, this framework offers a pathway for a more inclusive and sustainable future. The implications extend beyond nuclear energy, presenting a model for rethinking international cooperation in addressing shared vulnerabilities and advancing global sustainability goals.

2. The Case for a New Framework

The global nuclear governance system is at a critical crossroads. While the Nuclear Non-Proliferation Treaty (NPT) has historically provided a framework for managing nuclear energy and security, its limitations are increasingly evident in the face of urgent climate and energy challenges [5]. Persistent trust deficits, inequitable access to nuclear technology, and regional instabilities raise an important question: Is the current governance system adequately equipped to address these pressing issues?

For many developing nations, the potential of nuclear energy remains underutilized due to barriers to technology transfer, resource limitations, and geopolitical mistrust. Höffken and Ramana [2] underscore how existing governance structures often fail to address the environmental justice implications of nuclear energy distribution, perpetuating disparities in access. Similarly, Rublee and Cohen [6] observe that the current framework reinforces an implicit hierarchy of nuclear access, creating obstacles for nations striving to meet their energy and climate goals.

As Qudrat-Ullah [7] highlights, the NPT's structure was not designed with the aspirations or unique challenges of developing nations in mind. Consequently, the system often falls short in equitably addressing the needs of those most affected by climate vulnerabilities and energy insecurity. This gap underscores the need for a more inclusive governance model—one that aligns the imperatives of non-proliferation with the equitable distribution of nuclear energy technologies.

The Qudrat-Ullah Nuclear Peace and Trust (Q-NPT) framework seeks to address these gaps by reimagining nuclear governance through an equity-focused lens. It emphasizes granting developing nations greater agency in nuclear energy management under the oversight of the International Atomic Energy Agency (IAEA). By promoting trust-building, equitable technology sharing, and capacity development, the Q-NPT framework aims to create a balanced governance model that supports sustainable energy transitions while maintaining global security standards.

Achieving such a model is not merely a matter of fairness but a strategic necessity. As Beyer and McCauley [8] emphasize, governance systems must adapt to address the intertwined challenges of energy demand, decarbonization, and security. Nuclear energy's role as a scalable, low-carbon energy source, highlighted by Holechek et al. [9], makes it essential to global efforts to displace fossil fuels. Yet, the current governance structures continue to limit its accessibility to many nations that stand to benefit most.

The Q-NPT framework represents a forward-looking approach to these challenges. Rather than advocating for incremental changes, it envisions a comprehensive transformation of nuclear energy governance. By addressing inequities in access and fostering international collaboration, the Q-NPT framework seeks to redefine the role of nuclear energy as a tool for achieving shared global goals. While not without its challenges, this framework offers a pragmatic yet ambitious vision for creating a more inclusive, effective, and equitable nuclear governance system.

3. Pillars of the Q-NPT Framework

The Q-NPT framework is conceived as a normative vision grounded in the principles of equity, sustainability, and global collaboration. It integrates elements of conceptual modeling to guide governance practices and draws on theoretical insights to address systemic inequities in nuclear energy governance. The Q-NPT framework represents a strategic reimagining of nuclear governance that prioritizes transparency, inclusivity, and cooperation to address global energy and development challenges. In an era marked by the urgent need for energy transitions, the framework provides a pathway to foster trust between nuclear and non-nuclear states, ensuring equitable access to the benefits of nuclear energy. Figure 1 provides an overview of the framework, which emphasizes collaboration and mutual accountability as cornerstones for integrating nuclear technologies into the energy systems of developing nations. By doing so, it seeks to drive energy security, economic growth, and alignment with sustainable development goals [10,11]. Kühn [12] underscores that the realization of these objectives requires overcoming entrenched geopolitical, technical, and institutional barriers through trust-building measures, such as multilateral agreements and confidence-building mechanisms.

3.1. Equitable Technology Transfer

Equitable access to nuclear technology is a central tenet of the Q-NPT framework. Kühn [12] observes that restrictive technology transfer policies, often stemming from a lack of trust among states, contribute to global inequalities in nuclear energy access. To address these disparities, the Q-NPT framework seeks to facilitate the adoption of nuclear energy by developing nations for peaceful applications, such as electricity generation and medical advancements. This approach emphasizes reciprocal accountability and prioritizes collaborative partnerships rather than hierarchical or paternalistic models.

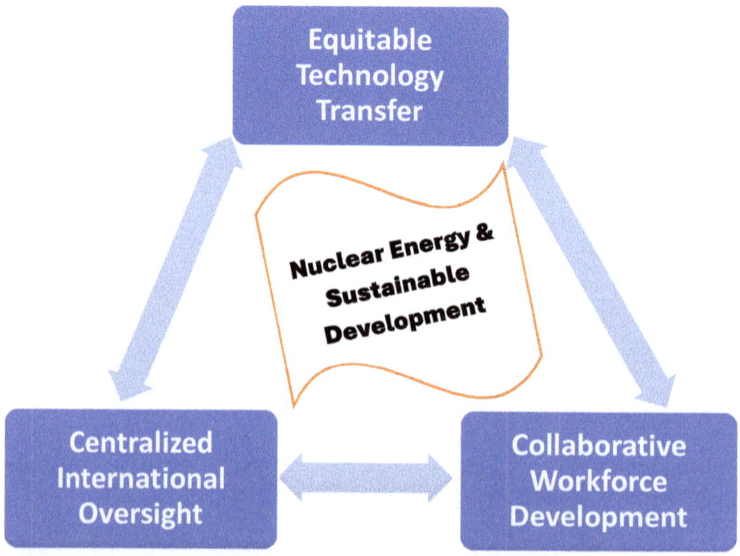

Figure 1. Core components and interactions within the Q-NPT framework.

Ruzicka and Wheeler [13] highlight that trust is a critical factor in the long-term effectiveness of multilateral nuclear agreements. Similarly, Dieguez Porras et al. [14] emphasize the role of capacity building in ensuring the secure and effective adoption of nuclear technologies. The Q-NPT framework integrates these perspectives, aiming to establish technology transfer mechanisms that are both equitable and underpinned by systematic trust-building processes.

3.2. *Centralized International Oversight*

The Q-NPT framework emphasizes centralized international oversight, designating the International Atomic Energy Agency (IAEA) as a key institution in global nuclear governance. Kühn [12] highlights the role of international institutions in promoting strategic stability by reducing uncertainty and mitigating risks associated with unilateral actions. Currently, the IAEA performs essential functions such as nuclear safeguards, safety monitoring, and the promotion of peaceful nuclear applications [15]. The Q-NPT framework proposes an expansion of the IAEA's mandate to include additional responsibilities, including capacity-building initiatives, facilitating equitable technology transfer, and establishing standardized frameworks for human resource development among member states.

In this expanded role, the IAEA would oversee collaborative workforce development programs, providing expertise and support tailored to the specific needs of developing countries, thereby enhancing trust and reducing proliferation risks. The framework also suggests the implementation of stricter, real-time monitoring mechanisms using advanced technologies such as blockchain and artificial intelligence to improve transparency and accountability in nuclear operations [16]. These measures would shift the IAEA's focus from primarily compliance to also actively supporting inclusive and sustainable nuclear energy governance.

Expanding the IAEA's responsibilities could strengthen international security measures and address gaps in access to and management of nuclear technology, particularly in developing regions. Centralized oversight aims to reduce risks related to the misuse of nuclear technologies and build global trust, contributing to a mechanism that supports safe and equitable adoption of nuclear energy in line with sustainable development objectives [10].

3.3. Collaborative Workforce Development

A third key component of the Q-NPT framework is its innovative approach to workforce development. Kühn [12] identifies workforce exchanges as vital trust-building measures that reduce misperceptions and enhance cooperation. Recognizing the expertise required to manage advanced nuclear technologies, the framework advocates establishing an international, rotational workforce. This approach builds capacity while reducing the risks of nuclear proliferation by promoting transparency and shared responsibility—critical elements in mitigating threats related to nuclear technology misuse.

Such collaborative strategies facilitate capacity building in emerging economies, foster international partnerships, and help mitigate proliferation risks. By encouraging cross-border cooperation, the Q-NPT framework ensures that nations collectively address challenges associated with nuclear energy integration, thereby supporting global energy transition efforts.

The effective integration of nuclear energy into sustainable energy frameworks requires not only technological advancements but also a comprehensive understanding of system dynamics, as noted by [7]. This underscores the need for a workforce skilled in both the technical and socio-political dimensions of nuclear energy systems.

Several international partnerships exemplify this collaborative workforce development approach. The 2008 India–United States civil nuclear agreement facilitated technology transfer and workforce training for civilian nuclear programs [17]. Similarly, the China–Pakistan partnership on the Chashma Nuclear Power Complex includes comprehensive technology transfer and training, enhancing Pakistan's nuclear capabilities [18]. Russia's support for Bangladesh's Rooppur Nuclear Power Plant incorporates extensive training programs for Bangladeshi engineers and technicians [19]. Additionally, the France–China collaboration on the Taishan Nuclear Power Plant involves significant technology transfer and joint workforce initiatives, notably through European pressurized reactor deployment [20]. Finally, the ITER project—a global fusion energy initiative—demonstrates how multinational partnerships promote joint workforce development and technological progress [21]. These examples highlight the critical importance of collaborative workforce development in building capacity, advancing nuclear energy technologies, and ensuring their safe and equitable integration into global energy systems.

3.4. Toward a Sustainable Nuclear Energy Future

The Q-NPT framework is based on three foundational pillars: equitable technology transfer, centralized international oversight, and collaborative workforce development. These components offer a structured approach to reforming nuclear governance in the 21st century. Implementation faces challenges such as geopolitical tensions and technological and institutional resistance to change. Addressing these challenges will require multilateral agreements and commitments from nuclear-armed states to promote transparency and trust in the international arena.

The framework emphasizes trust, transparency, and cooperation as essential elements to align nuclear governance with objectives of global equity and sustainability. It highlights the integration of nuclear and renewable energy sources, recognizing their complementary roles in reducing global carbon emissions.

The long-term effectiveness of the framework depends on its ability to adapt to emerging global challenges, including rapid technological innovation and evolving geopolitical contexts. An inclusive approach incorporating the perspectives of developing nations and non-nuclear states is necessary to maintain its relevance and effectiveness. Sustained cooperation and equitable practices among stakeholders are critical for the framework's contribution to global sustainable development.

4. A Dynamic Perspective of the Q-NPT

The Q-NPT framework signifies a paradigm shift in nuclear energy governance, moving beyond the limitations of traditional static models. Traditional frameworks often conceptualize governance as a linear and compartmentalized process, with fixed roles and responsibilities that lack flexibility to adapt to evolving global contexts. Such models typically emphasize compliance and control mechanisms without adequately addressing the dynamic interactions between various stakeholders and systems [22].

In contrast, the Q-NPT framework adopts a dynamic and interconnected perspective, emphasizing feedback loops and adaptive mechanisms to address the complex and interdependent challenges inherent in nuclear governance. This approach aligns with systems thinking principles, which advocate for understanding the interrelationships within complex systems rather than viewing components in isolation [7]. By incorporating adaptive strategies and continuous learning processes, the Q-NPT framework aims to enhance resilience and responsiveness in nuclear governance structures.

Figure 1 illustrates the dynamic governance framework of the Q-NPT framework, contrasting it with traditional static models. While conventional frameworks often treat nuclear governance as a linear and compartmentalized process, the Q-NPT framework acknowledges the multifaceted and evolving nature of global nuclear energy systems. Figure 1 underscores the paradigm shift required to address the complexities of nuclear energy governance effectively.

The Q-NPT framework's dynamic perspective advances global policymaking by emphasizing resilience and adaptability. It challenges stakeholders to move beyond siloed approaches, fostering an integrated governance model that prioritizes trust, equity, and security. This system-oriented perspective underscores the importance of dynamic policies that prioritize flexibility and adaptability, enabling more effective responses to the evolving challenges of nuclear governance.

The Q-NPT framework serves as both a theoretical construct and a practical roadmap, guiding policymakers and international organizations toward innovative strategies that account for feedback effects. By aligning nuclear energy governance with the principles of sustainability, equity, and collaboration, the framework provides a robust mechanism for fostering resilience and achieving long-term global development goals in the nuclear energy sector. Table 1 summarizes key differences between traditional nuclear non-proliferation frameworks and the proposed Q-NPT framework from a dynamic perspective.

Table 1. Comparative analysis of traditional nuclear non-proliferation frameworks and the Q-NPT framework.

Aspect	Traditional Nuclear Non-Proliferation Framework	Q-NPT Framework (Dynamic Perspective)
Core Focus	Static control and prevention of nuclear weapon spread	Dynamic governance balancing peaceful use and non-proliferation
Approach	Compliance-based, treaty enforcement	Systems thinking emphasizing feedback loops and adaptive policies

Table 1. *Cont.*

Aspect	Traditional Nuclear Non-Proliferation Framework	Q-NPT Framework (Dynamic Perspective)
Governance	Top-down, state-centric	Multi-level, inclusive of technology-sharing and capacity-building
Technological Consideration	Limited to static assessments of nuclear tech capabilities	Incorporates evolving technologies like SMRs, Gen IV reactors
Response to Change	Reactive, slow adaptation	Proactive, iterative learning and adaptation
Sustainability & Decarbonization	Peripheral or absent	Central role, integrating nuclear tech for decarbonization goals
Stakeholder Engagement	Limited, often state-only	Inclusive of developing countries, technology partners, and civil society
Capacity Building	Minimal emphasis	Core component to ensure effective technology adoption

5. Equitable Access to Nuclear Technology and Decarbonization

As the global community confronts the intertwined crises of energy inequity and climate change, the Q-NPT framework emerges as a transformative paradigm. At its core, the framework emphasizes equitable technology transfer, positioning nuclear energy as a means to address systemic disparities in energy access while advancing decarbonization objectives. By aligning nuclear technology sharing with the principles of energy justice, the Q-NPT framework highlights the urgent need for global partnerships that transcend traditional barriers, fostering a collective approach to sustainable development.

The dual nature of nuclear energy—its potential as a source of clean energy and the inherent risks of misuse—renders its equitable distribution both a necessity and a complex challenge. Qudrat-Ullah [23] underscores that a sustainable energy transition requires not only technical advancements but also governance frameworks that embed energy justice into global energy policies. By bridging the divide between nuclear-armed and non-nuclear nations, the Q-NPT framework offers a strategic pathway for a fair and inclusive transition toward low-carbon energy systems.

5.1. Technology Sharing: Bridging Inequities in Energy Access

Central to the Q-NPT framework is a vision for technology sharing that seeks to address historical disparities in nuclear energy access. By fostering partnerships between developed and developing nations, the framework aims to provide under-resourced countries with the expertise, infrastructure, and technical capacity necessary to integrate nuclear energy into their energy portfolios. This approach contributes to reducing dependence on fossil fuels while addressing principles of global energy justice.

Schneider and Ramana [24] highlight that equitable technology sharing extends beyond technical considerations and is fundamental to fairness in international energy policy. The Q-NPT framework builds on this principle by advocating for agreements that enable developing nations to utilize nuclear power for electricity generation, healthcare, and industrial applications. These agreements also incorporate safeguards to mitigate proliferation risks, supporting global security alongside energy equity.

Moreover, technology-sharing initiatives offer a means to reduce persistent inequalities in energy access. Developing countries often face resource constraints that limit their ability to independently invest in advanced technologies, thereby perpetuating energy poverty and economic vulnerabilities. Integrating nuclear technology into their energy strategies can provide these nations with access to clean and reliable power, thereby supporting broader social and economic development objectives.

Successful implementation of technology-sharing agreements also requires capacity-building efforts, including workforce development, regulatory support, and the establishment of robust governance structures. For instance, small modular reactors (SMRs) represent a viable technology for developing countries due to their scalability, cost-effectiveness, and compatibility with smaller energy grids. Similarly, Generation IV reactors, characterized by enhanced safety features and sustainable closed fuel cycles, provide long-term benefits in reducing nuclear waste and supporting decarbonization goals. These technologies directly address several challenges faced by developing nations:

- **Cost:** SMRs reduce upfront capital costs through modular construction, enabling countries with limited financial resources to invest in nuclear energy without the prohibitive expenses associated with traditional large-scale plants.
- **Infrastructure:** Unlike conventional nuclear power plants, SMRs have lower infrastructure requirements, allowing for deployment in remote or resource-limited regions where extensive grid networks may be absent or underdeveloped.
- **Safety:** Generation IV reactors incorporate advanced safety features such as passive cooling systems and fail-safe designs. These innovations not only mitigate the risks of nuclear accidents but also address long-standing public and environmental safety concerns, which are critical for fostering trust and adoption in new markets.

By integrating such advanced technologies into the Q-NPT framework, developing nations can harness nuclear energy effectively while addressing unique infrastructure and resource constraints. Without these elements, even well-intentioned technology transfers risk being underutilized or mismanaged.

Fostering trust and collaboration between partners becomes critical, not only for the immediate adoption of nuclear energy but also for ensuring its long-term sustainability in recipient nations. By embracing a holistic approach to technology sharing—one that includes advanced nuclear solutions like SMRs and Generation IV reactors—the Q-NPT framework not only contributes to reducing global emissions but also advances the moral imperative of creating an energy system that is equitable, inclusive, and resilient to future challenges.

5.2. A Case Study: Successful Technology Transfers in Developing Nations

The collaboration between the United Arab Emirates (UAE) and South Korea serves as a compelling example of the potential of international technology transfer to drive equitable energy development. Through the Barakah nuclear project, South Korea transferred not only advanced reactor technology but also the expertise and training necessary to build the UAE's capacity for sustainable nuclear energy. This partnership illustrates how strategic collaborations can help nations leapfrog traditional energy pathways, directly advancing their decarbonization efforts [15,25].

Beyond the infrastructure and expertise, the UAE–South Korea partnership highlights the broader benefits of fostering long-term relationships in energy development. It demonstrates how nations with established nuclear industries can support emerging nuclear states in achieving energy independence and contributing to global sustainability goals [7,10].

These collaborations also foster knowledge exchange and technical cooperation, enabling the recipient nation to strengthen its institutional capacity and regulatory frameworks, both of which are critical for sustaining a robust nuclear energy program.

This case underscores the importance of aligning technology transfer with tailored workforce development and localized expertise. By investing in human capital, the UAE has been able to operate and maintain its nuclear energy facilities while also cultivating a skilled workforce that contributes to the broader energy sector. Furthermore, the project serves as a model for integrating international safeguards and compliance with non-proliferation commitments, ensuring that the growth of nuclear energy remains aligned with global security standards.

Ultimately, the UAE–South Korea collaboration exemplifies how thoughtful, well-executed technology transfers can drive sustainable energy transitions in developing nations. This model can be replicated in other contexts, provided there is mutual trust, shared responsibility, and a commitment to equitable partnerships that prioritize long-term capacity building and environmental stewardship.

5.3. Nuclear Energy as a Pillar of Decarbonization

Nuclear energy's role in decarbonization extends far beyond its capacity to reduce carbon emissions. It offers a stable and scalable complement to renewable energy, addressing intermittency issues while enabling countries to meet ambitious climate targets. The integration of nuclear energy into diversified energy portfolios represents a critical strategy for achieving global sustainability goals.

Table 2 exemplifies nuclear energy's contribution to CO_2 reduction, highlighting its transformative potential in both established and emerging markets. For instance, France's reliance on nuclear power for 70% of its energy needs positions it as a global leader in low-carbon energy systems. Similarly, the U.S. and South Korea showcase the substantial emissions reductions achievable through nuclear power. Even emerging nuclear nations like the UAE illustrate the transformative potential of nuclear energy in achieving climate goals.

Table 2. The role of nuclear energy in achieving decarbonization goals.

Country	Nuclear Capacity (GW)	Share of Nuclear in Energy Mix (%)	CO_2 Emissions Reduction ($MtCO_2$ Annually)	Source
France	61.4	70	50	[26]
United States	95.5	20	200	[27]
South Korea	23.6	30	10	[25,28]
UAE (Barakah)	5.6	8	3.2	[28]

While nuclear energy alone cannot solve the climate crisis, its role as a low-carbon, high-capacity energy source makes it indispensable in the global energy transition. Its ability to produce consistent baseline power provides stability to grids increasingly reliant on intermittent renewable sources like solar and wind. This complementarity ensures a reliable energy supply, reducing the need for carbon-intensive backup systems. However, unlocking the full potential of nuclear energy demands addressing several challenges. These include managing public concerns over safety and waste disposal, resolving proliferation risks, and fostering international collaboration to support equitable technology transfer. Furthermore, equitable integration into global energy systems necessitates building trust among stakeholders and ensuring that nuclear technology benefits all nations, not just those with existing expertise or resources.

The Q-NPT framework offers a comprehensive pathway to overcome barriers in nuclear governance, positioning nuclear energy as a cornerstone of global decarbonization efforts. By emphasizing equitable technology sharing, the framework facilitates access to nuclear expertise and infrastructure for under-resourced nations, enabling them to bypass traditional, carbon-intensive energy development pathways. To foster trust between states, the Q-NPT framework incorporates specific initiatives designed to promote transparency, collaboration, and mutual accountability. These initiatives include the following:

1. **Establishing multilateral nuclear knowledge hubs**, regional centers of excellence where nuclear and non-nuclear states collaborate on research, training, and capacity-building programs to build mutual understanding and shared expertise;
2. **Implementing joint safeguard missions**, co-conducted by representatives from donor and recipient nations under the oversight of international bodies like the IAEA to ensure compliance with non-proliferation measures and build trust through shared accountability;
3. **Developing a transparency index**, an annual evaluation of participating nations' adherence to Q-NPT principles, including equitable technology sharing, non-proliferation compliance, and collaborative workforce development—this fosters trust by providing an objective and public measure of progress;
4. **Facilitating bilateral and multilateral confidence-building dialogues**, regular forums for open dialogue on concerns, progress, and opportunities, reducing geopolitical tensions and fostering long-term partnerships.

Additionally, the Q-NPT framework's robust safeguards and non-proliferation measures address security concerns, further strengthening global confidence in nuclear solutions. By harmonizing technological innovation, policy support, and equity, the framework provides a structured mechanism for fostering collaboration among nations, bridging gaps in resources and expertise, and aligning nuclear development with global sustainability goals. Through these concrete initiatives, nuclear power transcends its traditional limitations, fostering a sustainable, resilient, and inclusive energy future while establishing nuclear trust as a cornerstone of international collaboration.

6. Energy Policy Implications and Pathways for Implementation

The Q-NPT framework represents a paradigm shift in nuclear energy governance by addressing the interconnected challenges of equity, sustainability, and global collaboration. By prioritizing equitable technology sharing, environmental stewardship, and geopolitical stability, it provides a structured pathway for integrating nuclear energy into a decarbonized global energy system. Its focus on energy justice, especially for under-resourced nations, underscores its relevance to the sustainable development aspirations of the Global South.

6.1. National Energy Policies: Rethinking Integration in a Q-NPT Context

Realizing the vision of the Q-NPT framework requires adaptive national energy policies tailored to varying institutional contexts. Developed countries with established nuclear infrastructure, such as Canada, can focus on advancing safety and efficiency while supporting emerging nuclear states. Conversely, nations like India, which are expanding their nuclear programs, must prioritize aligning domestic energy strategies with international non-proliferation commitments to build trust and secure equitable technology access [29,30].

The Q-NPT framework's integration of nuclear energy into diversified energy portfolios highlights its complementarity with renewables, addressing energy resilience and inclusivity. Policymakers should adopt holistic approaches linking nuclear development with broader sustainability goals, transcending siloed strategies. For instance, Canada's leadership in reactor safety and India's focus on self-reliance present opportunities to align with Q-NPT principles. By framing nuclear energy as a global climate solution rather than a geopolitical tool, both nations can promote collaborative models bridging the Global North–South divide.

Collaborative initiatives, such as those spearheaded by the World Association of Nuclear Operators (WANO), have improved safety and operational performance globally. WANO's peer review mechanisms, along with the Nuclear Suppliers Group (NSG), foster trust by ensuring equitable access to nuclear materials and technologies. India's pursuit of NSG membership exemplifies its commitment to responsible nuclear governance and engagement in global non-proliferation frameworks [31,32].

6.2. Global Collaboration: A New Era of Shared Ambition

The Q-NPT framework envisions an era of collaboration-driven nuclear governance. Recent advancements, such as the U.S. Department of Energy's USD 800 million investment in Generation III+ SMRs, demonstrate the feasibility of this vision by enabling partnerships between utilities, vendors, and contractors to accelerate deployment [33]. Similarly, Canadian–Romanian collaborations, including a memorandum of understanding between the Canadian Nuclear Association and ROMATOM, highlight the importance of joint innovation in reactor development and SMR deployment [29]. These initiatives underscore the need for capacity building in emerging nuclear states. For instance, Canada's export financing of CAD 3 billion for Romania's nuclear projects reflects a commitment to shared energy security goals [34]. These collaborations foster mutual trust, equipping less-resourced nations with the tools to adopt nuclear energy responsibly while adhering to the principles of inclusivity and sustainability.

6.3. A Dynamic Roadmap for Implementation

The success of the Q-NPT framework hinges on its practical implementation, requiring a dynamic and iterative approach. Figure 2 outlines a roadmap where stakeholder engagement, capacity building, and continuous monitoring form the backbone of progress. This feedback loop is not merely a procedural step—it embodies the adaptive mindset necessary to navigate the complexities of nuclear energy governance.

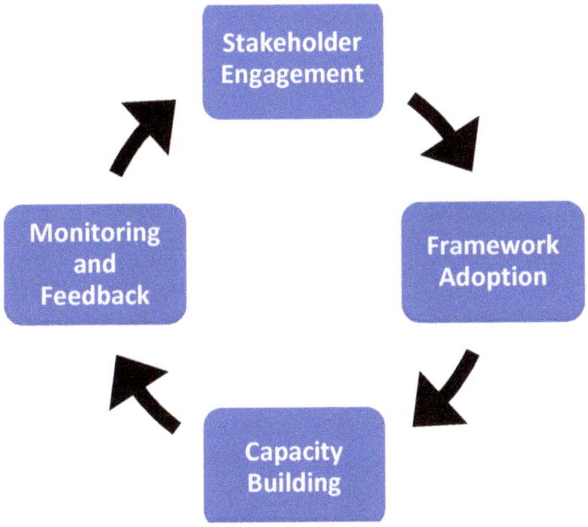

Figure 2. Roadmap for implementing the Q-NPT framework.

Operationalizing the Q-NPT framework requires collaboration between international bodies, such as the International Atomic Energy Agency (IAEA), and national governments.

For instance, the IAEA could establish a dedicated task force to oversee equitable technology transfer programs, while national governments might incorporate Q-NPT principles into bilateral and multilateral agreements. Pilot initiatives, such as regional nuclear energy hubs in developing nations, could serve as practical demonstrations of the framework's viability.

Key insights include the following:

- **Inclusive policy development**: Engaging diverse stakeholders ensures policies reflect the realities of specific contexts. The European Green Deal's adaptive strategies provide a model for recalibrating policies in response to challenges [26]. Workshops with regulatory bodies in developing nations could similarly address gaps in policy clarity, creating an enabling environment for nuclear energy adoption.
- **Addressing technological gaps**: Tailored capacity-building programs for emerging nuclear states, such as Canada's collaboration with Pakistan via the CANDU Owners Group, exemplify effective knowledge transfer [29]. Feedback mechanisms by agencies like the IAEA further guide the development of targeted training and resource allocation strategies.
- **Fostering public trust**: Transparent communication is vital. Romania's Doicești SMR project exemplifies how proactive measures, including IAEA Site and External Events Design reviews, build public confidence in nuclear initiatives [35].
- **Optimizing Resource Allocation**: Dynamic feedback loops enable efficient allocation of resources. For example, Romania's USD 4 billion SMR financing package illustrates how strategic investments can catalyze nuclear development [36].

6.4. Synthesis, Limitations, and Future Directions

The Q-NPT framework bridges policy, technology, and equity, offering a structured approach to advancing nuclear energy governance. It outlines pathways for integrating nuclear energy into global sustainability agendas while fostering collaboration among nations. While the framework provides a valuable perspective, significant challenges remain. Its success depends on adaptability to diverse national and regional contexts and its ability to address barriers such as public resistance, resource constraints, geopolitical tensions, and institutional inertia. Furthermore, the financial costs associated with implementing equitable governance structures and technology-sharing mechanisms may present additional obstacles. Future research should focus on addressing these limitations, particularly in politically unstable or resource-limited nations. Expanding case studies and incorporating lessons from other sectors, such as renewable energy, could enhance its applicability. By systematically addressing these challenges, the framework could support the development of a sustainable, inclusive, and resilient global energy system.

7. Conclusions: Reimagining Nuclear Energy Governance Through the Q-NPT Framework

The Q-NPT framework proposes a novel approach to nuclear energy governance, emphasizing trust, equity, and energy justice as central principles in the global energy transition. By addressing challenges related to accessibility, fairness, and inclusivity, the framework highlights the role of nuclear energy as both a technical solution and a means of fostering international collaboration toward sustainable development goals. This perspective positions the Q-NPT framework as a comprehensive model for enhancing cooperation among nations to tackle climate change and energy insecurity.

7.1. Q-NPT: A Pathway to Inclusive Nuclear Energy Governance

The Q-NPT framework proposes a model for integrating nuclear energy into strategies for decarbonization and energy justice. It prioritizes equitable access to nuclear technologies, particularly for developing nations, thereby enabling their participation in global decarbonization efforts [7,23,37]. This approach addresses historical constraints that have limited the Global South's engagement with nuclear energy adoption.

By incorporating trust-building measures and fostering collaborative oversight, the Q-NPT framework aligns nuclear energy governance with broader sustainable development goals. It emphasizes the integration of nuclear energy into a diversified low-carbon energy mix, highlighting its role as a complementary resource alongside renewable energy sources. The framework provides a practical and adaptable approach for nations seeking to balance energy security with environmental sustainability.

7.2. Q-NPT: A Blueprint for Equitable Energy Futures

The Q-NPT framework provides a foundation for immediate application while also highlighting critical areas for future exploration. To maximize its effectiveness, researchers and policymakers should focus on three key areas:

1. **Technology sharing and accessibility:** Strengthening mechanisms for equitable technology transfer is essential to empower nations in the Global South. Existing international frameworks, such as partnerships between Canada and Pakistan (CANDU operations) or technology-sharing initiatives between the United States and India, provide useful models. Future research should focus on developing scalable and replicable approaches to ensure widespread adoption.
2. **Integration with renewable energy systems:** Exploring the intersection of nuclear and renewable energy systems offers opportunities to create resilient, low-carbon energy systems tailored to global needs. Policymakers should prioritize frameworks that facilitate integration, fostering collaboration between nuclear and renewable energy sectors to enhance energy security and sustainability.
3. **Risk mitigation and systemic resilience:** Addressing systemic risks, including non-proliferation, safety, and public trust, is critical. Developing robust methodologies for monitoring and evaluation, supported by international cooperation, can help mitigate these risks. Adaptive governance structures will be necessary to align nuclear energy governance with the evolving nature of global energy challenges.

In conclusion, the Q-NPT framework represents a transformative model for governing nuclear energy systems, both existing and planned, by embedding principles of equity, resilience, and sustainability. Through the promotion of international collaboration, the integration of nuclear energy within diversified energy systems, and the mitigation of systemic risks, the framework repositions nuclear energy as a collaborative instrument for achieving global decarbonization goals rather than a source of geopolitical contention. The framework's ultimate success depends on the sustained commitment of policymakers and researchers to adopt and operationalize its principles, thereby ensuring that nuclear energy initiatives of today and the future advance inclusivity and sustainability within the global energy landscape.

While the Q-NPT framework offers a transformative vision for nuclear energy governance, its implementation may face significant challenges. These include political resistance from nuclear-armed states, the high costs of establishing equitable technology transfer mechanisms, and institutional inertia that can delay the adoption of new governance models. Acknowledging these limitations provides a realistic foundation for further discourse and refinement.

7.3. A Vision for the Future: Toward a Just and Decarbonized World

Amid the complexities of the climate crisis, the Q-NPT framework presents a compelling model for energy governance that prioritizes collaboration, justice, and sustainability. It underscores the need for collective efforts to dismantle entrenched barriers and positions nuclear energy as a unifying component of the global energy transition. By aligning nuclear governance with principles of equity and inclusivity, the Q-NPT framework has the potential to redefine the role of nuclear energy, shifting it from a contentious issue to a fundamental pillar of global progress.

The successful implementation of the Q-NPT framework hinges on the commitment of nations to adopt its principles and operationalize them effectively. This framework challenges global leaders, researchers, and practitioners to seize the opportunity to foster energy systems that are not only sustainable but also equitable and inclusive. Through sustained international collaboration and innovative approaches, the Q-NPT framework can provide a guiding vision for achieving a decarbonized, equitable, and resilient energy future.

Funding: This research received no external funding.

Conflicts of Interest: The author declares no conflict of interest.

References

1. Hendricks, C.; Ziegler, D. Nuclear Diplomacy and the Climate Crisis: A New Era of Nuclear Energy. *Glob. Policy* **2020**, *11*, 21–29.
2. Höffken, J.; Ramana, M.V. Nuclear power and environmental injustice. *Wiley Interdiscip. Rev. Energy Environ.* **2024**, *13*, e498. [CrossRef]
3. Duursma, A.; Bara, C.; Wilén, N.; Hellmüller, S.; Karlsrud, J.; Oksamytna, K.; Bruker, J.; Campbell, S.; Cusimano, S.; Donati, M.; et al. UN peacekeeping at 75: Achievements, challenges, and prospects. *Int. Peacekeep.* **2023**, *30*, 415–476. [CrossRef]
4. Bunn, M.; Wier, A. The Next Generation of Nuclear Disarmament. *J. Int. Aff.* **2018**, *71*, 45–58.
5. United Nations. *Treaty on the Non-Proliferation of Nuclear Weapons (NPT)*; United Nations Office for Disarmament Affairs: New York, NY, USA, 2020; Available online: https://www.un.org/disarmament/wmd/nuclear/npt/ (accessed on 30 April 2025).
6. Rublee, M.R.; Cohen, A. Nuclear norms in global governance: A progressive research agenda. *Contemp. Secur. Policy* **2018**, *39*, 317–340. [CrossRef]
7. Qudrat-Ullah, H. *Understanding the Dynamics of Nuclear Power and the Reduction of CO_2 Emissions—A System Dynamics Approach*; Springer: Cham, Switzerland, 2022. [CrossRef]
8. Beyer, D.; McCauley, K. Global Nuclear Governance and its Role in Energy Security. *J. Glob. Secur. Stud.* **2021**, *6*, 1–15.
9. Holechek, J.L.; Geli, H.M.E.; Sawalhah, M.N.; Valdez, R. A global assessment: Can renewable energy replace fossil fuels by 2050? *Sustainability* **2022**, *14*, 4792. [CrossRef]
10. Qudrat-Ullah, H. *Sustainable Energy: A myth or Reality*; Springer: Cham, Switzerland, 2024.
11. Qudrat-Ullah, H.; Karakul, M. Modeling for policy assessment in the electricity supply sector of Pakistan. *Int. J. Energy Sect. Manag.* **2007**, *3*, 240–256. [CrossRef]
12. Kühn, U. Strategic stability in the 21st century: An introduction. *J. Peace Nucl. Disarm.* **2023**, *6*, 1–8. [CrossRef]
13. Ruzicka, J.; Wheeler, N.J. The puzzle of trusting relationships in the nuclear non-proliferation treaty. *Int. Aff.* **2010**, *86*, 69–85. [CrossRef]
14. International Atomic Energy Agency. *International Conference on Nuclear Knowledge Management and Human Resources Development: Conference Programme*; IAEA: Vienna, Austria, 2024.
15. International Atomic Energy Agency. *IAEA Annual Report 2015*; IAEA: Vienna, Austria, 2015. Available online: https://www.iaea.org/publications/11093/iaea-annual-report-2015 (accessed on 30 March 2025).
16. Ecemis, I.N.; Ekinci, F.; Acici, K.; Guzel, M.S.; Medeni, I.T.; Asuroglu, T. Exploring blockchain for nuclear material tracking: A scoping review and innovative model proposal. *Energies* **2024**, *17*, 3028. [CrossRef]
17. U.S. Department of State. *United States-India Nuclear Cooperation Approval and Nonproliferation Enhancement Act*; U.S. Department of State: Washington, DC, USA, 2008. Available online: https://2001-2009.state.gov/p/sca/rls/fs/2008/109567.htm (accessed on 23 April 2025).
18. Reuters. *Pakistan and China to Expand Nuclear Cooperation with New Power Plant Projects*; Reuters: London, UK, 2023. Available online: https://www.reuters.com/business/energy/pakistan-china-sign-48-bln-nuclear-power-plant-deal-2023-06-20/ (accessed on 13 February 2025).
19. World Nuclear Association. *Rooppur Nuclear Power Plant: Enhancing Nuclear Capability in Bangladesh*; World Nuclear Association: London, UK, 2023. Available online: https://world-nuclear.org/information-library/country-profiles/countries-a-f/bangladesh (accessed on 11 April 2025).

20. EDF in China. Taishan Nuclear Power Plant: A Franco-Chinese Collaboration. Available online: https://www.edf.fr/sites/groupe/files/contrib/groupe-edf/espaces-dedies/espace-medias/dp/edf_in_china.pdf (accessed on 30 April 2025).
21. ITER Organization. About ITER: Bringing Nations Together For Fusion Energy. Available online: https://www.iter.org/few-lines (accessed on 30 April 2025).
22. International Atomic Energy Agency. *Fundamentals of Nuclear Safety (IAEA Safety Standards Series No. SF-1)*; International Atomic Energy Agency: Vienna, Austria, 2006; Available online: https://www.iaea.org/publications (accessed on 21 June 2024).
23. Qudrat-Ullah, H. *Exploring the Dynamics of Renewable Energy and Sustainable Development in Africa*; Springer: Cham, Switzerland, 2023. [CrossRef]
24. Schneider, M.; Ramana, M.V. Nuclear energy and the non-proliferation treaty: A retrospective examination. *J. Peace Nucl. Disarm.* **2023**, *6*, 165–174. [CrossRef]
25. World Nuclear Association. *Nuclear Power in South Korea*; World Nuclear Association: London, UK, 2022. Available online: https://world-nuclear.org/information-library/country-profiles/countries-o-s/south-korea (accessed on 12 March 2025).
26. European Commission. *The European Green Deal*; European Commission: Brussels, Belgium, 2020. Available online: https://ec.europa.eu/green-deal (accessed on 30 April 2025).
27. U.S. Energy Information Administration (EIA). *Nuclear Power and the Environment*; U.S. Energy Information Administration (EIA): Washington, DC, USA, 2023. Available online: https://www.eia.gov/energyexplained/nuclear/nuclear-power-and-the-environment.php (accessed on 30 April 2025).
28. Emirates Nuclear Energy Corporation. In the Year of Sustainability UAE Takes Significant Leap Towards Net Zero 2050 as Unit 3 of Barakah Nuclear Energy Plant Starts Commercial Operations. 2023. Available online: https://www.enec.gov.ae/news/latest-news/uae-takes-leap-towards-net-zero-2050-as-unit-3-plant-starts-commercial-operations (accessed on 30 April 2025).
29. Canadian Nuclear Association. New Partnership Formed Between Canadian and Romanian Nuclear Industry Groups to Promote Cooperation in the Field of Civil Nuclear Programs. 2021. Available online: https://cna.ca/2021/11/30/new-partnership-formed-between-canadian-and-romanian-nuclear-industry-groups-to-promote-cooperation-in-the-field-of-civil-nuclear-programs/ (accessed on 30 April 2025).
30. Canadian Nuclear Safety Commission. *Canadian National Report for the Convention on Nuclear Safety*; Canadian Nuclear Safety Commission: Ottawa, ON, Canada, 2023. Available online: https://www.cnsc-ccsn.gc.ca/eng/resources/publications/reports/cns/ninth/ (accessed on 30 April 2025).
31. WANO. *Peer Review*; World Association of Nuclear Operators: London, UK, 2024. Available online: https://www.wano.info/services-and-support/peer-review/ (accessed on 23 March 2023).
32. Vishwanathan, A. India and the Nuclear Suppliers Group: From estrangement to engagement? In *India in a Changing Global Nuclear Order*; Pentagon Press: New Delhi, India, 2009; pp. 189–207.
33. Reuters. *U.S. Opens Applications for $900 Million for Small Nuclear Reactors*; Reuters: London, UK, 2024. Available online: https://www.reuters.com/sustainability/climate-energy/us-opens-applications-900-million-small-nuclear-reactors-2024-10-16/ (accessed on 17 March 2025).
34. Government of Canada. *$3 Billion in Federal Export Finance to Support Canadian Businesses While Providing Clean Energy Security for Romania*; Government of Canada: Ottawa, ON, Canada, 2023. Available online: https://www.canada.ca/en/natural-resources-canada/news/2023/09/3-billion-in-federal-export-finance-to-support-canadian-businesses-while-providing-clean-energy-security-for-romania.html (accessed on 30 April 2025).
35. International Atomic Energy Agency. *IAEA SEED Follow-Up Mission Highlights Good Progress in SMR Site Selection in Romania*; International Atomic Energy Agency: Vienna, Austria, 2022. Available online: https://www.iaea.org/newscenter/news/iaea-seed-follow-up-mission-highlights-good-progress-in-smr-site-selection-in-romania (accessed on 30 April 2025).
36. Reuters. *Romania's Nuclearelectrica Sees Preliminary Decision on SMR Plant in 2025*; Reuters: London, UK, 2024. Available online: https://www.reuters.com/business/energy/romanias-nuclearelectrica-sees-preliminary-decision-smr-plant-2025-2024-03-18/ (accessed on 30 December 2024).
37. Pedro Dieguez Porras, J.; Sanz Gozalo, J.; Cizelj, L.; Ambrosini, W.; Lazaro Pavel, G. Sustaining regional nuclear human capacity building in Europe. *Nucl. Eng. Des.* **2024**, *419*, 112916. [CrossRef]

energies

Article

Sustainable Development of the European Electricity Sector: Investigating the Impact of Electricity Price, Market Liberalization and Energy Taxation on RES Deployment

George E. Halkos * and Apostolos S. Tsirivis

Department of Economics, University of Thessaly, 38333 Volos, Greece; atsirivis@uth.gr
* Correspondence: halkos@econ.uth.gr; Tel.: +30-2421074920

Abstract: Replacing conventional CO_2 intensive generation with green electricity from RES constitutes an essential prerequisite of sustainable development. Renewables play a vital role in achieving the UN's goals for clean low-cost energy production and the reverse of climate change process. Based on a comprehensive dataset including observations for 17 European countries between 2003 and 2020, the present research attempts to unveil the fundamental determinants of RES deployment. A panel FMOLS approach was utilized to provide a detailed analysis of the impact of electricity prices, energy taxes and competition level in both power generation and the retail electricity market on each country's RES percentage participation in electricity production fuel mix. The final econometric outcomes verified the strong statistical significance of all examined variables for the vast majority of the countries, constituting them crucial aspects of national energy strategies. However, both the actual effects as well as the impact size were found to differ significantly across Europe, signifying the complexity of the EU's task to develop a unified, autonomous and eco-friendly electricity market based on the principals of a fundamental energy strategy. Contributing to state authorities' and EU's colossal effort to deal with the crucial challenges of RES power generation, the paper proposes a series of targeted individual and groupwise policy implications.

Keywords: RES deployment; market liberalization; retail price; energy tax; SDG 7; SDG 13; energy strategy

1. Introduction

Extreme weather conditions and climate change are gradually becoming a reality for millions of people worldwide, giving rise to catastrophic consequences for the quality of life and economic prosperity. Air pollution, including the destruction of the ozone layers and the mass concentration of CO_2 into the Earth's atmosphere, is held responsible for triggering the greenhouse gas phenomenon and global warming. The last eight years were reported by [1] to be the warmest in history, since the beginning of official scientific recordings. Annual carbon emissions in 2022 continued to rise by nearly 2.1 ppm, while the average temperature exceeded that of the reference period, 1991–2000, by 0.3 °C despite the cooling La Niña effect, which was evident almost throughout the entire year for the third consecutive time. Putting Europe's environmental distortions under the microscope, Ref. [2] noted that the continent in 2022 experienced the second hottest year after 2020 and the summer with the worst heat wave ever recorded. The same report makes special reference to the fact that Europe was found to be the region in which average annual temperatures have increased at a higher pace than any other on the planet, reaching double the global average. It is indicative that the summer of 2022 in Western Europe was approximately 10 °C warmer than the typical seasonal maximum temperatures and together with the record sunshine duration, led to the largest glacier melts ever observed in European Alps and the second lowest river flow levels in history. Likewise, the combination of prolonged drought periods and extremely high temperatures in southern Europe created ideal conditions for some of the oldest forests in this geographical area to burst into flames. It is estimated that the emitted air pollutants due to the wildfires exceeded all previously documented levels in the last two decades, with the total burnt area being the second largest ever within a calendar year.

Households and industrial production of goods and services heavily rely on mass consumption and constant access to affordable electricity, the most essential energy source, which, according to [3], is accountable for roughly 40% of CO_2 emissions worldwide on year basis. This vital energy input is produced at a great cost of natural resources, inducing a series of environmental consequences. The United Nations (UN) could not have neglected the adverse effects of electricity generation in its effort to mobilize the global community in the joint cause against energy poverty and environmental degradation. The 2015 Paris Agreement and the sustainability goals for affordable clean energy production and tackling climate change (SDGs 7 and 13, respectively), which were endorsed by [4], put great emphasis on the development of a sustainable energy sector that is capable to provide citizens of the countries ratifying the terms of the agreement with green low-cost electricity. These parameters dictate an energy strategy that promotes a shift transition to a liberalized electricity market—in line with Refs. [5,6]—and towards an era where renewable energy sources (RES) will be the dominant energy source in the generation fuel mix. The decisive role of renewables in mitigating the carbon footprint of electricity generation in the nations comprising the EU and in major economies among with the highest GDPs was highlighted by [7], while [8] unveiled the beneficial effect of RES in moderating the upward price levels of European household electricity prices.

In line with its legal statutes to constantly enhance economic prosperity and its ethical mission to put environmental conditions as well as the quality of life of millions of citizens in the European region at the center of attention, the EU has become the frontrunner in the implementation of R&D and RES deployment strategies. In addition, the European Commission (EC), with its climate [9] and energy [10] directives, embraced the UN's ecological spirit and ideas expressed within the international environmental agreement. This newly established legal framework enabled the EU to align with its provisional commitments, while improving energy autonomy and diminishing import dependency. Nevertheless, the recent energy crisis, triggered by the geopolitical turmoil in Eastern Europe following the Russia–Ukraine conflict in 2022, brought to surface several structural weaknesses of EU's energy plan. The observed inability to effectively handle the energy price rally primarily stems from the lack of a diversified energy fuel mix and the strategic choice for overdependence on imported natural gas. The use of natural gas as the main intermediate fuel and basic instrument for achieving the full transition to green energy production magnified the impact of market shortages and limitations. High reliance on the more eco-friendly and relatively cheap Russian natural gas exposed EU countries to extreme price risk in cases of supply distortions, while increased complacency for technological advancements concerning RES efficiency and development of alternative energy sources, such as hydrogen.

In response to the latest extensive turmoil in the energy market and the lack of supply of natural resources from Russia, which is included in the top 10 countries with the largest natural resource reserves in the world [11], the EC, with its [12,13] directives, implemented several breakthrough energy policy adjustments. Theselegislative interventions and the RePowerEU plan brought fundamental changes—compared to the previously set goals in the Fit-for-55 scenario—, involving a rise of overall RES usage to 45%by 2030 and an increase in the incorporation of new RES technologies from 28% to 32%. Furthermore, specific targets with respect to the reduction of total demand for electricity were established, together with a revenue cap for power generators with lower costs than most expensive sources of electricity production. The chronicle of [14] about the progresses in the European electricity market in 2022 shows that during the last quarter of the year, overall demand for electricity within the EU diminished by 7.9% compared to 2021, with renewables accounting for 22% of total electricity production. This led to an all-time high for RES contribution, overtaking fossil fuel's participation in electricity generation fuel mix—which fell to 20%—yet the total share of fossil fuels still remained significantly large at 39%. However, despite all above actions, precautionary measures and achievements [15] support that the targets set by the RePowerEU plan concerning electricity generation and RES involvement in the transport sector are fairly unrealistic, unless EU member states accelerate the promotion of essential structural changes. Regarding the mitigation of CO_2 [15] underlines that fulfilling the EU's recently revised energy strategy demands the rapid expansion of RES electricity production, such that it approaches a 69% total share of the fuel mix. In order for these gaps to be filled, Ref. [16] proposed an ambitious overhaul of the EU's electricity market structure focusing on the reform of the current electricity pricing mechanism, which is currently mainly driven by short-term fossil fuel prices, in concurrence with strong financial incentives relative to RES deployment. Moreover, it is suggested for the liberalization process of the electricity sector in all member states to advance at a brisk pace, ensuring supply security, empowering consumers, and strengthening market transparency. Similarly, Ref. [17] suggests that strict provisions accompany the political agreement between EU countries about the reduction of the annual amount of emitted CO_2, while the EU is also advised to subsidize and finance the upgrade of the recharging network infrastructure for all types of electric-powered vehicles.

The EU's strategic energy planning, as described in [18], attempts to speed up procedures for a liberalized and integrated electricity market, which will be characterized by intense competition and will set control mechanisms to avoid the development of monopoly and oligopoly phenomena in electricity generation and retail market. The philosophy behind the EU's regulatory framework is to stimulate investments and market entries of new producers and distributors by lifting all legal, bureaucratic, and market barriers. The crucial influence of power sector's liberalization in RES deployment in developed western economies was highlighted by Refs. [19,20], with [21] putting the adverse effects of generation concentration in the epicenter. It is reasonable according to Refs. [22,23] for households in liberalized markets, with a high number of RES power producers and retailers, to choose providers with an ecological profile and at the same time be benefited by lower prices.

Electricity market openness and deregulation compose fundamental prerequisites for attracting and accumulating investments for additional RES projects, yet a series of related studies, among which [24], emphasize on the crucial impact of tax incentives and rational retail prices on maximizing the expected revenue of RES deployment schemes. It is noted by [25], that the announcement by governmental authorities in well-developed economies of RES related tax exemption incentives initiated a rapid response on behalf of private investors and firms, who distributed more of their available capital into projects for alternative electricity sources. Imposing energy taxes, and especially carbon emission taxes, on power production from conventional fossil fuel units was found by [26] to create an upward trend in RES investments.

Depending on the developmental stage, level of energy imports, abundance of natural resource reserves, degree of price stability in domestic electricity market and environmental quality, policymakers in individual countries are devising and modifying their energy strategies appropriately [27]. With respect to EU countries, Ref. [28] indicated that their shift towards renewable electricity varies mainly due to fundamental differences regarding their initial percentage RES usage as well as their dependency on foreign energy sources. The authors observed that European states with low installation capacities and high energy imports are capable of applying a more dynamic RES deployment strategy, since they tend to deal with less repercussions and hurdles from the power sector's stakeholders. Prior extended use of traditional carbon emitting energy sources was characterized by [29] as a critically detrimental factor obstructing the conversion of polluting electricity generation to RES. Several powerful and influential economic groups representing the gas, oil, and nuclear power industries can exert significant pressure on governments and force them to put a halt to their renewable energy strategies. The different paces by which EU member states are heading toward green electricity era were further observed by the recent analysis of [30].The authors highlighted the intense presence of lobbying, which hampers this already complicated procedure, while the opposite applies for stringent environmental rules and increased public awareness relative to the vital importance of RES.

Considering all previously discussed challenges, the present research attempts to clarify the influence of certain key determinants, representing the degree of power market deregulation, retail electricity prices, as well as environmental awareness and the provided investment incentives, onthe development of the European RES market. This research includes several novelties since the individual EU country characteristics are taken into account by utilizing the panel fully modified ordinary least square (FMOLS) econometric methodology. Furthermore, it is to the best of the authors' knowledge, the first time to investigate the effect of energy sector liberalization on RES share fuel mix participation deriving from electricity generation and the retail markets—of several European countries. For this purpose, this analysis will process a dataset comprised of the market shares of the largest electricity producer and the total number of market retail companies. Additionally, to provide a more spherical view of the driving forces of the EU's renewable market, it will be further scrutinized whether the amount of energy taxes imposed on fossil fuel power production and the level of household retail prices are sufficient enough to persuade new private investors and firms to enter the electricity market. These parameters will also be inspected for whether they motivate the existing market stakeholders to proceed to an increase in their RES installation projects. Based on the final outcomes of the econometric process, several suitable policy implications will be proposed accordingly for each country included in the study, in order to assist governmental authorities and the EC in their combined efforts for the transition to carbon neutral electricity generation leading to the alleviation of environmental degradation and the enhancement of the EU's energy autonomy.

Overall, this paper is organized as follows: Sections 1 and 2 contain the introduction and the literature review. Section 3 consists of the data and panel econometric methodology presentation. Sections 4 and 5 present a detailed commentary on the final empirical outcomes and corresponding policy recommendations. Lastly, Section 6 highlights the main findings and contributions of the paper.

2. State of the Art

Despite the plethora of papers in academia scrutinizing the potential implications on retail electricity prices from the delays, inconsistencies and procrastinations of central governments to stimulate the deregulation of the national energy markets, only a confined group of researchers focus on the mechanism upon which market openness affects the expansion of RES. After examining the data for OECD countries from the middle 1970s until the early 2010s, Ref. [31] detected that RES employment policies appear to be far more effective in deregulated electricity markets. Electricity market liberalization was concluded by [32,33] to enhance RES innovations, with [34] finding evidence of a quadratic relationship between the two factors, indicating that the intensity of the deregulation progression can be a decisive parameter of the impact magnitude. Interestingly, Ref. [21] processing long panel datasets on OECD countries, revealed that lifting market barriers for new entrants into the power sector was proved to be almost equally important in boosting RES deployment as consumers' personal available income and public awareness for tackling climate change and environmental depletion. The authors in the latter study further supported that whenever a variety of new actors was allowed to enter the electricity market, it diminished the concentration in power generation and abolished publicly owned monopolies, thereby triggering more progressive initiatives regarding RES sector's enlargement. Analyzing data for 25 developed OECD countries over a 30-year period—between 1985 and 2015—[20] claim that a competitive electricity market produces the necessary conditions under which RES development can flourish. Nevertheless, according to [35], the ability of RES to penetrate a deregulated energy market is not unlimited due to the evident weakness of highly competitive power industries to transparently deliver the proper market signals.

With respect to the impact of active governmental policy, involving tax incentives for the installation of carbon-free generation units and the simultaneous taxation of fossil-fuel-based power production, Ref. [36] underlines the importance of persistent commitment in support of a renewable-centered strategy combined with the willingness on behalf of regulatory authorities to intervene when necessary and impose essential measures (e.g., subsidies, taxes, etc.) to increase installed RES capacity. Exploring the Latin American energy market for a time frame between 2006 and 2015 [37] discovered that countries which adopt active promotion strategies regarding RES deployment accompanied by joint provisions for targeted tax incentives, are better able to cultivate a culture in favor of green electricity among their citizens, thereby increasing the overall contribution of renewables to total power production. By examining the Spanish power industry, Ref. [38] discovered that the most efficient way for the country's public administrators to reinforce RES's share involved establishing a policy mix of added taxes for non-renewable power production and tax discounts concerning investments in new zero-carbon electricity units, creating a competitive advantage over conventional polluting technologies. Likewise, conducting research on 27 EU countries [39] realized that a certain group was applied a dual-purpose tax strategy to intensify national RES usage, which involved financial motives that moderated initial investment budgets and allowed for competitively priced renewable electricity generation. In a similar study that focused on the entire EU electricity market from 2000 until 2015, Ref. [40] claim that imposed taxation on CO_2 emitting electricity production was proved a generally effective measure among all member states for incentivizing the expansion of nationally installed RES capacity. In the most recent relative research examining RES deployment in all EU member states, Ref. [25] highlights the vital significance for European governments to employ a tax incentive strategy that would allow them to achieve a more rapid switch from conventional fossil fuel to eco-friendly electricity production. Utilizing an extended panel dataset, containing information for 27 European countries and the vast majority of US states from 1990 until 2008, Ref. [41] validated the effectiveness of tax incentives to support the progress of green electricity in both regions. Solely referring to the US energy market [42,43] postulate that during the past three decades, tax credit extensions concerning RES electricity production enabled state governments to encourage investments in carbon-neutral power technologies and particularly wind farms. In contrast, the works of Refs. [26,44] examining the drivers behind eco-friendly power generation in China and 118 power markets respectively, determined that power market constraint policies fail to motivate RES investments—especially in regions with rich natural resources—while taxing fossil fuel electricity production results in an insignificant effect.

In spite of the extended number of academic papers found in the literature relative to the possible influence of RES on electricity prices, little research has been done to investigate the opposite effect. The work of [41] is one of the few studies to incorporate the price effect factor, underlying that feed-in tariff strategies increasing the final price received by green electricity producers proved capable of stimulating RES deployment within both the EU and the USA. In harmony with the evidence from the prior study, Ref. [40] recommend the implementation by European regulatory authorities of retail price levies as an effective measure to fund and promote renewable power production. Concentrating explicitly on the impact of electricity prices, Ref. [45] processing a large panel sample of well-developed OECD members unveiled that in countries with high percentage GDP growth, rising prices in the power market trigger an expansion of clean electricity consumption. Likewise, Ref. [46], based on data from 13 major economies from 2008 until 2018, argued that the impact of feed-in tariff policies, as well as the increasing price of electricity itself, is capable of stimulating investments in RES production units. utilizing evidence from 13 northeast US states [47] contend that the desire for higher financial returns generated by higher retail prices influenced investments in new solar photovoltaics, with [48] claiming that a decrease in electricity prices would indeed lead to a reduction of RES installations in the Spanish power market. Contrarily, Ref. [49] analyzing data for 38 of the most wealthy economies worldwide—including several European, OECD, and BRICS countries—from 1990 until 2010; they postulate that industrial electricity prices can negatively affect the rise of renewable electricity capacity, while the outcomes in the paper by [50] suggest that retail prices insignificantly contributed to the penetration of RES in the USA.

3. Methodology and Data

3.1. Data Presentation and Descricptive Statistics

Focusing on the key determinants of RES deployment within EU member states, the present research utilizes a balanced data sample containing yearly observations for a set of 17 countries, including the largest European power markets, during a time period between 2003 and 2020. The panel FMOLS econometric analysis that follows processes data for the total renewables' share in the electricity generation fuel mix [51], the average annual electricity retail price charged to household consumers (EUR/MWh) [52], the proportion of market concentration of the largest producer in the domestic power sector [53], the number of retail electricity providers [54] and the percentage of total national tax revenues corresponding solely to energy taxes [55]. The dataset was created with the use of the free online databases provided by the US Energy Information Administration and Eurostat. Table 1 contains brief and concentrated information of all variable details included in the analysis.

Table 1. Data sources and description.

Variable	Definition	Years	Data Sources
Renewables	Percentage RES participation into a country's electrcity generation fuel mix.	2003–2020	U.S. Energy Information Administration
Electricity Price	Cost of household electrcity per MWh.	2003–2020	Eurostat
Market_Share Largest Generator	Refers to percentage market concentration of the largest electrcity producing company.	2003–2020	Eurostat
Total Electricity Retailers	Refers to the total number of retail companies providing household electrcity.	2003–2020	Eurostat
Energy Taxes	Refers to the percentage share of total national tax revenues concerning exclusively energy production.	2003–2020	Eurostat

With respect to the selected variable representing RES's impact, the study followed the recommendation of [49] due to the fact that energy policymaking involves specific targets for RES participation in the generation scheme, while climate change preventive initiatives set certain goals for the increase of RES usage at the expense of carbon intensive technologies. Moreover, the energy tax variable is comprised of all taxes imposed on energy production. These include taxes on all types of fossil fuels used for electricity generation, taxes on energy product stocks, carbon emission taxes, as well as any other tax referring to greenhouse emissions.

In Tables 2 and 3, a detailed analysis of the entire panel and individual country descriptive statistics has been provided. Table 2 generates a broad but representative view of the energy markets in the examined EU member states and the energy sector in the European region as a whole. The statistical values for renewables' share reveal persistent high reliance on fossil fuels. A roughly 30% RES participation is reported on average, which is well under the EC's anticipated usage rate during the investigated period and way poorer than the necessary level of RES deployment set by IEA to successfully carry out the recently established RePowerEU plan. What is more, the minimum and maximum values for percentage RES contribution in national electricity production show that within the same dataset countries co-exist with recorded values from nearly 0% up to 99.47%. On the other hand, the nearly equal mean and median values regarding electricity prices indicate a strong sign of coherence among national energy strategies and compliance with the EC's proposed initiatives, targeting European power market integration. Nonetheless, the almost double than average maximum electricity price of 197.6 EUR/MWh is indicative of the relative instability that was present even prior to the recent energy crisis, possibly because of the EU's lack of natural resource autonomy and the insufficient development of RES. Focusing on electricity generation deregulation, it is evident that in a high number of the sample's countries there is absence of market competition. In most cases, one main utility company holds a market share exceeding 40%, indicating oligopolistic conditions. Conversely, the retail electricity market reflects a high degree of liberalization and economic antagonism; as a result, European consumers are benefited by the existence of several electricity providers. Exploring the taxation of energy production, there is an obvious policy convergence between EU country members with a common tax rate that accounts for approximately 5% of overall government tax revenues. Lastly, results for skewness, kurtosis, and the Jarque–Bera normality test in Table 2 reject the normality hypothesis for the dependent variables, as well as all explanatory variables.

Table 2. Descriptive statistics of total panel (Years: 2003–2020).

	Renewables (% Total Fuel Mix)	Electricity Price (EUR/MWh)	Market_Share Largest Generator (% Total)	Total Electricity Retailers	Energy Taxes (% Total Taxes)
Mean	34.09	112.09	49.09	190.48	5.36
Median	27.60	109.3	41.85	100	5.04
Std. dev.	25.96	26.98	24.94	282.37	1.52
Minimum	0.98	53.5	10.27	1	2.77
Maximum	99.47	197.6	100	1485	9.93
Skewness	0.90	0.74	0.44	2.63	0.69
Kurtosis	3.09	3.58	1.82	10.17	2.94
Jarque–Bera	41.9 ***	32.08 ***	27.56 ***	1009 ***	24.46 ***

*** Denotes significance at 1%.

Table 3 shows the main statistical values for each of the 17 nations, unveiling a broad profile of their domestic power market. Interestingly, Scandinavian countries appear to be the protagonists in green electricity generation, with all four of them being in the leading places of that particular category. Specifically, Norway achieved an average 98.22% annual electricity production through RES between 2003 and 2020. This strategic energy planning resulted in less volatile and relatively lower electricity prices. Among with the highest electricity prices lie heavily industrialized economies, including Germany, Italy, Spain and Belgium, showing a tendency for both high prices and increased instability. Scandinavian countries, together with Poland, Germany, and Spain, have the lowest generation concentration, while the most pluralistic retail electricity markets appear to be present within the largest European economies. Finally, all of the dataset's countries seem to have an aligned policy relative to energy taxes.

Table 3. Descriptive statistics of total panel (Years: 2003–2020).

	Renewables (% Total Fuel Mix)				Electricity Price (EUR/MWh)				Market_Share Largest Generator (% Total)				Total Electricity Retailers				Energy Taxes (% Total Taxes)			
	Mean	**Std.dev.**	**Min**	**Max**	**Mean**	**Std.dev.**	**Min**	**Max**	**Mean**	**Std.dev.**	**Min**	**Max**	**Mean**	**Std.dev.**	**Min**	**Max**	**Mean**	**Std.dev.**	**Min**	**Max**
Belgium	12.86	9.50	1.43	27.90	155.35	30.04	110.1	197.6	64.64	19.79	38.36	92	43.56	12.15	23	60	3.71	0.38	3.22	4.31
Croatia	58.44	11.28	37.81	74.61	94.81	11.66	72.5	110	84.47	2.93	80.01	92	5	3.24	1	9	6.16	0.81	4.81	7.21
Czech Republic	8.29	3.97	2.27	13.51	113.52	21.80	67.80	137.7	67.25	6.32	54.34	74.19	347.11	56.32	238	419	5.90	0.53	5	6.59
Denmark	46.36	22.45	17.49	82.76	101.72	10.88	86.90	126.2	39.40	7.50	31.87	56	50.89	20.07	33	113	4.91	0.67	3.53	6.16
Finland	40.89	6.35	29.61	52.94	99.96	13.83	77.60	120.5	21.79	4.36	15.77	27	98.33	4.01	86	100	4.42	0.26	3.82	4.79
France	15.37	4.35	9.86	25.10	104.27	13.16	89	129.2	80.31	9.32	63.87	90.20	172.33	7.62	160	185	3.76	0.31	3.33	4.28
Germany	23.41	12.20	7.55	46.52	137.43	6.37	125.4	148.9	27.84	2.78	22	32	1144.8	203.18	940	1485	4.61	0.73	3.53	5.97
Greece	20.13	10.22	7.33	41.86	102.36	17.98	72.20	127.8	76.36	19.59	50.02	100	10.5	8.15	2	26	6.21	1.85	3.91	8.34
Hungary	8.72	4.48	0.98	17.88	98.95	18.86	72.80	132	36.96	7.09	23.66	47.10	33	13.56	12	52	5.05	0.38	4.59	5.83
Italy	29.85	11.07	15.48	43.58	144.63	7.86	132.6	162.6	30.35	6.70	23	46.30	490	153.72	268	775	5.98	0.59	4.80	6.85
Latvia	59.32	6.92	50.17	72.50	92.94	16.40	58.40	114.4	88.16	2.74	85.47	95	10.72	8.57	1	26	8.27	1.24	6.05	9.93
Lithuania	33.37	31.77	1.76	89.65	81.22	15.09	53.5	104.8	49.36	20.46	24.9	79.70	16.5	7.75	7	27	5.84	0.42	4.98	6.88
Norway	98.22	1.11	95.73	99.47	117.11	13.96	92.70	138.8	30.78	1.46	27.4	33.60	194.67	16.47	163	226	3.33	0.27	2.77	4.08
Poland	9.06	5.69	1.50	19.44	97.41	13.09	73.70	119.5	14.90	3.49	10.27	19.20	162.44	36.52	118	265	6.80	0.27	6.26	7.30
Slovenia	28.56	4.48	22.28	38.52	104.91	12.61	85.20	119.2	55.40	6.96	50.10	82	15.67	4.69	7	23	7.46	1.07	5.98	8.81
Spain	30.78	9.47	15.2	44.62	144.52	37.05	87.20	194.7	26.62	5.92	17.92	39.10	300.39	94.63	121	459	4.31	0.33	3.72	4.79
Sweden	55.98	6.46	43.38	69.81	114.38	18.55	80.60	134.5	39.98	5.68	30.1	47	142.28	31.29	75	193	4.44	0.58	3.55	5.16

3.2. Causality Analysis

3.2.1. Pearson Correlation Test

The Pearson's correlation coefficients illustrated in Table 4 signify the considerably low correlation between the five investigated variables. In detail, the most commentworthy relationships include that of the electricity prices with generation concentration and total electricity retailers, which appear to be statistically significant at 1% with positive correlation coefficients (0.3118 and 0.3886, respectively). These relationships basically reflect the propensity of electricity prices to rise alongside the market share of the largest stakeholders in electricity production, while higher prices is reasonable to attract more electricity providers that are willing to enter the retail market. Likewise, a higher number of electricity retail companies seem to be negatively correlated with generation concentration. This strongly statistically significant connection suggests that a competitive electricity retail market influences a more liberalized power sector.

Table 4. Pearson's correlation coefficients.

Variable	Renewables	Electricity Price	Market_Share Largest Generator	Total Electricity Retailers	Energy Taxes
Renewables	1.0000				
Electricity Price	−0.0079 (0.8900)	1.0000			
Market_Share Largest Generator	−0.1110 * (0.0524)	0.3118 *** (0.0000)	1.0000		
Total Electricity Retailers	−0.1024 * (0.0737)	0.3886 *** (0.0000)	−0.3571 *** (0.0000)	1.0000	
Energy Taxes	−0.0857 ** (0.1349)	−0.2335 *** (0.0000)	0.2143 *** (0.0002)	−0.2021 *** (0.0004)	1.0000

*** Denotes significance at 1%, ** at 5%, and * at 10% level, respectively. Numbers in parentheses show the test corresponding p-values.

3.2.2. Dumitrescu–Hurlin (2012) Causality Test

Conducting a causality analysis among the dependent and explanatory variables is a vital prerequisite for avoiding the econometric process of a misspecified model and the misinterpretation of the potential effects between the investigated parameters of interest. The present paper employs the Dumitrescu–Hurlin (2012) [56] causality test, which is widely accepted in the academic community as a fairly reliable test that is commonly applied in the research field of economics and energy. Table 5 exclusively portrays the statistically significant causal connections concerning the five parameters included in the data sample with the implementation of a 2-period lag. The outcomes in Table 5 reveal two bidirectional relationships, one among RES and generation concentration and another among generation concentration and energy taxes. These findings suggest that investments in renewable electricity and energy taxes can affect the competition level of power production. Furthermore, there is a unidirectional causal effect running from electricity price towards RES and total retailers, meaning that electricity prices can trigger RES deployment as well as the entry of extra electricity sellers. Finally, from the one-way effect of RES to energy taxes and from generation concentration to total retailers it can be comprehended that state policies—up to an extent—involve the taxation of green electricity production, while electricity generators can intervene in the retail market.

Table 5. Dumitrescu–Hurlin (2012) causality testing, Lag Order: 2.

Null Hypothesis (H_0)	Obs	Test-Statistic	p-Value
Electricity Price does not Granger cause **Renewables**	302	**2.9575**	**0.0031**
Renewables does not Granger Cause **Market_Share**	302	**2.7704**	**0.0056**
Market_Share does not Granger Cause **Renewables**	302	**3.2157**	**0.0013**

Table 5. *Cont.*

Null Hypothesis (H_0)	Obs	Test-Statistic	p-Value
Renewables does not Granger Cause **Energy Taxes**	302	**5.7545**	**0.0000**
Electricity Price does not Granger Cause **Total Retailers**	302	**3.4564**	**0.0005**
Market_Share does not Granger Cause **Total Retailers**	302	**2.0952**	**0.0362**
Market_Share does not Granger Cause **Energy Taxes**	302	**3.7058**	**0.0002**
Energy Taxes does not Granger Cause **Market_Share**	302	**2.1929**	**0.0283**

Note: For the estimation of Dumitrescu–Hurlin (2012) [56] causality test, the analysis used the *xtgcause* command with 2 lags of "STATA" software (version 15.0).

3.3. Model Specification and Diagnostic Tests

The Dumitrescu–Hurlin (2012) [56] test validated a variety of interconnections among the dataset's examined variables. As a result, the econometric model that will be utilized to assess the impact of electricity sector liberalization, household retail prices, and energy tax policy on RES deployment in the 17 European states is formed as follows:

Main model econometric representation:

$$\begin{aligned} Renewables = \ & \beta_0 + \beta_1 Electricity\ Price_{i,t} + \beta_2 Market\_Share\_Largest\_Generator_{i,t} \\ & + \beta_3 Total\_Retailers_{i,t} + \beta_4 Energy\ Tax_{i,t} + \varepsilon_{i,t} \end{aligned} \quad (1)$$

The existence of a strong statistical connection between the component variables is an important first step for correct model specification. Nevertheless, to ensure the robustness and reliability of the model's econometric analysis through the selected panel methodological procedures, it necessitates a series of statistical diagnostic checks. The confirmation of cross-sectional dependence in the error term ($\varepsilon_{i,t}$) might be responsible for poor estimation of both the variables' coefficients and standard errors, which may in turn result in misleading final outcomes and unsound generalizations, in case the proper regression estimator is not applied. Tables 6 and 7 report the findings for time-series and panel cross-sectional dependence, respectively. Apparently, the null hypotheses (H_0) for both time-series and panel cross-sectional independence are emphatically rejected by all implemented tests, suggesting that the examined countries act under a central, longstanding EU energy strategy and that any shocks related to the deregulation of the electricity market, retail prices, and tax policy are spread from one or more countries to all the rest. The size and persistence of this effect may differ from one country to another and might be subject to regional and local characteristics.

Table 6. Cross-section dependence of panel time-series.

Variable	Pesaran (2004) CD_{test}	Correlation	Absolute Correlation	Pesaran (2015) Weak CD_{test}
Renewables	30.39 *** (0.0000)	0.614	0.714	46.638 *** (0.0000)
Electricity Price	24.72 *** (0.0000)	0.500	0.541	48.930 *** (0.0000)
Market_Share Largest Generator	21.25 *** (0.0000)	0.429	0.493	48.452 *** (0.0000)
Total Electricity Retailers	9.96 *** (0.0000)	0.201	0.454	45.098 *** (0.0000)
Energy Taxes	5.65 *** (0.0000)	0.114	0.448	48.801 *** (0.0000)

Note: *** Denotes significance at 1%. Numbers in parentheses show the test corresponding p-values. The null hypothesis (H_0) of Pesaran (2004) [57] CD test assumes strict cross-sectional independence. The null hypothesis (H_0) of Pesaran (2015) [58] CD test assumes weak cross-sectional independence. For the Pesaran (2004) CD [57] and Pesaran (2015) [56] CD tests, the *xtcd* and the *xtcd2* commands of "STATA" software were utilized. Correlation and Absolute (correlation) are the average (absolute) value of the off-diagonal elements of the cross-sectional correlation matrix of residuals.

Table 7. Cross-section dependence among groups.

Pesaran's test of cross-sectional independence	1.848 *	
	(0.064)	
Friedman's test of cross-sectional independence	26.684 **	
	(0.045)	
Frees' test of cross sectional independence	3.201	
Critical values from Frees' Q distribution:	**Alpha** = 0.10	0.1438
	Alpha = 0.05	0.1888
	Alpha = 0.01	0.2763

** Denotes significance at 5% and * at 10% level, respectively. Numbers in parentheses show the test corresponding *p*-values. For the Pesaran, Friedman [59], and Frees [60,61], group cross-sectional dependence tests, the *xtcsd pesaran abs*, *friedman xtcsd*, and *frees xtcsd* post commands after *xtreg* POLS regression in "STATA" software were utilized.

The appropriate implementation of panel FMOLS approach presupposes co-integrated variables, as well as the elimination of the possibility of potential existence of unit roots in the processed data sample. With respect to the dataset's stationarity, the analysis makes use of the traditional LLC test [62], which manages to generate fairly consistent results when applied to panel data with a long time-series, along with the ADF–Fisher test [63–67] with relaxed conditions relative to the allowed lag length across units. In addition, the analysis further utilizes the CIPS stationarity test developed by Pesaran (2007) [68], a second generation panel unit-root test able to avoid the restriction of the first two tests by relying on a standard type of distribution to produce the corresponding P-values, which is proved in the literature to outperform all other unit-root tests in the presence of cross-sectional dependence. Similarly, the validation of co-integration hypothesis is examined by Pedroni (1999) [65] and Pedroni (2004) [66] tests, as well as the second generation Westerlund (2005) [67] co-integration test. The latter is based on an error-correction pattern, allowing it to remain robust regardless of the potential existence of cross-sectional dependence. Table 8, Table 9 andTable 10 summarize the outcomes of the statistical tests by examining the previously discussed econometric parameters. In Table 8, it is shown that the unit root null hypothesis (Ho) is rejected at first difference by all three tests at 1% statistical significance, including both intercept and trend options. This outcome signifies that the dataset's components are integrated and are stationary at order one I(1). Tables 9 and 10 highlight the rejection of non-co-integration null hypothesis (H_0) by all three tests at 1% and 5% significance levels respectively.

Finalizing the necessary statistical diagnostic procedure, which aims to select appropriate and more robust panel econometric techniques and estimators, includes checking the investigated dataset for the presence of heteroskedasticity and serial correlation. Table 11 portrays the results of a variety of widely acceptable heteroskedasticity tests, together with the Breusch–Godfrey/Wooldridge (2010) [68] test for serial correlation. From the table, it is made clear that within the processed data sample both statistical phenomena are present, while Table A1 verifies the absence of multicollinearity in the data.

Table 8. Unit root tests.

	Level						First-Difference					
	Intercept			Intercept and Trend			Intercept			Intercept and Trend		
Variable	ADF–Fisher	LLC	CIPS	ADF–Fisher	LLC	CIPS	ADF–Fisher	LLC	CIPS	ADF–Fisher	LLC	CIPS
Renewables	13.183 (0.999)	−4.597 (1.000)	−2.772 ***	31.464 (0.592)	0.324 (0.627)	−3.082 ***	130.289 *** (0.000)	−6.487 *** (0.000)	−4.020 ***	113.655 *** (0.000)	−3.716 *** (0.000)	−4.065 ***
Electricity Price	39.598 (0.234)	−5.985 *** (0.000)	−2.518 ***	27.450 (0.779)	−0.941 (0.173)	−2.451	65.527 *** (0.000)	−5.624 *** (0.000)	−3.816 ***	64.129 *** (0.001)	−3.194 *** (0.000)	−3.922 ***
Market_Share Largest Generator	36.508 (0.352)	−0.960 (0.168)	2.886 ***	44.473 (0.107)	−4.413 *** (0.000)	−3.034 ***	166.756 *** (0.000)	−9.122 *** (0.000)	−4.339 ***	122.529 *** (0.000)	−6.802 *** (0.000)	−4.341 ***
Total Electricity Retailers	35.046 (0.177)	−0.141 (0.443)	−2.200 *	28.311 (0.742)	0.286 (0.612)	−3.104 ***	99.104 *** (0.000)	−14.269 *** (0.000)	−4.344 ***	33.107 *** (0.511)	−5.165 *** (0.000)	−4.459 ***
Energy Taxes	32.620 (0.535)	1.948 (0.974)	−2.074	30.402 (0.992)	1.07 0 (0.857)	−2.629 *	109.784 *** (0.000)	−5.924 *** (0.000)	−3.751 ***	66.435 *** (0.000)	−5.949 *** (0.000)	−3.838 ***

Note: *** Denotes significance at 1% and * at 10% level. Numbers in parentheses show the test's corresponding P-values. The null hypotheses (H_0) of the tests assume non-stationary variables. For the ADF–Fisher, LLC, and CIPS Refs. [62–64] unit root tests, the *xtunitroot* and *xtcips* commands of "STATA" software were utilized. Critical values for the CIPS test of Pesaran (2007) [68] are −2.1 (10%), −2.21 (5%), and −2.4 (1%) for constant, and −2.63 (10%), −2.73 (5%), and −2.92 (1%) for trend, respectively. The optimal lag selection was made based on the Akaike Information Criterion, while the Bartlett kernel was selected with the maximum number of lags being determined by the Newey and West bandwidth selection algorithm.

Table 9. Pedroni's (1999, 2004) panel co-integration tests.

Pedroni (2001)	Panel v-Statistic	Panel ho-Statistic	Panel t-Statistic	Panel ADF-Statistic	Group rho-Statistic	Group t-Statistic	Group ADF-Statistic
Test-Statistics	−1.33	1.803	−8.002	2.603	3.39	−8.525	3.823
Pedroni (2004)						**Statistic**	***p*-value**
Modified Phillips-Perron-t						3.7523	0.0001
Phillips-Perron-t						−8.9462	0.0000
Augmented Dickey-Fuller-t						−21.2287	0.0000

Note: For the Pedroni (1999, 2001) [65,66] panel co-integreation test, the *xtpedroni* command of "STATA" software was utilized with *trend* option. The optimal lag selection was made based on the Akaike Information Criterion, while the Bartlett kernel was selected with the maximum number of lags being determined by the Newey and West bandwidth selection algorithm. The null hypothesis (H_0) of the test assumes no co-integration in the examined models while the alternative hypothesis (Ha) assumes that all panels are co-integrated. All test statistics are distributed N(0,1) and diverge to negative infinity except for panel v, in which the test statistic diverges to positive infinity. The null hypothesis (H_0) assumes no co-integration of the models' variables. The probability upon which it is decided whether to reject or accept the H_0 is estimated based on the z-score of the values of the test stastistics. Additionally, for the Pedroni (1999, 2004) Refs. [63,64] panel co-integreation test, the *xtcointtest pedroni* command of "STATA" software was utilized, with *kernel (bartlett), trend,* and *demean** options. The optimal lag length was selected automatically based on the Akaike Information Criterion (AIC), and all other bandwidth orders are set according to the rule 4(T/100)2/9 ≈ 3. The null hypothesis (Ho) of the test assumes no co-integration in the examined models while the alternative hypothesis (Ha) assumes that all panels are co-integrated. The null hypothesis (H_0) of the test assumes no co-integration in the examined models while the alternative hypothesis (Ha) assumes that all panels are co-integrated. *Demean option: Stata computes the mean of the series across panels and subtracts this mean from the series. Ref. [62] suggests this procedure to mitigate the impact of cross-sectional dependence.

Table 10. Westerlund panel co-integration test.

	Statistic	*p*-Value
Variance ratio	−1.9407	0.0262

Note: The null hypothesis (H_0) of the Westerlund (2005) [65] panel co-integration test assumes no co-integration, while the alternative hypothesis (H_1) assumes co-integration in all panels. The Westerlund (2005) [67] tests were estimated by using the xtcointtest westerlund command of "STATA" software, including the *trend* and *demean** options. *Demean option: Stata computes the mean of the series across panels and subtracts this mean from the series. Ref. [62] suggests this procedure to mitigate the impact of cross-sectional dependence.

Table 11. Heteroskedasticity and serial correlation tests.

	Statistic	*p*-Value
Breusch–Pagan Heteroskedasticity test	45.94	0.0000
Glejser Heteroskedasticity test	28.04	0.0000
Harvey Heteroskedasticity test	10.07	0.0000
White Heteroskedasticity test	12.06	0.0000
Breusch–Godfrey/Wooldridge Serial Correlation test	113.98	0.0000

Note: The null hypothesis (H_0) of the Breusch–Pagan (1979) [69], Glejser (1969) [70], Harvey (1976) [71] and White (1980) [72] tests assume no heteroskedasticity in the models. Similarly, the null hypothesis (H_0) of the Breusch-Godfrey/Wooldridge (2010) [68] test assumes no serial correlation, (pbgtest {plm} from "R" software, version 4.3.1).

4. Empirical Analysis and Results

The recent studies from Refs. [27–30] unveiled the different rhythm that European nations seem to adopt in promoting RES deployment strategies due to structural differences in energy taxation and in the deregulation process of power generation and retail markets. In harmony with the findings and recommendations of the aforementioned papers, the present research attempts to shine a spotlight on individual electricity markets and country-specific characteristics that may affect the transition of certain EU countries to renewable electricity. For this purpose, the analysis incorporates the FMOLS econometric methodology, which enables the generation of endogeneity and serial-correlation robust error-terms (even though no instrumental or synthetic variable has been utilized), as well as separate regression coefficients for each panel of the 17 examined countries. These essential merits have been praised by a series of academic papers. The FMOLS model is recommended by Refs. [73–75] as a fairly suitable and appropriate model for processing extended economic, energy, and RES panel datasets. The FMOLS approach was preferred from other panel methodologies, such as difference and system GMM, since the relative econometric literature, among which Refs. [76,77], proposes that when the number of time-series (T) in the panel is larger than the number of total panels (N) (i.e., T > N), then FMOLS and PDOLS methodologies provide more accurate and consistent estimates. The FMOLS estimator was initially developed by [78] with the goal of overcoming the inconsistencies stemming from the long-run correlation between the co-integrating equation and the stochastic regressor innovations. The typical FMOLS estimator, which was originally intended for time-series econometric processing, was further modified into the widely applied pooled FMOLS by Refs. [79,80].

Table 12 contains the final outcomes of the panel regression analysis that was conducted with the use of the FMOLS model and was suitably adjusted for the aims and scope of this study. Interestingly, the findings for the individual country panels fully verify the conclusions of all previous studies referring to the exhibited disparities concerning the sustainable development of the energy sector, even between countries belonging to the same region and actively participate in international country coalitions, such as the EU. Household electricity prices proved to be a highly statistically significant parameter of RES contribution in the production scheme for 14 out of the 17 countries, with the vast majority of the panels revealing a slightly negative effect of the variable—except for the cases of Greece, France, and Hungary. In all other countries, for every 1% increase in retail prices, there has been a recorded negative impact on renewable electricity production, varying from −0.02% in Italy and Spain to −0.53% in Croatia and −1.1% in Lithuania ceteris paribus. Similarly, lack of competition and monopolistic or oligopolistic conditions in power generation constitutes a detrimental factor for the transition to eco-friendly electricity in 13 countries. A potential increase in the percentage market share of the largest power company by 1% can approximately negatively affect RES contribution by −0.59% in Spain, −0.82% in Germany, and -0.94% in Latvia—when holding constant all other explanatory parameters. Surprisingly, such an increase in Norway, Sweden, Italy, and Croatia seems to benefit renewable electricity production, implying that the stakeholder's power production activity in these countries is mainly supported by zero-carbon emitting technologies. In contrast with generation concentration, an increase in the number of total electricity retail sellers triggers a rather complicated effect to the overall contribution of RES into the various energy systems' fuel mixes. Specifically, for 9 of the countries, the entry of 1% additional electricity providers boosts RES development, with that influence varying from nearly 0.05% for Czech Republic and France up to 2.17% for Croatia. For 5 of the countries, increasing the number of total electricity retail companies causes the exact opposite effect; particularly in Slovenia an increase of 1% reduces the proportion of renewable electricity by 0.66%. For Norway, Sweden, and Poland, this parameter is found to be statistically insignificant. Finally, in 9 countries the intensification of the imposed energy taxes led to the development of an upward trend in RES installation projects. It is noteworthy that for 6 of these countries, a 1% growth of domestic energy taxation increases RES share in total electricity generation multiple times. Indicatively, this increase reaches roughly 4.72% in Croatia, 6.78% in Spain, and 7.39% in Sweden. The opposite, however, seems to apply for France, Germany, and Greece, while for Belgium, Poland, and Latvia, where energy taxes compose a statically insignificant factor.

Table 12. Panel FMOLS coefficients for the 17 European markets for the period of 2003–2020.

	Panel FMOLS			
Country	**Electricity Price**	**Market_Share Largest Generator**	**Total Electricity Retailers**	**Energy Taxes**
Belgium	0.000 (0.05)	−0.210 *** (−14.65)	0.080 *** (9.09)	−0.370 (−0.71)
Croatia	−0.53 *** (−21.71)	1.77 *** (49.25)	2.170 *** (26.75)	4.720 *** (21.53)
Czech Republic	−0.000 (−0.67)	−0.080 *** (−2.36)	0.040 *** (11.64)	0.51 * (1.90)
Denmark	−0.120 *** (−4.19)	−0.380 *** (−8.40)	0.150 *** (5.36)	3.12 *** (3.75)
Finland	−0.060 *** (−9.88)	−0.10 *** (−6.05)	−0.160 *** (−21.70)	2.70 *** (24.44)
France	0.070 ** (2.55)	−0.48 *** (−8.27)	0.050 *** (2.97)	−1.21 *** (−2.59)
Germany	−0.080 *** (−5.96)	−0.820 *** (−19.34)	0.010 *** (6.89)	−3.69 *** (−6.99)
Greece	0.110 ** (2.38)	−0.280 *** (−6.40)	0.300 *** (4.15)	−0.930 *** (−3.56)
Hungary	0.010 * (2.47)	−0.010 (−1.45)	−0.060 *** (−10.14)	0.900 *** (7.16)
Italy	−0.02 *** (−3.57)	1.120 *** (73.19)	−0.030 *** (−67.15)	3.400 *** (50.96)
Latvia	−0.39 *** (−5.29)	−0.94 ** (−2.56)	0.270 * (1.92)	−0.120 (−0.20)
Lithuania	−1.100 *** (−25.01)	−0.620 *** (−13.17)	0.410 *** (4.01)	2.370 ** (2.34)
Norway	−0.040 *** (−10.76)	0.440 *** (16.81)	−0.000 (−1.25)	0.760 *** (3.32)
Poland	0.010 (1.37)	−0.110 ** (−2.26)	−0.000 (−0.21)	−0.510 (−1.40)
Slovenia	−0.080 *** (−4.43)	−0.060 *** (−5.74)	−0.660 *** (−16.54)	0.980 *** (8.38)
Spain	−0.020 *** (−9.46)	−0.590 *** (−35.59)	−0.020 *** (−31.71)	6.780 *** (38.42)
Sweden	−0.140 *** (−10.71)	0.59 *** (4.87)	−0.020 (−1.37)	7.390 *** (7.18)

Note: ***, **, and * indicate statistical significance at 1%, 5%, and 10%, respectively. For the estimation of panel FMOLS, the *xtcointreg* command of "STATA" software was utilized along with suitable options for *kernel*, *bmeth*, *eqtrend*, *xtrend*, and *stage*.

5. Discussion and Policy Implications

By applying a sophisticated panel econometric approach, the research revealed the complex interactions and underlying causality relationships between RES deployment and electricity retail market, power generation, and energy taxation in Europe. The econometric analysis highlighted the unique characteristics of each country's energy sector as well as the individual effects of the four explanatory variables on RES share electricity production. Focusing on the impact of retail electricity charges, in 11 out of 17 countries price increases constituted a detrimental driver of new RES investment plans. Investors are expecting household consumers to negatively react in potential rises of electricity bills by reducing their annual consumption in the near future, leading to lower marginal revenues and profits for electricity generation companies. As a result, it would be wise for public administrators in these countries to incorporate into their national energy strategy a flexible feed-in tariff policy supporting RES producers whenever electricity consumption falls below a certain level. A similar feed-in tariff policy is also suggested for Greece, France, and Hungary, with a price subsidy clause being activated whenever electricity prices drop and cross the point which green electricity production units are profitable and viable long term investment.

Furthermore, the degree of liberalization in power production reflected by generation concentration proved to be a considerably adverse determinant of RES development for 14 of the countries, indicating that the lack of competition between generators makes them more reluctant to replace conventional fossil fuel plants with RES electricity production units. Hence, regulators in these countries should build an energy strategy based on two main pillars. The first pillar, in collaboration with the EC, is advisable to involve subsidies that target the amortization of RES installation costs concerning the initial investment. In this way, according to [24], potential private investors will enjoy higher total revenues, incentivizing them to prioritize RES investment plans in their portfolios. Such subsidy policies are capable according to [81] of guaranteeing an extensive generation contribution by clean electricity technologies, regardless the degree of market deregulation. In addition to the possible financial aid in the form of European and state subsidies, central governments are advised to further enable the funding of RES projects by relaxing the credit conditions set by banks and other financial institutions in order for private investors to gain access to affordable loans. Nevertheless, Ref. [82] alleges that such funding policies necessitate the complete reform of the current financial sector so that credit availability regarding carbon-free generation units is hierarchized at the top of loan request lists. Clean energy funds are reported by [54] to have increased market penetration for all types of alternative energy production. On the other hand, the second pillar is proposed to include the reduction of market concentration of publicly owned generation companies through gradual privatizations. With regard to the countries where generation concentration composes an encouraging driving force of RES deployment, the dominant power generation stakeholder, which is most probably a public utility company, seems to have already invested vast amounts of capital in RES in the previous years. Among others, an essential benefit of RES, based on Refs. [83,84], lies in their ability to enhance a country's energy self-sufficiency. Hence, these generation companies most likely follow the central government's inducements for a more eco-friendly strategic energy planning, so that it complies with the EC's initiatives about certain climate change goals and future prospects for accomplishing energy autonomy.

With respect to the influence of the number of electricity retailers, the mixed econometric outcomes dictate governmental authorities in countries where a higher number of providers enhance RES deployment to proceed on the establishment of a liberalized legal framework which will effectively remove any entry barriers to the retail market. Conversely, for a group of countries including Italy, Spain, Finland, Hungary, and Slovenia, in which constantly increasing the total number of electricity sellers triggers an unfavorable effect on clean electricity production, certain control measures of the retail market should be implemented.

Lastly, the outcomes for energy taxes show a controversial impact of this factor on the development of an eco-friendly energy system, which depends on the different countries' power sector characteristics. Energy taxes appear to considerably increase RES electricity production in 7 out of the 17 countries, particularly Spain, Sweden, Italy, and Croatia, while it shows an analogous but more moderate effect in another 3 countries. Apparently, the imposed energy taxes in these countries either solely concern fossil fuel energy conversion or the ratio between them and RES taxes is comparatively well uneven. Therefore, central governments are recommended to follow a dual strategy involving the continuation of the current tax policy narrowing fossil fuel electricity, while further increase energy tax rates to narrow fossil fuel electricity production and accumulate essential capital for financing new RES projects. Conversely, it would be wise for Germany, France, and Greece to relax energy taxation and proceed to incorporate tax reductions of RES electricity production as soon as the figures in their national budgets allows for it.

6. Conclusions

Replacing conventional CO_2 intensive electricity generation with green energy from RES constitutes an essential prerequisite of sustainable development. In order to cope with the challenges of the EU's newly established REPowerEU plan, as well as to fulfill the provisional commitments relative to the UN's SDG 7 and 13referring to clean low-cost energy production and tackling climate change, European governments need to shed light on the fundamental determinants driving RES deployment. The present research, utilizing a panel FMOLS econometric methodology, provides a detailed analysis of the impact of electricity prices, energy taxes, and the competition level in both power generation and the retail market respectively, in each of the 17 European countries included in the processed data sample. The different outcomes from country to country signify the complexity of effectively implementing a common European energy strategy that is obligatory for all EU member states. The main problem behind this compound task lies in the multiple conflicts of interest due to the different socioeconomic characteristics of each country, hence a series of groupwise policy implications are proposed with respect to the four explanatory variables. Considering the causal interactions between RES and generation concentration, as well as the outcomes of the FMOLS model, policymakers are advised to rapidly respond to the dynamic effect of the examined parameters. Green electricity production might have a broader impact that can spread to other activity sectors such as transport, considering the explosive growth in electric vehicles which constantly gain higher market shares. In addition, both individual states and the EU are recommended to fund acts and an extensive pro-environmental campaign that will mobilize European public opinion to actively participate in the promotion of clean electricity consumption, even when this requires putting aside personal interest and supporting RES producers at the expense of a price premium. In economically developed and sophisticated societies [85,86] observed that consumers show a condensed willingness to pay for even slightly more expensive RES-generated electricity. Reversing this quite worrying phenomenon requires a modern education policy emphasizing on environmental awareness and ecological training of its citizens, such that to embrace post-materialistic values and choose the wellbeing and the innumerable benefits of a healthy natural environment over personal wealth. Finally, dealing with the main limitation of the study, at a future time it would be wise for the dynamic aspect of the dataset to be taken into account by additionally employing the PDOLS methodology and examine whether the latter confirms the current study's results. Likewise, as a further extension of the current paper and prospect for future research, it would be rather interesting to scrutinize if splitting the same dataset into groups based on each country's energy import dependency, as advocated by Refs. [28,30], would cause alterations in the identified relationships as well as their size effect.

Author Contributions: G.E.H. and A.S.T. conceived and designed the analysis. Both G.E.H. and A.S.T. wrote the manuscript and contributed to the final version of the manuscript. G.E.H. supervised the paper, provided critical feedback, and helped shape the research. All authors have read and agreed to the published version of the manuscript.

Funding: This research received no external funding.

Data Availability Statement: Dataset derived from public domain resources. Full dataset is available upon request from the authors.

Conflicts of Interest: The authors declare no conflict of interest.

Abbreviations

RES	Renewable Energy Sources
EU	European Union
EC	European Commission
GDP	Gross Domestic Products
MWh	Mega Watt hour
UN	United Nations
SDG	Sustainable Development Goal

Appendix A

Table A1. Variance Inflation Factor Test.

Variable	VIF	1/VIF
Electricity Price	1.25	0.7974
Market_Share Largest Generator	1.21	0.8240
Total Electricity Retailers	1.28	0.7812
Energy Taxes	1.09	0.9162

Note: For the variance inflation factor test the *estat vif* command of "STATA" software was utilized, which calculates the centered variance inflation factors (VIFs) for the independent variables specified in a linear regression model.

References

1. World Meteorological Organization. State of the Global Climate 2022. WMO-No. 1316. 2023. Available online: https://library.wmo.int/doc_num.php?explnum_id=11593 (accessed on 15 June 2023).
2. European Union's Copernicus Climate Change Service. European Centre for Medium-Range Weather Forecasts (ECMWF): Global Climate Highlights 2022. 2022. Available online: https://climate.copernicus.eu/copernicus-2022-was-year-climate-extremes-record-high-temperatures-and-rising-concentrations (accessed on 15 June 2023).
3. United Nations Economic Commission for Europe Carbon. Neutrality in the UNECE Region: Integrated Life-Cycle Assessment of Electricity Sources. March 2022. Available online: https://unece.org/sed/documents/2021/10/reports/life-cycle-assessment-electricity-generation-options (accessed on 15 June 2023).
4. United Nations (UN). General Assembly, Transforming Our World: The 2030 Agenda for Sustainable Development, 21 October 2015 A/RES/70/1. 2015. Available online: https://www.refworld.org/docid/57b6e3e44.html (accessed on 15 June 2023).
5. Cai, Y.; Xu, J.; Ahmad, P.; Anwar, A. What drives carbon emissions in the long-run? The role of renewable energy and agriculture in achieving the sustainable development goals. *Econ. Res. Ekon. Istraživanja* **2022**, *35*, 4603–4624. [CrossRef]
6. Al-Barakani, A.; Bin, L.; Zhang, X.; Saeed, M.; Qahtan, A.; Ghallab, H. Spatial analysis of financial development's effect on the ecological footprint of belt and road initiative countries: Mitigation options through renewable energy consumption and institutional quality. *J. Clean. Prod.* **2022**, *366*, 132696. [CrossRef]
7. Halkos, G.E.; Tsirivis, A.S. Electricity Production and Sustainable Development: The Role of Renewable Energy Sources and Specific Socioeconomic Factors. *Energies* **2023**, *16*, 721. [CrossRef]
8. Halkos, G.E.; Tsirivis, A.S. Electricity Prices in the European Union Region: The Role of Renewable Energy Sources, Key Economic Factors and Market Liberalization. *Energies* **2023**, *16*, 2540. [CrossRef]
9. European Parliament. Regulation (EU) 2021/1119 of the European Parliament and of the Council of 30 June 2021 Establishing the Framework for Achieving Climate Neutrality and Amending Regulations (EC) No 401/2009 and (EU) 2018/1999 ('European Climate Law'). 2020. Available online: https://eur-lex.europa.eu/legal-content/EN/TXT/?uri=CELEX:32021R1119 (accessed on 15 June 2023).
10. European Commission. Directorate-General for Communication, Decarbonising our Energy System to Meet Our Climate Goals, Publications Office of the European Union. 2021. Available online: https://data.europa.eu/doi/10.2775/582723 (accessed on 15 June 2023).
11. Statista. Leading Countries Worldwide Based on Natural Resource Value as of 2021. 2023. Available online: https://www.statista.com/statistics/748223/leading-countries-based-on-natural-resource-value/?fbclid=IwAR0-6mASnJ3778BLKYIDTqjee0IrPrwGTg6b0AlXdms5Dz-8rqawdSyN_Jo (accessed on 15 June 2023).
12. European Commission. REPowerEU: Joint European Action for More Affordable, Secure and Sustainable Energy, 8 March 2022 COM 108 Final. 2022. Available online: https://eur-lex.europa.eu/legal-content/EN/TXT/?uri=COM:2022:108:FIN (accessed on 15 June 2023).

13. Council of the European Union. Council Regulation (EU) 2022/1854 on an Emergency Intervention to Address High Energy Prices. Official Journal of the European Union. 2022. Available online: https://eur-lex.europa.eu/eli/reg/2022/1854/oj (accessed on 15 June 2023).
14. EMBER. European Electricity Review 2023. 2023. Available online: https://ember-climate.org/app/uploads/2023/01/Report-European-Electricity-Review-2023.pdf (accessed on 15 June 2023).
15. IEA. Is the European Union on Track to Meet Its REPowerEUgoals? 2022. Available online: https://www.iea.org/reports/is-the-european-union-on-track-to-meet-its-repowereu-goals (accessed on 15 June 2023).
16. European Parliamentary Research Service (EPRS). Reforming the EU Electricity Market. 2023. Available online: https://www.europarl.europa.eu/RegData/etudes/BRIE/2023/739374/EPRS_BRI(2023)739374_EN.pdf (accessed on 15 June 2023).
17. Council of the European Union. Proposal for a Regulation of the European Parliament and of the Council on the Deployment of Alternative Fuels Infrastructure, and Repealing Directive 2014/94/EU of the European Parliament and of the Council. Official Journal of the European Union. 2022. Available online: https://data.consilium.europa.eu/doc/document/ST-9111-2022-INIT/en/pdf (accessed on 15 June 2023).
18. European Parliament. Regulation (EU) 2019/943 of the European Parliament and of the Council of 5 June 2019 on the Internal Market for Electricity. 2019. Available online: https://eur-lex.europa.eu/legal-content/EN/TXT/?uri=CELEX:32019R0943 (accessed on 15 June 2023).
19. Llobet, G.; Padilla, J. Conventional Power Plants in Liberalized Electricity Markets with Renewable Entry. *Energy J.* **2018**, *39*, 69–92. Available online: https://www.jstor.org/stable/26534443 (accessed on 15 June 2023). [CrossRef]
20. Agyeman, S.D.; Lin, B. Electricity industry (de)regulation and innovation in negative-emission technologies: How do market liberalization influences climate change mitigation? *Energy* **2023**, *270*, 126863. [CrossRef]
21. Nicolli, F.; Vona, F. Energy market liberalization and renewable energy policies in OECD countries. *Energy Policy* **2019**, *128*, 853–867. [CrossRef]
22. Hirth, L. The market value of variable renewables: The effect of solar wind power variability on their relative price. *Energy Econ.* **2013**, *38*, 218–236. [CrossRef]
23. Erni, F.; Pietsch, D.; Paraschiv, R. The impact of renewable energies on EEX day-ahead electricity prices. *Energy Policy* **2014**, *73*, 196–210. [CrossRef]
24. D'Adamo, I.; Mammetti, M.; Ottaviani, D.; Ozturk, I. Photovoltaic systems and sustainable communities: New social models for ecological transition. The impact of incentive policies in profitability analyses. *Renew. Energy* **2023**, *202*, 1291–1304. [CrossRef]
25. Bölük, G.; Kaplan, R. Effectiveness of renewable energy incentives on sustainability: Evidence from dynamic panel data analysis for the EU countries and Turkey. *Environ. Sci. Pollut. Res.* **2022**, *29*, 26613–26630. [CrossRef] [PubMed]
26. Hao, F.; Shao, W. What really drives the deployment of renewable energy? A global assessment of 118 countries. *Energy Res. Soc. Sci.* **2021**, *72*, 101880. [CrossRef]
27. Batool, Z.; Raza, S.M.F.; Ali, S.; Ul Abidin, S.Z. ICT, renewable energy, financial development, and CO2 emissions in developing countries of East and South Asia. *Environ. Sci. Pollut. Res.* **2022**, *29*, 35025–35035. [CrossRef]
28. Kahia, M.; Aïssa, M.S.B.; Lanouar, C. Renewable and non-renewable energy use—Economic growth nexus: The case of MENA Net Oil Importing Countries. *Renew. Sustain. Energy Rev.* **2017**, *71*, 127–140. [CrossRef]
29. Marques, A.C.; Fuinhas, J.A. Drivers promoting renewable energy: A dynamic panel approach. *Renew. Sustain. Energy Rev.* **2011**, *15*, 1601–1608. [CrossRef]
30. Marra, A.; Colantonio, E. The institutional and socio-technical determinants of renewable energy production in the EU: Implications for policy. *J. Ind. Bus. Econ.* **2022**, *49*, 267–299. [CrossRef]
31. Nesta, L.; Vona, F.; Nicolli, F. Environmental policies, competition and innovation in renewable energy. *J. Environ. Econ. Manag.* **2014**, *67*, 396–411. [CrossRef]
32. Jamasb, T.; Pollitt, M.G. Electricity sector liberalisation and innovation: An analysis of the UK's patenting activities. *Res. Policy* **2011**, *40*, 309–324. [CrossRef]
33. Wang, N.; Mogi, G. Deregulation, market competition, and innovation of utilities: Evidence from Japanese electric sector. *Energy Policy* **2017**, *111*, 403–413. [CrossRef]
34. Marino, M.; Parrotta, P.; Valletta, G. Electricity (de)regulation and innovation. *Res. Policy* **2018**, *48*, 748–758. [CrossRef]
35. Blazquez, J.; Fuentes-Bracamontes, R.; Bollino, C.A.; Nezamuddin, N. The renewable energy policy Paradox. *Renew. Sustain. Energy Rev.* **2018**, *82*, 1–5. [CrossRef]
36. Schallenberg-Rodriguez, J. Renewable electricity support systems: Are feed-in systems taking the lead? *Renew. Sustain. Energy Rev.* **2017**, *76*, 1422–1439. [CrossRef]
37. Washburn, C.; Pablo-Romero, M. Measures to promote renewable energies for electricity generation in Latin American countries. *Energy Policy* **2019**, *128*, 212–222. [CrossRef]
38. Langarita, R.; Cazcarro, I.; Sánchez-Chóliz, J.; Sarasa, C. The role of fiscal measures in promoting renewable electricity in Spain. *Energy Convers. Manag.* **2021**, *244*, 114480. [CrossRef]
39. Cansino, J.M.; Pablo-Romero, M.d.P.; Román, R.; Yñiguez, R. Promoting renewable energy sources for heating and cooling in EU-27 countries. *Energy Policy* **2011**, *39*, 3803–3812. [CrossRef]
40. Mac Domhnaill, C.; Ryan, L. Towards renewable electricity in Europe: Revisiting the determinants of renewable electricity in the European Union. *Renew. Energy* **2020**, *154*, 955–965. [CrossRef]

41. Kilinc-Ata, N. The evaluation of renewable energy policies across EU countries and US states: An econometric approach. *Energy Sustain. Dev.* **2016**, *31*, 83–90. [CrossRef]
42. Shrimali, G.; Lynes, M.; Indvik, J. Wind energy deployment in the U.S.: An empirical analysis of the role of federal and state policies. *Renew. Sustain. Energy Rev.* **2015**, *43*, 796–806. [CrossRef]
43. Mai, T.; Cole, W.; Lantz, E.; Marcy, C.; Sigrin, B. *Impacts of Federal Tax Credit Extensions on Renewable Deployment and Power Sector Emissions*; U.S. Department of Energy: Washington, DC, USA, 2016. [CrossRef]
44. Hu, X.; Guo, Y.; Zheng, Y.; Liu, L.; Yu, S. Which types of policies better promote the development of renewable energy? Evidence from China's provincial data. *Renew. Energy* **2022**, *198*, 1373–1382. [CrossRef]
45. Chang, T.-H.; Huang, C.-M.; Lee, M.-C. Threshold effect of the economic growth rate on the renewable energy development from a change in energy price: Evidence from OECD countries. *Energy Policy* **2009**, *37*, 5796–5802. [CrossRef]
46. Azhgaliyeva, D.; Beirne, J.; Mishra, R. What matters for private investment in renewable energy? *Clim. Policy* **2023**, *23*, 71–87. [CrossRef]
47. Gelabert, L.; Labandeira, X.; Linares, P. An ex-post analysis of the effect of renewables and cogeneration on Spanish electricity prices. *Energy Econ.* **2011**, *33*, S59–S65. [CrossRef]
48. Crago, C.L.; Koegler, E. Drivers of growth in commercial-scale solar PV capacity. *Energy Policy* **2018**, *120*, 481–491. [CrossRef]
49. Aguirre, M.; Ibikunle, G. Determinants of renewable energy growth: A global sample analysis. *Energy Policy* **2014**, *69*, 374–384. [CrossRef]
50. Shrimali, G.; Kniefel, J. Are government policies effective in promoting deployment of renewable electricity resources? *Energy Policy* **2011**, *39*, 4726–4741. [CrossRef]
51. U.S. Energy Information Administration. Electricity Generation from Renewables (GWh), Interantional/Electricity. 2023. Available online: https://www.eia.gov/international/data/world/electricity/electricity-generation (accessed on 15 June 2023).
52. Eurostat. Electricity Prices for Household Consumers, (nrg_pc_204). 2023. Available online: https://ec.europa.eu/eurostat/databrowser/view/nrg_pc_204/default/table?lang=en (accessed on 15 June 2023).
53. Eurostat. Market Share of the Largest Generator in the Electricity Market, (ten00119). 2023. Available online: http://data.europa.eu/88u/dataset/vk1YAqU00TW2OBXK8KqR0A (accessed on 15 June 2023).
54. Eurostat. Total Number of Electricity Retailers, (nrg_ind_market, [ERTL]). 2023. Available online: https://ec.europa.eu/eurostat/databrowser/view/nrg_ind_market/default/table?lang=en (accessed on 15 June 2023).
55. Eurostat. Energy Tax Revenues (ENV_AC_TAX). 2023. Available online: https://ec.europa.eu/eurostat/databrowser/view/env_ac_tax/default/table?lang=en (accessed on 15 June 2023).
56. Dumitrescu, E.I.; Hurlin, C. Testing for Granger non-causality in heterogeneous panels. *Econ. Model.* **2012**, *29*, 1450–1460. [CrossRef]
57. Pesaran, M.H. *General Diagnostic Tests for Cross Section Dependence in Panels*; Cambridge Working Papers No: 0435 in Economics; Faculty of Economics, University of Cambridge: Cambridge, UK, 2004. [CrossRef]
58. Pesaran, M.H. Testing Weak Cross-Sectional Dependence in Large Panels. *Econom. Rev.* **2015**, *34*, 1089–1117. [CrossRef]
59. Friedman, M. The use of ranks to avoid the assumption of normality implicit in the analysis of variance. *J. Am. Stat. Assoc.* **1937**, *32*, 675–701. [CrossRef]
60. Frees, E.W. Assessing cross-sectional correlation in panel data. *J. Econom.* **1995**, *69*, 393–414. [CrossRef]
61. Frees, E.W. *Longitudinal and Panel Data: Analysis and Applications in the Social Sciences*; Cambridge University Press: Cambridge, UK, 2004. [CrossRef]
62. Levin, A.; Lin, C.F.; Chu, C.S.J. Unit root tests in panel data: Asymptotic and finite-sample properties. *J. Econom.* **2002**, *108*, 1–24. [CrossRef]
63. Maddala, G.S.; Wu, S. A Comparative Study of Unit Root Tests with Panel Data and a New Simple Test. *Oxf. Bull. Econ. Stat.* **1999**, *61*, 631–652. [CrossRef]
64. Pesaran, M.H. A Simple Panel Unit Root Test in the Presence of Cross-Section Dependence. *J. Appl. Econom.* **2007**, *22*, 265–312. Available online: http://www.jstor.org/stable/25146517 (accessed on 15 June 2023). [CrossRef]
65. Pedroni, P. Critical values for cointegration tests in heterogeneous panels with multiple regressors. *Oxf. Bull. Econ. Stat.* **1999**, *61*, 653–670. [CrossRef]
66. Pedroni, P. Panel co-integration: Asymptotic and finite sample properties of pooled time series tests with an applicaton to the PPP hypothesis. *Econom. Theory* **2004**, *20*, 597–625. [CrossRef]
67. Westerlund, J. New Simple Tests for Panel Cointegration. *Econom. Rev.* **2005**, *24*, 297–316. [CrossRef]
68. Wooldridge, J.M. *Econometric Analysis of Cross Section and Panel Data*; MIT Press: Cambridge, MA, USA, 2010; Available online: http://www.jstor.org/stable/j.ctt5hhcfr (accessed on 15 June 2023).
69. Breusch, T.S.; Pagan, A.R. A Simple Test for Heteroscedasticity and Random Coefficient Variation. *Econometrica* **1979**, *47*, 1287–1294. [CrossRef]
70. Glejser, H. A New Test for Heteroskedasticity. *J. Am. Stat. Assoc.* **1969**, *64*, 316–323. [CrossRef]
71. Harvey, A.C. Estimating Regression Models with Multiplicative Heteroscedasticity. *Econometrica* **1976**, *44*, 461–465. [CrossRef]
72. White, H. A heteroscedasticity Consistent Covariance Matrix Estimator and a Direct Test of Heteroscedasticity. *Econometrica* **1980**, *48*, 817–838. [CrossRef]
73. Neal, T. Panel Cointegration Analysis with Xtpedroni. *Stata J.* **2014**, *14*, 684–692. [CrossRef]

74. Liddle, B. The importance of energy quality in energy intensive manufacturing: Evidence from panel cointegration and panel FMOLS. *Energy Econ.* **2012**, *34*, 1819–1825. [CrossRef]
75. Damette, O.; Marques, A.C. Renewable energy drivers: A panel cointegration approach. *Appl. Econ.* **2018**, *51*, 2793–2806. [CrossRef]
76. Neal, T. Using Panel Co-Integration Methods To Understand Rising Top Income Shares. *Econ. Rec.* **2013**, *89*, 83–98. [CrossRef]
77. Chudik, A.; Pesaran, M.H.; Smith, R.P. *Pooled Bewley Estimator of Long-Run Relationships in Dynamic Heterogenous Panels*; Globalization Institute Working Papers, 409; Federal Reserve Bank of Dallas: Dallas, TX, USA, 2021. [CrossRef]
78. Phillips, P.; Hansen, B. Statistical Inference in Instrumental Variables Regression with I(1) Processes. *Rev. Econ. Stud.* **1990**, *57*, 99–125. [CrossRef]
79. Phillips, P.; Moon, H. Linear Regression Limit Theory for Nonstationary Panel Data. *Econometrica* **1999**, *67*, 1057–1111. [CrossRef]
80. Pedroni, P. Fully Modified OLS for Heterogeneous Cointegrated Panels. In *Nonstationary Panels, Panel Cointegration, and Dynamic Panels, Advances in Econometrics*; Baltagi, B.H., Fomby, T.B., Hill, R.C., Eds.; Emerald Group Publishing Limited: Bingley, UK, 2001; Volume 15, pp. 93–130. [CrossRef]
81. Shayegh, S.; Sanchez, D.L. Impact of market design on cost-effectiveness of renewable portfolio standards. *Renew. Sustain. Energy Rev.* **2021**, *136*, 110397. [CrossRef]
82. Bashira, M.F.; Ma, B.; Bashirb, M.A.; Radulescu, M.; Shahzade, U. Investigating the role of environmental taxes and regulations for renewable energy consumption: Evidence from developed economies. *Econ. Res.* **2021**, *35*, 1262–1284. [CrossRef]
83. Schumacher, K.; Krones, F.; McKenna, R.; Schultmann, F. Public acceptance of renewable energies and energy autonomy: A comparative study in the French, German and Swiss Upper Rhine region. *Energy Policy* **2019**, *126*, 315–332. [CrossRef]
84. Ayala-Chauvin, M.; Sanmartí, G.R.; Riba, C.; Lara, P. Evaluation of the energy autonomy of urban areas as an instrument to promote the energy transition. *Energy Sources* **2022**, *17*, 2053897. [CrossRef]
85. Mulder, M.; Zomer, S.P.E. Contribution of green labels in electricity retail markets to fostering renewable energy. *Energy Policy* **2016**, *99*, 100–109. [CrossRef]
86. Ndebele, T. Assessing the potential for consumer-driven renewable energy development in deregulated electricity markets dominated by renewables. *Energy Policy* **2020**, *136*, 111057. [CrossRef]

MDPI AG
Grosspeteranlage 5
4052 Basel
Switzerland
Tel.: +41 61 683 77 34

Energies Editorial Office
E-mail: energies@mdpi.com
www.mdpi.com/journal/energies

www.ingramcontent.com/pod-product-compliance
Lightning Source LLC
Chambersburg PA
CBHW061848210926
51850CB00006B/999